APPLIED MATRIX AND TENSOR ANALYSIS

APPLIED MATRIX AND TENSOR ANALYSIS

BY

JOHN A. EISELE

AND

ROBERT M. MASON

Naval Research Laboratory
Washington, D.C.

WILEY-INTERSCIENCE

a Division of John Wiley & Sons, Inc.

NEW YORK · LONDON · SYDNEY · TORONTO

Library of Congress Catalogue Card Number: 71-114916

SBN 471 23465 6

Printed in the United States of America

10 9 8 7 6 5 4 3 2 1

We know that intellectual food is sometimes more easily digested if not taken in the most condensed form.

F. CAJORI, *History of Mathematical Notations*

Preface

This book is based on a series of lectures on mathematical and theoretical physics given to graduate students under the Science Education Program of the Naval Research Laboratory in cooperation with the Physics Department of the University of Maryland.

In order to present topics of recognized importance and relevance that unfortunately are neither universally taught nor well understood, we have attempted to fill in the gaps in coverage of other recent books. By depending on sufficient detail, illustrative examples, drawings, and mnemonic devices, we hope to evoke such responses as "Why, I never knew that before," if not "At last, I finally understand what it all means."

The purpose of this book would not be realized without applications to physics. Even a grade school student knows, for example, that it does little good to hold an opaque object up to the light to inspect it; a better plan is to turn the object around under the light, to feel it, and to look at it from different angles. Few if any of the numerous concepts of quantum mechanics and relativity—the foundation stones of modern physics, which is now a 70-year-old structure—really can be said to be transparent. The details shown here, however, may enable the reader to turn these opaque conceptual objects around, to grasp them firmly, and to perceive them with a fresh perspective. When it presents a novel aspect, even an old fact sometimes becomes mentally exciting.

Such are the considerations that have guided our efforts. We hope that the resulting portfolio of interesting vignettes will save the reader much time and lead to an earlier commencement of active research for the student.

We wish to thank Reginald Larson for reading the manuscript and making many useful suggestions.

December 1969
Oxon Hill, Maryland
Temple Hills, Maryland

John A. Eisele
Robert M. Mason

Contents

APPLIED MATRIX AND TENSOR ANALYSIS

Part I

Vector and Matrix Algebra

1 Vector Algebra

A. Scalars and Vectors

Two important formulas describe the stationary aspects of any vector **V**, directed from the origin in Euclidean three-space, and having scalar components x, y, z along the three coordinate axes. The first formula is for the length:

$$\text{length of } \mathbf{V} = (x^2 + y^2 + z^2)^{1/2} \tag{1}$$

and the second for the cosine of the angle ϕ between two such vectors $v_1 = (x_1, y_1, z_1)$ and $v_2 = (x_2, y_2, z_2)$:

$$\cos\phi = \frac{x_1x_2 + y_1y_2 + z_1z_2}{(x_1{}^2 + y_1{}^2 + z_1{}^2)^{1/2}(x_2{}^2 + y_2{}^2 + z_2{}^2)^{1/2}}. \tag{2}$$

The length of a vector is denoted by $\|\mathbf{V}\|$ or sometimes by $|\mathbf{V}|$, although the second symbol should be reserved for the absolute value of a scalar.

A vector **V** is said to be "n-dimensional" if and only if it has n components. The inner product of two real n-dimensional vectors $\mathbf{x} = (x_1, x_2, \cdots, x_n)$ and $\mathbf{y} = (y_1, y_2, \cdots, y_n)$ is

$$\mathbf{x} \cdot \mathbf{y} = x_1y_1 + x_2y_2 + \cdots + x_ny_n. \tag{3}$$

Hence $\cos\phi$ becomes

$$\cos\phi = \frac{\mathbf{x} \cdot \mathbf{y}}{|\mathbf{x}|\,|\mathbf{y}|}. \tag{4}$$

EXERCISE 1

Prove that $|\cos\phi| \leq 1$. It may be helpful to read what several other textbooks have to say about the classical inequalities of Hölder, Minkowski, and Schwarz.

Start with two column vectors:

$$\mathbf{A} = \begin{pmatrix} A_1 \\ A_2 \\ \cdot \\ \cdot \\ \cdot \\ A_n \end{pmatrix} \quad \text{and} \quad \mathbf{B} = \begin{pmatrix} B_1 \\ B_2 \\ \cdot \\ \cdot \\ B_n \end{pmatrix}. \tag{5}$$

The "inner" or "scalar product" of these two vectors is

$$\mathbf{A}\cdot\mathbf{B} = \tilde{A}B = (A_1, A_2, \cdots, A_n)\begin{pmatrix} B_1 \\ B_2 \\ \cdot \\ \cdot \\ \cdot \\ B_n \end{pmatrix} = A_1B_1 + A_2B_2 + \cdots + A_nB_n. \quad (6)$$

The "outer," "direct," or "tensor product" of these two vectors is

$$\mathbf{A}\times\mathbf{B} = A\tilde{B} = \begin{pmatrix} A_1 \\ A_2 \\ \cdot \\ \cdot \\ \cdot \\ A_n \end{pmatrix} (B_1, B_2, \cdots, B_n) = \begin{pmatrix} A_1B_1 & A_1B_2 & \cdots & A_1B_n \\ A_2B_1 & A_2B_2 & \cdots & A_2B_n \\ \cdot & \cdot & \cdot & \cdot \\ \cdot & \cdot & \cdot & \cdot \\ \cdot & \cdot & \cdot & \cdot \\ A_nB_1 & A_nB_2 & \cdots & A_nB_n \end{pmatrix}$$

$$= C = \begin{pmatrix} C_{11} & C_{12} & \cdots & C_{1n} \\ C_{21} & C_{22} & \cdots & C_{2n} \\ \cdot & \cdot & \cdot & \cdot \\ \cdot & \cdot & \cdot & \cdot \\ \cdot & \cdot & \cdot & \cdot \\ C_{n1} & C_{n2} & \cdots & C_{nn} \end{pmatrix}. \quad (7)$$

Although the inner product is simply a scalar, it may be thought of as a one-element matrix; on the other hand, the outer product is a square array of numbers, which may be thought of as a matrix or tensor operator of higher rank.

$$\begin{aligned} \mathbf{A}\cdot\mathbf{B} &= \sum_{i=1}^{n} A_iB_i = A_iB_i, \\ \mathbf{A}\times\mathbf{B} &= (A_iB_j) = (C_{ij}) = C. \end{aligned} \quad (8)$$

These equations can be specialized to any number of dimensions, n. Thus for three dimensions definition (6) becomes

$$\mathbf{A}\cdot\mathbf{B} = (A_1, A_2, A_3)\begin{pmatrix} B_1 \\ B_2 \\ B_3 \end{pmatrix} = A_1B_1 + A_2B_2 + A_3B_3 \quad (9)$$

and definition (7) becomes

$$\mathbf{A} \times \mathbf{B} = \begin{pmatrix} A_1 \\ A_2 \\ A_3 \end{pmatrix} (B_1, B_2, B_3) = \begin{pmatrix} A_1B_1 & A_1B_2 & A_1B_3 \\ A_2B_1 & A_2B_2 & A_2B_3 \\ A_3B_1 & A_3B_2 & A_3B_3 \end{pmatrix} \tag{10}$$

$$= \begin{pmatrix} C_{11} & C_{12} & C_{13} \\ C_{21} & C_{22} & C_{23} \\ C_{31} & C_{32} & C_{33} \end{pmatrix} = C.$$

The outer product can be extended to produce an oblong array as follows:

$$\mathbf{A} \times \mathbf{B} = \begin{vmatrix} A_1B_1 & A_1B_2 & \cdots & A_1B_n \\ A_2B_1 & A_2B_2 & \cdots & A_2B_n \\ \cdot & \cdot & \cdot & \cdot \\ \cdot & \cdot & \cdot & \cdot \\ \cdot & \cdot & \cdot & \cdot \\ A_mB_1 & A_mB_2 & \cdots & A_mB_n \end{vmatrix} \tag{11}$$

provided $m \neq n$. In general, any rectangular array of individual elements enclosed in parentheses is called a "matrix." The entries may be scalars, polynomials, functions, or even matrices. A row vector is readily seen to be a 1 by n matrix and a column vector, an m by 1 matrix, where m and n are determined by inspection.

EXERCISE 2

To check on the dimensional balance or physical correctness of applied matrix equations, it is sometimes convenient to incorporate physical units within the scalar components of vectors and matrices. As an actual *cookbook* example, employ vector multiplication to determine the total cost of ingredients needed to bake a pound cake according to the following recipe*: "1 lb. cake flour, 1 lb. butter, 1 lb. sugar, 1 lb. eggs in shell (about 9–10), 2 tablespoons vanilla extract or brandy."

*Mrs. Simon Kander, *The Settlement Cookbook*, 13th ed., New York; Simon and Schuster, 1951.

B. Orthogonal Coordinate Systems and Transformations

Returning to the two-dimensional case, the relationships between cartesian and polar coordinates are as follows:

$$x = r\cos\theta, \qquad r = (x^2 + y^2)^{1/2},$$
$$y = r\sin\theta, \qquad \theta = \arctan\frac{y}{x}. \tag{1}$$

Set up unit* vectors $\hat{\mathbf{i}}$ and $\hat{\mathbf{j}}$ in the x, y system and unit vectors $\hat{\mathbf{e}}_r$ and $\hat{\mathbf{e}}_\theta$ in the r, θ system, as illustrated:

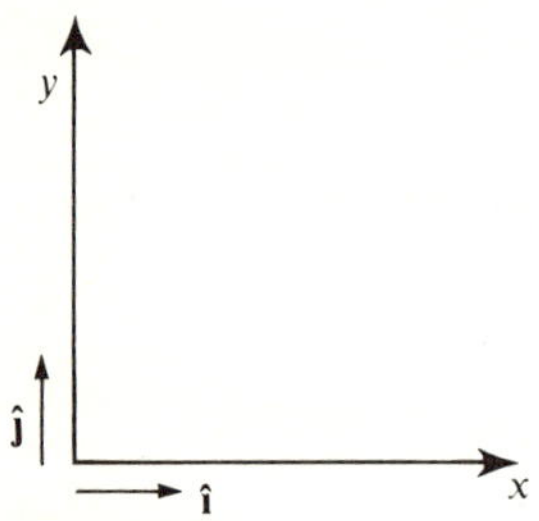

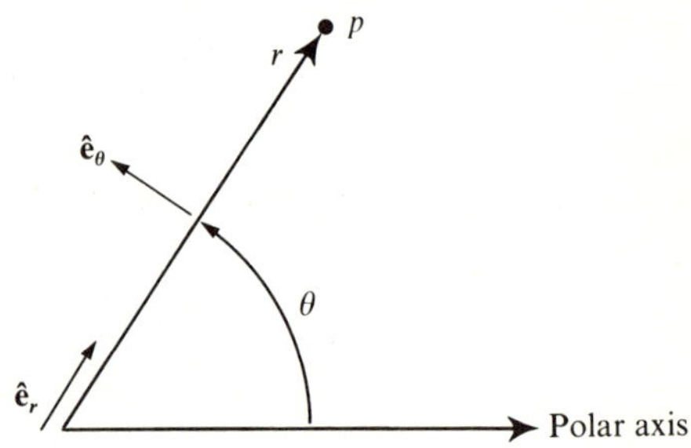

The "position vector" $\mathbf{r}$ of the point p is given by

$$\mathbf{r} = \hat{\mathbf{i}}x + \hat{\mathbf{j}}y = \hat{\mathbf{e}}_r r = \hat{\mathbf{i}}(r\cos\theta) + \hat{\mathbf{j}}(r\sin\theta). \tag{2}$$

Thus

$$\hat{\mathbf{e}}_r = (\hat{\mathbf{i}}\cos\theta + \hat{\mathbf{j}}\sin\theta). \tag{3}$$

The differential of $\mathbf{r}$ is given by

$$d\mathbf{r} = \hat{\mathbf{i}}\,dx + \hat{\mathbf{j}}\,dy = \frac{\partial\mathbf{r}}{\partial r}dr + \frac{\partial\mathbf{r}}{\partial\theta}d\theta = \mathbf{a}_r\,dr + \mathbf{a}_\theta\,d\theta. \tag{4}$$

Thus

$$\mathbf{a}_r = \frac{\partial\mathbf{r}}{\partial r}, \qquad \mathbf{a}_\theta = \frac{\partial\mathbf{r}}{\partial\theta}. \tag{5}$$

Here $\mathbf{a}_r$ and $\mathbf{a}_\theta$ form a set of "base or basis vectors," that is, a set of independent vectors that span the space. More will be said about such vectors later.

$$\mathbf{a}_r = \frac{\partial\mathbf{r}}{\partial r} = \hat{\mathbf{i}}\cos\theta + \hat{\mathbf{j}}\sin\theta \equiv \hat{\mathbf{e}}_r,$$

$$\mathbf{a}_\theta = \frac{\partial\mathbf{r}}{\partial\theta} = \hat{\mathbf{i}}(-r\sin\theta) + \hat{\mathbf{j}}(r\cos\theta) = r(-\hat{\mathbf{i}}\sin\theta + \hat{\mathbf{j}}\cos\theta) \neq \hat{\mathbf{e}}_\theta. \tag{6}$$

*The caret (^) appearing over a letter is common way of signifying a unit vector.

Notice that $\mathbf{a}_\theta$ is not the same as the unit vector in either its direction or its physical dimensions. To imbed the polar coordinates r and θ in a cylindrical coordinate system, introduce the z-coordinate and the unit vector $\mathbf{k} \equiv \mathbf{a}_z \equiv \hat{\mathbf{e}}_z$.

EXERCISE 3

A set of vectors $\{\mathbf{e}_1, \mathbf{e}_2, \ldots, \mathbf{e}_n\}$ is called a "basis on" or "for" a finite-dimensional vector space V if and only if (a) every vector $\mathbf{x}$ in V can be written as a linear combination of the $\mathbf{e}_j$, that is,

$$\mathbf{x} = \sum_{j=1}^{n} \alpha_j \mathbf{e}_j.$$

Briefly, this means the $\mathbf{e}$ vectors span the space, (b) the set $\{\mathbf{e}_1, \mathbf{e}_2, \cdots, \mathbf{e}_n\}$ must be independent.

Verify that

$$\mathbf{e}_1 = \begin{pmatrix} 1 \\ 0 \\ \cdot \\ \cdot \\ \cdot \\ 0 \end{pmatrix}, \quad \mathbf{e}_2 = \begin{pmatrix} 0 \\ 1 \\ \cdot \\ \cdot \\ \cdot \\ 0 \end{pmatrix}, \quad \cdots, \quad \mathbf{e}_n = \begin{pmatrix} 0 \\ 0 \\ \cdot \\ \cdot \\ \cdot \\ 1 \end{pmatrix}$$

are a basis for complex Euclidean n-space.

Quite often it is expedient to transform a vector from one coordinate system to another, by using certain transformation equations, such as $x = r\cos\theta$ and $y = r\sin\theta$, that relate the new and old sets of coordinates. In most familiar situations these basic equations relate two orthogonal systems, although in the general case neither the old nor the new system is required to be orthogonal.

Two steps are involved in changing coordinates:

Step A. The components of each vector in the first system must be converted into components in the second system.

Step B. The original coordinates remaining in the new components then must be replaced by new coordinates.

Step B is carried out by straightforward substitution. Step A is ideally an exercise in matrix algebra, but it can be done by substitution also. If only one of the two steps is carried out, the result is hard to use and difficult to interpret.

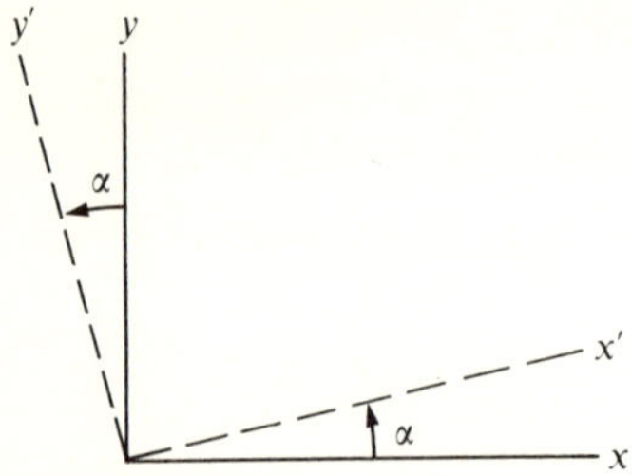

Figure 1. Rotation of the xy-plane about the origin.

The fortuitous coincidence that makes a cartesian coordinate indistinguishable from a cartesian component lulls many students into a false sense of security. The transformation from one three-dimensional cartesian system to another is deceptive in its simplicity and misleading because the coordinates x, y, z *may* be thought of as components of the position vector $\mathbf{r}$:

$$\mathbf{r} = x\hat{\mathbf{i}} + y\hat{\mathbf{j}} + z\hat{\mathbf{k}},$$

with length: (7)

$$\mathbf{r} = (x^2 + y^2 + z^2)^{1/2}.$$

Consider, for example, rotation of two-dimensional cartesian coordinates as shown in Fig. 1. The coordinate transformation called for in Fig. 1 is expressed by either one of the following systems of equations:

$$\begin{cases} x' = x\cos\alpha + y\sin\alpha, \\ y' = -x\sin\alpha + y\cos\alpha \end{cases} \tag{8}$$

or

$$\begin{cases} x = x'\cos\alpha - y'\sin\alpha, \\ y = x'\sin\alpha + y'\cos\alpha. \end{cases} \tag{9}$$

These equations are needed in any straightforward substitution as in step B.

If

$$\mathbf{r} = \{r_x, r_y\} = \{x, y\} \tag{10}$$

is thought of as a vector, then the preceding systems of equations may be written in matrix notation as follows:

$$\mathbf{r}' = \begin{pmatrix} x' \\ y' \end{pmatrix} = \begin{pmatrix} \cos\alpha & \sin\alpha \\ -\sin\alpha & \cos\alpha \end{pmatrix}\begin{pmatrix} x \\ y \end{pmatrix} = M\mathbf{r} \tag{11}$$

and

$$\mathbf{r} = \begin{pmatrix} x \\ y \end{pmatrix} = \begin{pmatrix} \cos\alpha & -\sin\alpha \\ \sin\alpha & \cos\alpha \end{pmatrix}\begin{pmatrix} x' \\ y' \end{pmatrix} = M^{-1}r'. \tag{12}$$

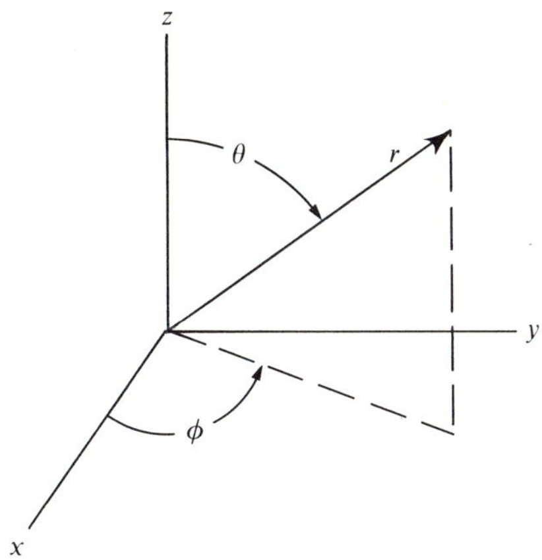

Figure 2. Three-dimensional coordinate system.

It is easy to verify that

$$MM^{-1} = M^{-1}M = I \tag{13}$$

where I symbolizes the so-called "identity matrix" by multiplying out:

$$\begin{pmatrix} \cos\alpha & \sin\alpha \\ -\sin\alpha & \cos\alpha \end{pmatrix}\begin{pmatrix} \cos\alpha & -\sin\alpha \\ \sin\alpha & \cos\alpha \end{pmatrix} = \begin{pmatrix} \cos^2\alpha + \sin^2\alpha & 0 \\ 0 & \sin^2\alpha + \cos^2\alpha \end{pmatrix}$$

$$= \begin{pmatrix} \cos\alpha & -\sin\alpha \\ \sin\alpha & \cos\alpha \end{pmatrix}\begin{pmatrix} \cos\alpha & \sin\alpha \\ -\sin\alpha & \cos\alpha \end{pmatrix} = \begin{pmatrix} 1 & 0 \\ 0 & 1 \end{pmatrix}. \tag{14}$$

By analogy with equation (12) any vector $\mathbf{V}$ can be expressed by

$$\mathbf{V} = \begin{pmatrix} V_x \\ V_y \end{pmatrix} = \begin{pmatrix} \cos\alpha & -\sin\alpha \\ \sin\alpha & \cos\alpha \end{pmatrix}\begin{pmatrix} V'_x \\ V'_y \end{pmatrix} = M^{-1}\mathbf{V}' \tag{15}$$

and any vector $\mathbf{V}'$ by

$$\mathbf{V}' = M\mathbf{V}. \tag{16}$$

Now consider the three-dimensional spherical coordinates r, θ, ϕ shown in Fig. 2. The transformation equations are

$$\left.\begin{aligned} x &= r\cos\phi\sin\theta \\ y &= r\sin\phi\sin\theta \\ z &= r\cos\theta \end{aligned}\right\} \quad \text{or} \quad \left\{\begin{aligned} r &= \sqrt{x^2 + y^2 + z^2} \\ \phi &= \tan^{-1}\left(\frac{y}{x}\right) \\ \theta &= \cos^{-1}\left(\frac{z}{\sqrt{x^2 + y^2 + z^2}}\right). \end{aligned}\right. \tag{17}$$

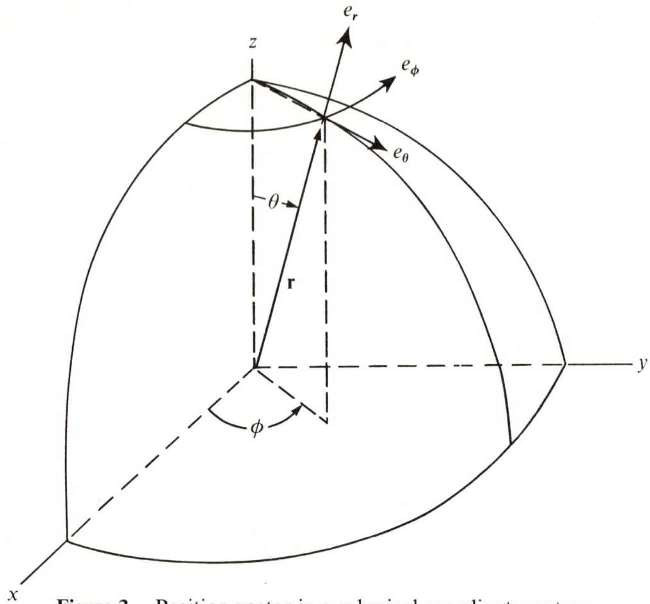

Figure 3. Position vector in a spherical coordinate system.

Here, of course, r, θ, and ϕ are not the components of the position vector $\mathbf{r}$. Transformation equations (17) may be used in straightforward substitution, but they do not yield a matrix of transformation M.

A linear transformation relationship between linear cartesian coordinates and curvilinear coordinates is globally impossible. *Note:* It is still an important requirement that the coordinate systems being dealt with are *orthogonal.*

In relating cartesian coordinates x, y, z to spherical coordinates r, θ, ϕ (see Fig. 3) a set of three orthogonal unit vectors $\hat{\mathbf{i}}$, $\hat{\mathbf{j}}$, $\mathbf{k}$ must be similarly related to a second set of three orthogonal unit vectors $\hat{\mathbf{e}}_r$, $\hat{\mathbf{e}}_\theta$, $\hat{\mathbf{e}}_\phi$, rotated in a complicated way with respect to the original cartesian vectors by the spherical coordinates. This rule is a key point to remember in searching for a matrix transformation approach.

Now, start with a three-dimensional vector $\mathbf{V}$:

$$\mathbf{V} = V_x\hat{\mathbf{i}} + V_y\hat{\mathbf{j}} + V_z\hat{\mathbf{k}} = \{V_x, V_y, V_z\}. \tag{18}$$

For simplicity, first study the problem in sections:

$$\begin{cases} x = r \sin\theta \cos\phi, \\ y = r \sin\theta \sin\phi, \\ z = r \cos\theta. \end{cases} \tag{19}$$

If $\mathbf{V} \equiv \{V_r, 0, 0\}$, then

$$\begin{cases} V_x = V_r \sin\theta \cos\phi, \\ V_y = V_r \sin\theta \sin\phi, \\ V_z = V_r \cos\theta. \end{cases} \tag{20}$$

And in matrix form:

$$\begin{pmatrix} V_x \\ V_y \\ V_z \end{pmatrix} = \begin{pmatrix} \sin\theta\cos\phi & A & B \\ \sin\theta\sin\phi & C & D \\ \cos\theta & E & F \end{pmatrix} \begin{pmatrix} V_r \\ 0 \\ 0 \end{pmatrix}, \tag{21}$$

where A, B, C, D, E, F are arbitrary for this transformation. If $\mathbf{V} \equiv \{0, 0, V_\phi\}$, then

$$\begin{cases} V_x = -V_\phi \sin\phi, \\ V_y = V_\phi \cos\phi, \\ V_z = 0, \end{cases} \tag{22}$$

and

$$\begin{pmatrix} V_x \\ V_y \\ V_z \end{pmatrix} = \begin{pmatrix} F & A & -\sin\phi \\ G & C & \cos\phi \\ H & E & 0 \end{pmatrix} \begin{pmatrix} 0 \\ 0 \\ V_\phi \end{pmatrix}, \tag{23}$$

where, once again, A, C, E, F, H are arbitrary. Then, if

$$\mathbf{V} \equiv \{V_r, 0, V_\phi\},$$

it follows that

$$\begin{pmatrix} V_x \\ V_y \\ V_z \end{pmatrix} = \begin{pmatrix} \sin\theta\cos\phi & A & -\sin\phi \\ \sin\theta\sin\phi & C & \cos\phi \\ \cos\theta & E & 0 \end{pmatrix} \begin{pmatrix} V_r \\ 0 \\ V_\phi \end{pmatrix}, \tag{24}$$

where now only A, C, and E are arbitrary. Finally consider

$$\mathbf{V} \equiv \{0, V_\theta, 0\}$$

for which two projections are required (as shown in Fig. 4). Then

$$\begin{cases} V_x = V_\theta \cos\theta \cos\phi, \\ V_y = V_\theta \cos\theta \sin\phi, \\ V_z = -V_\theta \sin\theta, \end{cases} \tag{25}$$

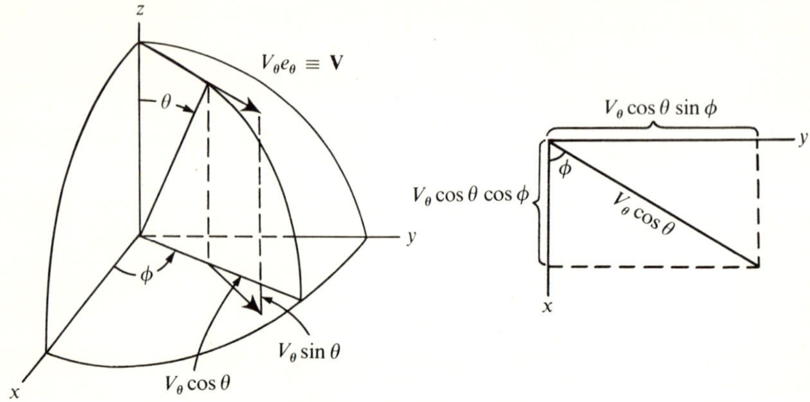

Figure 4. Projections of **V**.

so that

$$\begin{pmatrix} V_x \\ V_y \\ V_z \end{pmatrix} = \begin{pmatrix} E & \cos\theta\cos\phi & B \\ G & \cos\theta\sin\phi & D \\ H & -\sin\theta & F \end{pmatrix} \begin{pmatrix} 0 \\ V_\theta \\ 0 \end{pmatrix}, \tag{26}$$

with B, D, E, F, G, H arbitrary. Then in the most general case,

$$\mathbf{V} \equiv \{V_r, V_\theta, V_\phi\}$$

and

$$\begin{pmatrix} V_x \\ V_y \\ V_z \end{pmatrix} = \begin{pmatrix} \sin\theta\cos\phi & \cos\theta\cos\phi & -\sin\phi \\ \sin\theta\sin\phi & \cos\theta\sin\phi & \cos\phi \\ \cos\theta & -\sin\theta & 0 \end{pmatrix} \begin{pmatrix} V_r \\ V_\theta \\ V_\phi \end{pmatrix}, \tag{27}$$

or

$$\begin{pmatrix} V_x \\ V_y \\ V_z \end{pmatrix} = M \begin{pmatrix} V_r \\ V_\theta \\ V_\phi \end{pmatrix}. \tag{28}$$

Inversion is easy, because the matrix M satisfies the orthogonality condition $M^T = M^{-1}$. Therefore, using M transpose to get M inverse,

$$\begin{pmatrix} V_r \\ V_\theta \\ V_\phi \end{pmatrix} = \begin{pmatrix} \sin\theta\cos\phi & \sin\theta\sin\phi & \cos\theta \\ \cos\theta\cos\phi & \cos\theta\sin\phi & -\sin\theta \\ -\sin\theta & \cos\phi & 0 \end{pmatrix} \begin{pmatrix} V_x \\ V_y \\ V_z \end{pmatrix}, \tag{29}$$

or

$$\begin{pmatrix} V_r \\ V_\theta \\ V_\phi \end{pmatrix} = M^{-1} \begin{pmatrix} V_x \\ V_y \\ V_z \end{pmatrix}. \tag{30}$$

It is again easy to show that

$$MM^{-1} = M^{-1}M = I, \tag{31}$$

but this is not the complete story. In the limit as $\theta \to 0$, $\phi \to 0$, M should be the identity matrix; however, it is jumbled in its present form, as seen in the following:

$$\lim_{\substack{\theta \to 0 \\ \phi \to 0}} M = \begin{pmatrix} \sin 0 \cos 0 & \cos 0 \cos 0 & -\sin 0 \\ \sin 0 \sin 0 & \cos 0 \sin 0 & \cos 0 \\ \cos 0 & -\sin 0 & 0 \end{pmatrix} = \begin{pmatrix} 0 & 1 & 0 \\ 0 & 0 & 1 \\ 1 & 0 & 0 \end{pmatrix} = P \neq I. \tag{32}$$

Matrices like P and its powers P^2 and P^3 are called "permutation matrices." Note that P is an identity matrix of sorts, so luckily a cyclic permutation of the columns of M and elements of $\{V_r, V_\theta, V_\phi\}$ leads to a natural order of the components, as follows:

$$\begin{pmatrix} V_x \\ V_y \\ V_z \end{pmatrix} = \begin{pmatrix} \cos\theta\cos\phi & -\sin\phi & \sin\theta\cos\phi \\ \cos\theta\sin\phi & \cos\phi & \sin\theta\sin\phi \\ -\sin\theta & 0 & \cos\theta \end{pmatrix} \begin{pmatrix} V_\theta \\ V_\phi \\ V_r \end{pmatrix}$$

$$\xrightarrow[\theta \to 0]{} \begin{pmatrix} \cos\phi & -\sin\phi & 0 \\ \sin\phi & \cos\phi & 0 \\ 0 & 0 & 1 \end{pmatrix} \begin{pmatrix} V_{\theta \to 0} \\ V_\phi \\ V_r \end{pmatrix} \xrightarrow[\phi \to 0]{} \begin{pmatrix} 1 & 0 & 0 \\ 0 & 1 & 0 \\ 0 & 0 & 1 \end{pmatrix} \begin{pmatrix} V_{\theta \to 0} \\ V_{\phi \to 0} \\ V_r \end{pmatrix}. \tag{33}$$

Thus

$$M \to I \quad \text{as} \quad \theta \text{ and } \phi \to 0. \tag{34}$$

It is difficult to go further with the straightforward substitution of the coordinates in Step B without knowing the explicit form of V_x, V_y, and V_z *or* V_r, V_ϕ, and V_θ, thus the calculation must end with the determination of the transformation matrix.

EXERCISES

4. A similar transformation problem is encountered for Euler angles in the rotation of bodies in mechanics. Explain.

5. (*a*) Generate the two-dimensional transformation between two spherical coordinate systems ϕ, θ and ϕ', θ' that are inclined to each other by a fixed set of angles α, β.

(*b*) Apply this to the problem of converting the position of a celestial object from the alt azimuth coordinate system to right ascension and declination.

(*c*) The angle α can be related to the observer's latitude. Make the angle β time dependent to describe the rotation of the Earth on its axis. Assume a spherical Earth.

A vector may be transformed from one orthogonal system of basis vectors to another by using a more general procedure than Step *A*. This general transformation then may be applied to either a simple, two-dimensional rotation of axes or a complicated transformation from cartesian coordinates to spherical coordinates.

Consider a transformation that takes the vector $\mathbf{V}$ from an original system with the base vectors $\hat{\mathbf{e}}_1$, $\hat{\mathbf{e}}_2$, $\hat{\mathbf{e}}_3$ into the vector $\mathbf{V}'$ in a "new" system with base vectors $\hat{\mathbf{e}}'_1$, $\hat{\mathbf{e}}'_2$, $\hat{\mathbf{e}}'_3$. This transformation, in matrix notation, may be written $\mathbf{V}' = M\mathbf{V}$, where M is again to be determined (see Fig. 5). The components of $\mathbf{V}'$ entail contributions from each component of $\mathbf{V}$, as follows:

$$
\begin{aligned}
V'_1 &= V_1\hat{\mathbf{e}}_1\cdot\hat{\mathbf{e}}'_1 + V_2\hat{\mathbf{e}}_2\cdot\hat{\mathbf{e}}'_1 + V_3\hat{\mathbf{e}}_3\cdot\hat{\mathbf{e}}'_1,\\
V'_2 &= V_1\hat{\mathbf{e}}_1\cdot\hat{\mathbf{e}}'_2 + V_2\hat{\mathbf{e}}_2\cdot\hat{\mathbf{e}}'_2 + V_3\hat{\mathbf{e}}_3\cdot\hat{\mathbf{e}}'_2,\\
V'_3 &= V_1\hat{\mathbf{e}}_1\cdot\hat{\mathbf{e}}'_3 + V_2\hat{\mathbf{e}}_2\cdot\hat{\mathbf{e}}'_3 + V_3\hat{\mathbf{e}}_3\cdot\hat{\mathbf{e}}'_3.
\end{aligned}
\tag{35}
$$

Or, in matrix form:

$$
\mathbf{V}' = \begin{pmatrix} V'_1 \\ V'_2 \\ V'_3 \end{pmatrix} = \begin{pmatrix} \hat{\mathbf{e}}_1\cdot\hat{\mathbf{e}}'_1 & \hat{\mathbf{e}}_1\cdot\hat{\mathbf{e}}'_2 & \hat{\mathbf{e}}_1\cdot\hat{\mathbf{e}}'_3 \\ \hat{\mathbf{e}}_2\cdot\hat{\mathbf{e}}'_1 & \hat{\mathbf{e}}_2\cdot\hat{\mathbf{e}}'_2 & \hat{\mathbf{e}}_2\cdot\hat{\mathbf{e}}'_3 \\ \hat{\mathbf{e}}_3\cdot\hat{\mathbf{e}}'_1 & \hat{\mathbf{e}}_3\cdot\hat{\mathbf{e}}'_2 & \hat{\mathbf{e}}_3\cdot\hat{\mathbf{e}}'_3 \end{pmatrix} \begin{pmatrix} V_1 \\ V_2 \\ V_3 \end{pmatrix}. \tag{36}
$$

If the z-axis is held fixed, then for a counterclockwise rotation of a cartesian system through an angle α, the inner products of the base vectors become

$$
\begin{aligned}
&\hat{\mathbf{e}}_1\cdot\hat{\mathbf{e}}'_1 = \cos\alpha, &&\hat{\mathbf{e}}_2\cdot\hat{\mathbf{e}}'_1 = \sin\alpha, &&\hat{\mathbf{e}}_3\cdot\hat{\mathbf{e}}'_1 = \cos 90^\circ = 0,\\
&\hat{\mathbf{e}}_1\cdot\hat{\mathbf{e}}'_2 = -\sin\alpha, &&\hat{\mathbf{e}}_2\cdot\hat{\mathbf{e}}'_2 = \cos\alpha, &&\hat{\mathbf{e}}_3\cdot\hat{\mathbf{e}}'_2 = \cos 90^\circ = 0,\\
&\hat{\mathbf{e}}_1\cdot\hat{\mathbf{e}}'_3 = \cos 90^\circ = 0, &&\hat{\mathbf{e}}_2\cdot\hat{\mathbf{e}}'_3 = \cos 90^\circ = 0, \quad\text{and} &&\hat{\mathbf{e}}_3\cdot\hat{\mathbf{e}}'_3 = \cos 0^\circ = 1.
\end{aligned}
\tag{37}
$$

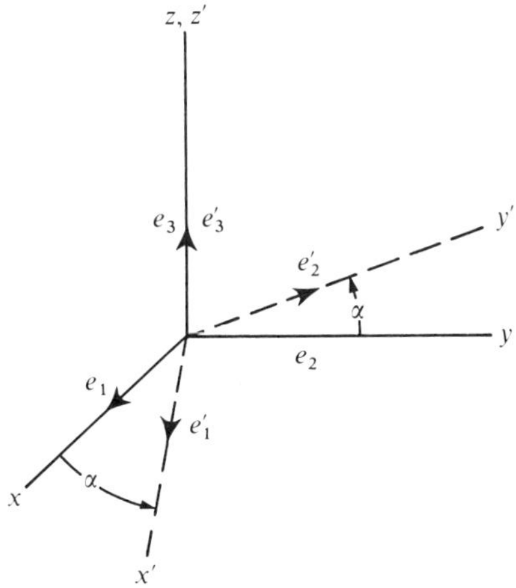

Figure 5. Simple rotation of cartesian coordinates.

Thus

$$M = \begin{pmatrix} \cos\alpha & \sin\alpha & 0 \\ -\sin\alpha & \cos\alpha & 0 \\ 0 & 0 & 1 \end{pmatrix}. \tag{38}$$

To revert to the original coordinate system, that is, to perform a clockwise rotation through an angle α holding the z-axis fixed, M^T, the transpose of M, is used since for an orthogonal matrix it is the same as the inverse. To transform a vector $\mathbf{V}$ from a cartesian system to a spherical system:

$$\begin{array}{llll} V_1 = V_x & V_1' = V_r & \hat{\mathbf{e}}_1 = \hat{\mathbf{i}} & \hat{\mathbf{e}}_1' = \hat{\mathbf{e}}_r \\ V_2 = V_y & V_2' = V_\theta & \hat{\mathbf{e}}_2 = \hat{\mathbf{j}} & \hat{\mathbf{e}}_2' = \hat{\mathbf{e}}_\theta \\ V_3 = V_z & V_3' = V_\phi & \hat{\mathbf{e}}_3 = \hat{\mathbf{k}} & \hat{\mathbf{e}}_3' = \hat{\mathbf{e}}_\phi. \end{array} \tag{39}$$

The matrix elements for the transformation $\mathbf{V}' = M\mathbf{V}$ are evaluated in the same way as before:

$$\begin{aligned} \hat{\mathbf{e}}_1 \cdot \hat{\mathbf{e}}_1' &= \hat{\mathbf{i}} \cdot \hat{\mathbf{e}}_r &&= \sin\theta\cos\phi, \\ \hat{\mathbf{e}}_1 \cdot \hat{\mathbf{e}}_2' &= \hat{\mathbf{i}} \cdot \hat{\mathbf{e}}_\theta &&= \cos\theta\cos\phi, \\ \hat{\mathbf{e}}_1 \cdot \hat{\mathbf{e}}_3' &= \hat{\mathbf{i}} \cdot \hat{\mathbf{e}}_\phi &&= -\sin\phi; \\ \\ \hat{\mathbf{e}}_2 \cdot \hat{\mathbf{e}}_1' &= \hat{\mathbf{j}} \cdot \hat{\mathbf{e}}_r &&= \sin\theta\sin\phi, \\ \hat{\mathbf{e}}_2 \cdot \hat{\mathbf{e}}_2' &= \hat{\mathbf{j}} \cdot \hat{\mathbf{e}}_\theta &&= \cos\theta\sin\phi, \\ \hat{\mathbf{e}}_2 \cdot \hat{\mathbf{e}}_3' &= \hat{\mathbf{j}} \cdot \hat{\mathbf{e}}_\phi &&= \cos\phi; \end{aligned} \tag{40}$$

and

$$\hat{\mathbf{e}}_3 \cdot \hat{\mathbf{e}}'_1 = \hat{\mathbf{k}} \cdot \hat{\mathbf{e}}_r = \cos\theta,$$
$$\hat{\mathbf{e}}_3 \cdot \hat{\mathbf{e}}'_2 = \hat{\mathbf{k}} \cdot \hat{\mathbf{e}}_\theta = -\sin\theta,$$
$$\hat{\mathbf{e}}_3 \cdot \hat{\mathbf{e}}'_3 = \hat{\mathbf{k}} \cdot \hat{\mathbf{e}}_\phi = 0.$$

Thus

$$M = \begin{pmatrix} \sin\theta\cos\phi & \sin\theta\sin\phi & \cos\theta \\ \cos\theta\cos\phi & \cos\theta\sin\phi & -\sin\theta \\ -\sin\phi & \cos\phi & 0 \end{pmatrix}. \tag{41}$$

C. Orthogonality and Normalization by the Gram-Schmidt Method

Given a set of nonorthogonal basis vectors or coordinate axes, it is always possible to generate an orthogonal set. Let the given nonorthogonal set $\mathbf{e}_i$ be normalized to the set $\hat{\mathbf{e}}_i$. As shown in Fig. 6, vector $\hat{\mathbf{e}}_1$ can be used unchanged in the orthogonal set of coordinates.

To obtain the projection shown in Fig. 6c, form the scalar or dot product of $\hat{\mathbf{e}}_1$ and $\hat{\mathbf{e}}_2$:

$$\hat{\mathbf{e}}_2 \cdot \hat{\mathbf{e}}_1 = \hat{\mathbf{e}}_1 \cdot \hat{\mathbf{e}}_2 = \cos\theta. \tag{1}$$

Now multiply expression (1) by the unit vector $\hat{\mathbf{e}}_1$:

$$[\hat{\mathbf{e}}_1 \cdot \hat{\mathbf{e}}_2]\hat{\mathbf{e}}_1 = \cos\theta\, \hat{\mathbf{e}}_1, \tag{2}$$

a vector of length $\cos\theta$ in the $\hat{\mathbf{e}}_1$ direction. By vector subtraction:

$$\mathbf{V} = \hat{\mathbf{e}}_2 - \cos\theta\, \hat{\mathbf{e}}_1 = \hat{\mathbf{e}}_2 - (\hat{\mathbf{e}}_2 \cdot \hat{\mathbf{e}}_1)\hat{\mathbf{e}}_1. \tag{3}$$

Since $\mathbf{V}$ is perpendicular to $\hat{\mathbf{e}}_1$:

$$\mathbf{V} \cdot \hat{\mathbf{e}}_1 = \hat{\mathbf{e}}_2 \cdot \hat{\mathbf{e}}_1 - (\hat{\mathbf{e}}_2 \cdot \hat{\mathbf{e}}_1)\hat{\mathbf{e}}_1 \cdot \hat{\mathbf{e}}_1 = 0. \tag{4}$$

Thus an acceptable system of base vectors is $\hat{\mathbf{e}}_1$ and $\mathbf{V}$. To normalize $\mathbf{V}$ requires the magnitude or absolute value of $\mathbf{V}$, which is found as follows:

$$\begin{aligned} |\mathbf{V}|^2 = \mathbf{V} \cdot \mathbf{V} = V^2 &= [\hat{\mathbf{e}}_2 - (\hat{\mathbf{e}}_2 \cdot \hat{\mathbf{e}}_1)\hat{\mathbf{e}}_1] \cdot [\hat{\mathbf{e}}_2 - (\hat{\mathbf{e}}_2 \cdot \hat{\mathbf{e}}_1)\hat{\mathbf{e}}_1] \\ &= \hat{\mathbf{e}}_2 \cdot \hat{\mathbf{e}}_2 + (\hat{\mathbf{e}}_2 \cdot \hat{\mathbf{e}}_1)^2 \hat{\mathbf{e}}_1 \cdot \hat{\mathbf{e}}_1 - 2(\hat{\mathbf{e}}_2 \cdot \hat{\mathbf{e}}_1)\hat{\mathbf{e}}_1 \cdot \hat{\mathbf{e}}_2 \\ &= [1 - (\hat{\mathbf{e}}_2 \cdot \hat{\mathbf{e}}_1)^2] = [1 - \cos^2\theta] = \sin^2\theta. \end{aligned} \tag{3}$$

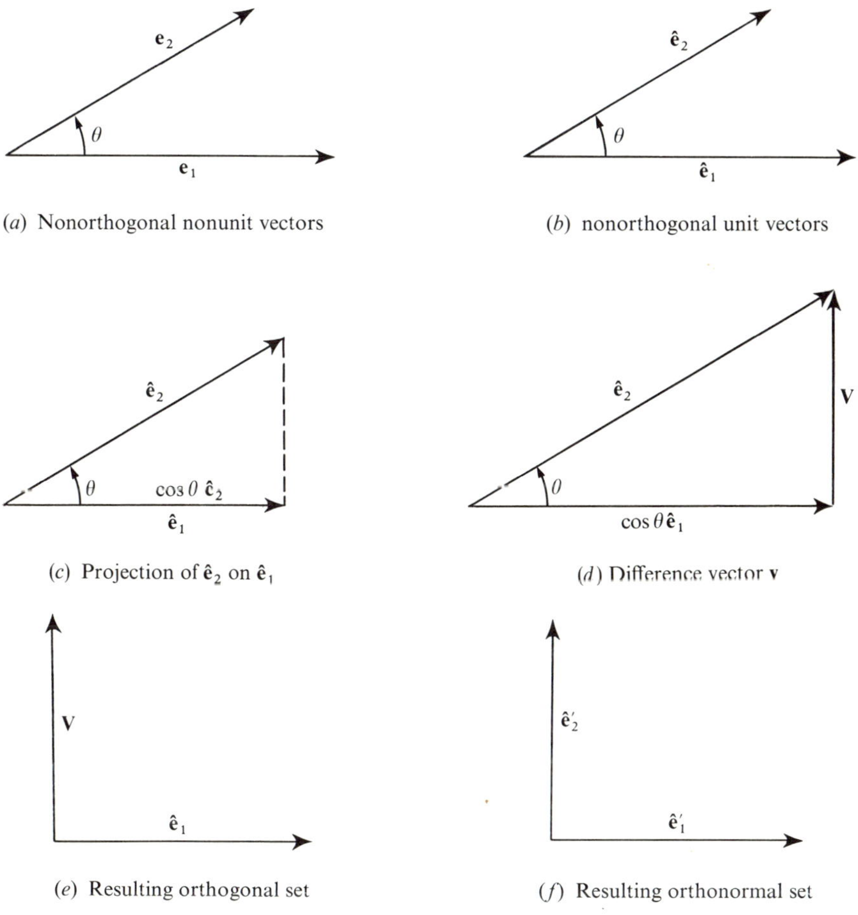

Figure 6. Two-dimensional orthogonalization.

Then:

$$|\mathbf{V}| = \sqrt{1 - (\hat{\mathbf{e}}_2 \cdot \hat{\mathbf{e}}_1)^2} = \sin\theta, \tag{6}$$

and the vector

$$\frac{\mathbf{V}}{|\mathbf{V}|} = \frac{\hat{\mathbf{e}}_2 - (\hat{\mathbf{e}}_2 \cdot \hat{\mathbf{e}}_1)\hat{\mathbf{e}}_1}{\sqrt{1 - (\hat{\mathbf{e}}_2 \cdot \hat{\mathbf{e}}_1)^2}} = \frac{\hat{\mathbf{e}}_2 - \cos\theta\hat{\mathbf{e}}_1}{\sin\theta} = \hat{\mathbf{e}}_2' \tag{7}$$

is a unit vector. Thus $\hat{\mathbf{e}}_1$ and $\hat{\mathbf{e}}_2'$ form an "orthonormal" (orthogonal + normal) set of vectors.

Then use

$$\hat{\mathbf{e}}_1' = \hat{\mathbf{e}}_1 \qquad \hat{\mathbf{e}}_2' = \frac{\hat{\mathbf{e}}_2 - (\hat{\mathbf{e}}_2 \cdot \hat{\mathbf{e}}_1)\hat{\mathbf{e}}_1}{\sqrt{1 - (\hat{\mathbf{e}}_2 \cdot \hat{\mathbf{e}}_1)^2}}. \tag{8}$$

The extension to three or more coordinates is straight-forward. Assume that a third vector $\hat{\mathbf{e}}_3$ is given; then form

$$\mathbf{W} = \hat{\mathbf{e}}_3 - (\hat{\mathbf{e}}_3 \cdot \hat{\mathbf{e}}_2)\hat{\mathbf{e}}_2 - (\hat{\mathbf{e}}_3 \cdot \hat{\mathbf{e}}_1)\hat{\mathbf{e}}_1, \tag{9}$$

a vector orthogonal to both $\hat{\mathbf{e}}_1$ and $\hat{\mathbf{e}}_2$. Next form

$$\mathbf{W} \cdot \hat{\mathbf{e}}_1 = \hat{\mathbf{e}}_3 \cdot \hat{\mathbf{e}}_1 - (\hat{\mathbf{e}}_3 \cdot \hat{\mathbf{e}}_1)\hat{\mathbf{e}}_1 \cdot \hat{\mathbf{e}}_1 = 0 \tag{10}$$

and

$$\mathbf{W} \cdot \hat{\mathbf{e}}_2 = \hat{\mathbf{e}}_3 \cdot \hat{\mathbf{e}}_2 - (\hat{\mathbf{e}}_3 \cdot \hat{\mathbf{e}}_1)\hat{\mathbf{e}}_2 \cdot \hat{\mathbf{e}}_2 = 0. \tag{11}$$

The normalizing factor, as before, is obtained by using the length of the vector as a divisor:

$$\begin{aligned}
|\mathbf{W}|^2 &= \mathbf{W} \cdot \mathbf{W} \\
&= [\hat{\mathbf{e}}_3 - (\hat{\mathbf{e}}_3 \cdot \hat{\mathbf{e}}_2)\hat{\mathbf{e}}_2 - (\hat{\mathbf{e}}_3 \cdot \hat{\mathbf{e}}_1)\hat{\mathbf{e}}_1] \cdot [\hat{\mathbf{e}}_3 - (\hat{\mathbf{e}}_3 \cdot \hat{\mathbf{e}}_2)\hat{\mathbf{e}}_2 - (\hat{\mathbf{e}}_3 \cdot \hat{\mathbf{e}}_1)\hat{\mathbf{e}}_1] \\
&= \hat{\mathbf{e}}_3 \cdot \hat{\mathbf{e}}_3 + (\hat{\mathbf{e}}_3 \cdot \hat{\mathbf{e}}_2) \cdot (\hat{\mathbf{e}}_3 \cdot \hat{\mathbf{e}}_2) + (\hat{\mathbf{e}}_3 \cdot \hat{\mathbf{e}}_1)(\hat{\mathbf{e}}_3 \cdot \hat{\mathbf{e}}_1) \\
&\quad + 2(\hat{\mathbf{e}}_3 \cdot \hat{\mathbf{e}}_2)(\hat{\mathbf{e}}_3 \cdot \hat{\mathbf{e}}_1)\hat{\mathbf{e}}_2 \cdot \hat{\mathbf{e}}_1 - 2(\hat{\mathbf{e}}_3 \cdot \hat{\mathbf{e}}_2)\hat{\mathbf{e}}_3 \cdot \hat{\mathbf{e}}_2 - 2(\hat{\mathbf{e}}_3 \cdot \hat{\mathbf{e}}_1)\hat{\mathbf{e}}_3 \cdot \hat{\mathbf{e}}_1 \\
&= [1 - (\hat{\mathbf{e}}_3 \cdot \hat{\mathbf{e}}_2)^2 - (\hat{\mathbf{e}}_3 \cdot \hat{\mathbf{e}}_1)^2].
\end{aligned} \tag{12}$$

Thus the third unit vector becomes

$$\hat{\mathbf{e}}_3' = \frac{\hat{\mathbf{e}}_3 - (\hat{\mathbf{e}}_3 \cdot \hat{\mathbf{e}}_2)\hat{\mathbf{e}}_2 - (\hat{\mathbf{e}}_3 \cdot \hat{\mathbf{e}}_1)\hat{\mathbf{e}}_1}{\sqrt{1 - (\hat{\mathbf{e}}_3 \cdot \hat{\mathbf{e}}_2)^2 - (\hat{\mathbf{e}}_3 \cdot \hat{\mathbf{e}}_1)^2}}. \tag{13}$$

Although complicated in appearance, the process thus far has been basically the following:

$$\begin{aligned}
\hat{\mathbf{e}}_3' &= A\hat{\mathbf{e}}_3 + B\hat{\mathbf{e}}_2 + C\hat{\mathbf{e}}_1 \\
\hat{\mathbf{e}}_2' &= \phantom{A\hat{\mathbf{e}}_3 + {}} D\hat{\mathbf{e}}_2 + E\hat{\mathbf{e}}_1 \\
\hat{\mathbf{e}}_1' &= \phantom{A\hat{\mathbf{e}}_3 + D\hat{\mathbf{e}}_2 + {}} \hat{\mathbf{e}}_1.
\end{aligned} \tag{14}$$

Finally, this may be written

$$\begin{matrix} \hat{\mathbf{e}}_3' \\ \hat{\mathbf{e}}_2' \\ \hat{\mathbf{e}}_1' \end{matrix} = \begin{pmatrix} A & B & C \\ 0 & D & E \\ 0 & 0 & 1 \end{pmatrix} \begin{pmatrix} \hat{\mathbf{e}}_3 \\ \hat{\mathbf{e}}_2 \\ \hat{\mathbf{e}}_1 \end{pmatrix} \tag{15}$$

or

$$\hat{\mathbf{e}}' = M\hat{\mathbf{e}}. \tag{16}$$

Thus the Gram-Schmidt orthogonalization method is equivalent to a matrix transformation between orthogonal and nonorthogonal coordinates.

EXERCISES

6. The determinant of an n-square matrix A, denoted by various authors as $d(A)$, $D(A)$, $|A|$, or $\Delta(A)$, is defined by

$$\det A = \sum \operatorname{sgn} \sigma \prod_{i=1}^{n} a_{i\sigma(i)},$$

where σ runs over all permutations of the integers $1, 2, \cdots, n$ and the function $\operatorname{sgn} \sigma$ is 1 if the permutation σ is even, and -1 otherwise. Thus an n-square determinant is the sum of n-factorial products, each product containing n matrix elements as factors.

Prove that in a rectangular xyz-coordinate system

$$\begin{vmatrix} x & y & z & 1 \\ x_1 & y_1 & z_1 & 1 \\ x_2 & y_2 & z_2 & 1 \\ x_3 & y_3 & z_3 & 1 \end{vmatrix} = 0$$

is the equation of a plane through the points (x_1, y_1, z_1), (x_2, y_2, z_2), and (x_3, y_3, z_3). Write this equation in expanded form.

7. The following linear transformation is used to study the distribution of the sample variance in statistical sampling from a normal population:

$$\begin{cases} y_1 = \dfrac{1}{\sqrt{1\cdot 2}}(x_1 - x_2), \\ y_2 = \dfrac{1}{\sqrt{2\cdot 3}}(x_1 + x_2 - x_3), \\ \quad\vdots \qquad\qquad \vdots \\ y_i = \dfrac{1}{\sqrt{i(i+1)}}(x_1 + \cdots + x_i - x_{i+1}),\ i = 1, 2, \cdots, n-1, \\ \quad\vdots \qquad\qquad \vdots \\ y_n = \dfrac{1}{\sqrt{n}}(x_1 + x_2 + \cdots + x_n) \end{cases}$$

a. Determine the elements of the matrix $A = (a_{ij})$ such that $\mathbf{y} = A\mathbf{x}$.
b. What is det A? (*Hint:* First remove a scalar factor.)
c. Do the rows of A form a system of orthogonal vectors?

A set of vectors $\{x_j\}, j = 1, 2, \cdots, n$ is said to be "linearly independent" if no possible choice of scalars α_j, not all zero, causes the linear combination $\sum \alpha_j x_j = \alpha_1 x_1 + \alpha_2 x_2 + \cdots + \alpha_n x_n$ to vanish.* Let $x_1, x_r, \cdots, x_2$ be a set of linearly independent vectors. Then there exist constants a_{ij} in a lower triangular array:

$$\begin{array}{llll} a_{11} & & & \\ a_{21} & a_{22} & & \\ a_{31} & a_{32} & a_{33} & \\ \cdot & \cdot & \cdot & \\ \cdot & \cdot & \cdot & \\ \cdot & \cdot & \cdot & \\ a_{n1} & a_{n2} & a_3 & \cdots a_{nn}, \end{array} \tag{17}$$

*In Euclidean three-space the vectors

$$\begin{pmatrix} 3 \\ 2 \\ -7 \end{pmatrix} \qquad \begin{pmatrix} 6 \\ 4 \\ -14 \end{pmatrix}$$

are dependent since

$$2\begin{pmatrix} 3 \\ 2 \\ -7 \end{pmatrix} - 1.\begin{pmatrix} 6 \\ 4 \\ -14 \end{pmatrix} = 0;$$

the vectors

$$\begin{pmatrix} 1 \\ 0 \\ 0 \end{pmatrix} \text{ and } \begin{pmatrix} 0 \\ 1 \\ 0 \end{pmatrix}$$

are independent since

$$\alpha_1\begin{pmatrix} 1 \\ 0 \\ 0 \end{pmatrix} + \alpha_2\begin{pmatrix} 0 \\ 1 \\ 0 \end{pmatrix} = \begin{pmatrix} \alpha_1 \\ \alpha_2 \\ 0 \end{pmatrix}$$

can be zero only if both α_1 and α_2 equal zero.

such that the set of vectors

$$\begin{aligned}
\mathbf{x}_1^* &= a_{11}\mathbf{x}_1 \\
\mathbf{x}_2^* &= a_{21}\mathbf{x}_1 + a_{22}\mathbf{x}_2 \\
\mathbf{x}_3^* &= a_{31}\mathbf{x}_1 + a_{32}\mathbf{x}_2 + a_{33}\mathbf{x}_3 \\
&\cdots \\
\mathbf{x}_n^* &= a_{n1}\mathbf{x}_1 + a_{n2}\mathbf{x}_2 + a_{n3}\mathbf{x}_3 + \cdots + a_{nn}\mathbf{x}_n
\end{aligned}$$

forms an orthogonal system.

The appropriate explicit representation of the Gram-Schmidt process, geared for computer use,* employs so-called "Gram determinants." In an n-dimensional Euclidean vector space E_n, composed of vectors

$$\mathbf{x}_1 : (x_{11}, x_{12}, \cdots, x_{1n}), \quad \mathbf{x}_2 : (x_{21}, x_{22}, \cdots, x_{2n}),$$

and so forth, the determinant

$$G - G(\mathbf{x}_1, \mathbf{x}_2, \cdots, \mathbf{x}_n) = |xx^T| - \begin{vmatrix} \mathbf{x}_1 \cdot \mathbf{x}_1 & \mathbf{x}_1 \cdot \mathbf{x}_2 \cdots & \mathbf{x}_1 \cdot \mathbf{x}_n \\ \mathbf{x}_2 \cdot \mathbf{x}_1 & \mathbf{x}_2 \cdot \mathbf{x}_2 \cdots & \mathbf{x}_2 \cdot \mathbf{x}_n \\ \cdot & \cdot & \cdot \\ \cdot & \cdot & \cdot \\ \cdot & \cdot & \cdot \\ \mathbf{x}_n \cdot \mathbf{x}_1 & \mathbf{x}_n \cdot \mathbf{x}_2 \cdots & \mathbf{x}_n \cdot \mathbf{x}_n \end{vmatrix} - |x|^2 \geq 0, \tag{19}$$

where $|x|$ is the value of the determinant:

$$|x| = \begin{vmatrix} x_{11} & x_{12} \cdots & x_{1n} \\ x_{21} & x_{22} \cdots & x_{2n} \\ \cdot & \cdot & \cdot \\ \cdot & \cdot & \cdot \\ \cdot & \cdot & \cdot \\ x_{n1} & x_{n2} \cdots & x_{nn} \end{vmatrix} \tag{20}$$

called the "Gram determinant" or "Gramian."

8. What concrete meaning has the Gramian for a parallelopiped whose edges are formed by the vectors $\mathbf{x}_1$ to $\mathbf{x}_n$?

*P. Davis, "Orthonormalization Codes in Numerical Analysis," *Survey of Numerical Analysis* (J. Todd, editor), McGraw-Hill (1962); John R. Rice, "Experiments on Gram-Schmidt Orthonormalization," *Math. Comp.*, **20,** 325–328 (1966).

9. Prove the following theorem: The vectors $\mathbf{x}_1, \mathbf{x}_2, \cdots, \mathbf{x}_n$ are linearly dependent if and only if $G = 0$.

10. Given the vectors $(4, -1, 2)$ and $(7, 3, 1)$, find a vector normal to their plane.

11. Express the vector $(3, 1, 4)$ as a linear combination of the vectors $(1, 2, 1)$, $(0, 3, 2)$, and $(2, 5, 3)$.

12. Let V be a vector space spanned by the vectors

$$\mathbf{x}_1 = \begin{pmatrix} 1 \\ -1 \\ 0 \\ 1 \\ 2 \end{pmatrix}, \quad \mathbf{x}_2 = \begin{pmatrix} 2 \\ 0 \\ 4 \\ 4 \\ 4 \end{pmatrix}, \quad \mathbf{x}_3 = \begin{pmatrix} 1 \\ -2 \\ -2 \\ 0 \\ 2 \end{pmatrix}, \quad \text{and} \quad \mathbf{x}_4 = \begin{pmatrix} 0 \\ 5 \\ -1 \\ 3 \\ 1 \end{pmatrix}.$$

Apply the Gram-Schmidt method to $\mathbf{x}_1, \mathbf{x}_2, \mathbf{x}_3, \mathbf{x}_4$ in this order to obtain an orthonormal basis for V.

2 Matrix Algebra

A. Simultaneous Linear Algebraic Equations

Since the right-hand side is zero, the matrix equation

$$A\mathbf{x} = 0, \tag{1}$$

where A is an n-square matrix and $\mathbf{x}$ is the unknown n by 1 column vector, is termed a "homogeneous equation." Its treatment is illustrated by letting A be a general 2 by 2 matrix:

$$A = \begin{pmatrix} a & b \\ c & d \end{pmatrix}, \tag{2}$$

with elements a, b, c, d taken from the real field.

Case 1. $a = b = c = d = 0$.

$$\begin{pmatrix} 0 & 0 \\ 0 & 0 \end{pmatrix}\begin{pmatrix} x_1 \\ x_2 \end{pmatrix} = 0 \qquad \begin{cases} 0x_1 + 0x_2 = 0 \\ 0x_1 + 0x_2 = 0. \end{cases} \tag{3}$$

Consequently any $(x_1, x_2)^T$ is a solution of equation (1).

Case 2. $\det A = ad - bc \neq 0$.

$$\begin{pmatrix} a & b \\ c & d \end{pmatrix}\begin{pmatrix} x_1 \\ x_2 \end{pmatrix} = 0 \qquad \begin{cases} ax_1 + bx_2 = 0 \\ cx_1 + dx_2 = 0. \end{cases} \tag{4}$$

Clearly a and c are not both zero. With no loss of generality, assume that $a \neq 0$:

$$\begin{pmatrix} a & b \\ c & d \end{pmatrix} \to \begin{pmatrix} 1 & \frac{b}{a} \\ c & d \end{pmatrix} \to \begin{pmatrix} 1 & \frac{b}{a} \\ 0 & d - \frac{cb}{a} \end{pmatrix} \to \begin{pmatrix} 1 & \frac{b}{a} \\ 0 & 1 \end{pmatrix} \to \begin{pmatrix} 1 & 0 \\ 0 & 1 \end{pmatrix}$$

$$\to \begin{pmatrix} 1 & 0 \\ 0 & 1 \end{pmatrix}\begin{pmatrix} x_1 \\ x_2 \end{pmatrix} = \begin{pmatrix} 0 \\ 0 \end{pmatrix} \tag{5}$$

Consequently, $\mathbf{x} = (0, 0)^T$.

Case 3. $\det A = ad - bc = 0$.

$$\begin{pmatrix} a & b \\ c & d \end{pmatrix} \to \begin{pmatrix} 1 & \dfrac{b}{c} \\ 0 & \dfrac{ad - bc}{a} \end{pmatrix} \to \begin{pmatrix} 1 & \dfrac{b}{a} \\ 0 & 0 \end{pmatrix} \tag{6}$$

or

$$\begin{aligned} x_1 + \frac{b}{a}x_2 &= 0 \\ x_1 &= -\frac{b}{a}x_2. \end{aligned} \tag{7}$$

Consequently, $\mathbf{x} = \left(-\dfrac{b}{a}k, k\right)^T$.

EXERCISES

1. Prove that, given integers $n > m$, a homogeneous system of m simultaneous linear equations in n unknowns always has a nontrivial solution.

2. State and prove Cramer's rule for solving a system of simultaneous linear algebraic equations.

3. Solve the following system of equations:

$$\begin{cases} w + 2x + 7y + z = 13, \\ 2w + x - y + 2z = 2, \\ w + x + 2y + z = 5. \end{cases}$$

4. (*a*) List and illustrate six properties of a determinant.
(*b*) Prove these properties.

5. (*a*) State the conditions under which the system $A\mathbf{x} = \mathbf{b}$ of m simultaneous linear equations in n unknowns has a unique solution.
(*b*) Obtain the answer assuming that these conditions are met.
(*c*) Prove that this answer is a solution.
(*d*) Show that there are no other solutions.

If the right-hand side of the matrix equation

$$A\mathbf{x} = \mathbf{b} \tag{8}$$

is not the zero vector, then equation (8) is termed "nonhomogeneous." A process known as "Standard Gaussian elimination" often is used to reduce the number of unknowns. Numerical analysts now admit that this invention of Gauss can hardly be improved upon as a basic computer algorithm.

His treatment can be illustrated by solving equation (8) for a general 2-square matrix with real elements:

$$\begin{pmatrix} a & b \\ c & d \end{pmatrix} \begin{pmatrix} x \\ y \end{pmatrix} = \begin{pmatrix} p \\ g \end{pmatrix}. \tag{9}$$

This becomes

$$\begin{cases} ax + by = p, \\ cx + dy = q. \end{cases} \tag{10}$$

To solve system (10) for y, first multiply the top equation by c/a and then subtract the result from the bottom equation, as follows:

$$\left(c - \frac{c}{a} \cdot a\right) x + \left(d - \frac{c}{a} \cdot b\right) y = q - \frac{c}{a} p, \tag{11}$$

so

$$y = \frac{q - \dfrac{c}{a} p}{d - \dfrac{c}{a} b} = \frac{aq - cp}{ad - bc} = \frac{\begin{vmatrix} a & p \\ c & q \end{vmatrix}}{\begin{vmatrix} a & b \\ c & d \end{vmatrix}}. \tag{12}$$

Working numerically, it is easy to solve system (10) for x by back substitution once y is known, to obtain a similar quantity:

$$x = \frac{\begin{vmatrix} p & b \\ q & d \end{vmatrix}}{\begin{vmatrix} a & b \\ c & d \end{vmatrix}}. \tag{13}$$

Three cases arise in handling system (8) when A is of size m by n, which corresponds to a system of m linear equations in n unknowns:

1. *Inconsistent:* There is no possible solution.
2. *Determined:* There is a unique solution.
3. *Undetermined:* There is more than one solution.

Three subprocesses are entailed in Gaussian elimination:

1. *Triangularization.* Operate on rows of A.
2. *Transformation:* Operate on vector $\mathbf{b}$.
3. *Back-substitution:* Solve the resulting triangularized system.

EXERCISES

6. Show that

$$\begin{vmatrix} a_{11} & a_{12} & a_{13} \\ a_{21} & a_{22} & a_{23} \\ a_{31} & a_{32} & a_{33} \end{vmatrix} = a_{11}\begin{vmatrix} a_{22} & a_{23} \\ a_{32} & a_{33} \end{vmatrix} + a_{13}\begin{vmatrix} a_{21} & a_{22} \\ a_{31} & a_{32} \end{vmatrix} - a_{22}\begin{vmatrix} a_{11} & a_{13} \\ a_{31} & a_{33} \end{vmatrix}$$

$$+ a_{31}\begin{vmatrix} a_{12} & a_{13} \\ a_{22} & a_{23} \end{vmatrix} + a_{33}\begin{vmatrix} a_{11} & a_{12} \\ a_{21} & a_{22} \end{vmatrix}$$

Discover a similar expansion for a 5-square determinant.

7. Evaluate

$$\det\begin{pmatrix} 0 & a & 0 & b & 0 & c & 0 \\ d & 0 & e & 0 & f & 0 & g \\ 0 & h & 0 & j & 0 & k & 0 \\ l & 0 & m & 0 & n & 0 & p \\ 0 & q & 0 & r & 0 & s & 0 \\ t & 0 & u & 0 & v & 0 & w \\ 0 & x & 0 & y & 0 & z & 0 \end{pmatrix}.$$

8. (*a*) Evaluate

$$\begin{vmatrix} 1 & -1 & 0 & 1 \\ -1 & 0 & 1 & 7 \\ 1 & 2 & 1 & -2 \\ 2 & 1 & 1 & -1 \end{vmatrix}$$

using the Laplace expansion in first-order determinants.

(*b*) Evaluate

$$\begin{vmatrix} 1 & 1 & 2 & 0 \\ -2 & -3 & 1 & 4 \\ -1 & 0 & 3 & 2 \\ 5 & 1 & 2 & 1 \end{vmatrix}$$

using the Laplace expansion in second-order determinants.

9. (*a*) The "Vandermonde matrix" is defined as the m by n matrix:

$$X_{mn} = \begin{pmatrix} 1 & x_1 & {x_1}^2 & \cdots & {x_1}^{n-1} \\ 1 & x_2 & {x_2}^2 & \cdots & {x_2}^{n-1} \\ \cdot & \cdot & \cdot & \cdot & \cdot \\ \cdot & \cdot & \cdot & \cdot & \cdot \\ \cdot & \cdot & \cdot & \cdot & \cdot \\ 1 & x_m & {x_m}^2 & \cdots & {x_m}^{n-1} \end{pmatrix}.$$

Prove that

$$\det X_n = \prod_{\substack{i,\, j=1 \\ i>j}}^{n} (x_i - x_j).$$

10. If known, the inverse Vandermonde matrix X^{-1} can be used to evaluate the coefficients a_i in a Lagrangian interpolating polynomial:

$$y = a_0 + a_1 x + \cdots + a_{n-1} x^{n-1},$$

of degree $n-1$ that fits a given function $f(x)$ exactly at the n distinct points $x_1, x_2, \cdots, x_n$ of the independent variable x and serves as an approximation to $f(x)$ elsewhere. These coefficients are expressed formally as the vector

$$\mathbf{a} = X^{-1}\mathbf{y},$$

where

$$\mathbf{y} = [y(x_1), y(x_2), \cdots, y(x_n)]^T.$$

(*a*) Let

$$\frac{d}{dx} = N\left(\frac{\partial}{\partial x_1}, \frac{\partial}{\partial x_2}, \cdots, \frac{\partial}{\partial x_n}\right);$$

compute

$$X' = \frac{d}{dx} X = XM,$$

where M is the matrix of some interest. (See page 33 for notation.)

(*b*) Similarly, compute $\int X\, dx$.

(*c*) Derive the matrix

$$\begin{pmatrix} 0 & 0 & 0 \\ \frac{-5}{8} & \frac{-12}{8} & \frac{-1}{12} \\ \frac{-1}{3} & \frac{-4}{3} & \frac{-1}{3} \end{pmatrix},$$

the elements of whose rows pertain to Simpson's quadrature rules.

(*d*) Generalize this derivation to n-point Newton-Cotes formulas.

B. Matrices, Rank, and Dimensionality

Consider the system of "simultaneous linear algebraic equations" defined by

$$\begin{cases} a_{11}x_1 + a_{12}x_2 + \cdots + a_{1n}x_n = b_1, \\ a_{21}x_1 + a_{22}x_2 + \cdots + a_{2n}x_n = b_2, \\ \quad \cdot \quad \cdot \quad \cdot \\ a_{m1}x_1 + a_{m2}x_2 + \cdots + a_{mn}x_n = b_m. \end{cases} \tag{1}$$

The rectangular array of coefficients of the n variables $x_1, x_2, \cdots, x_n$ in this system is called a "matrix":

$$A = (a_{ij}) = \begin{pmatrix} a_{11} & a_{12} \cdots a_{1n} \\ a_{21} & a_{22} \cdots a_{2n} \\ \cdot & \cdot \quad \cdot \quad \cdot \\ \cdot & \cdot \quad \cdot \quad \cdot \\ \cdot & \cdot \quad \cdot \quad \cdot \\ a_{m1} & a_{m2} \cdots a_{mn} \end{pmatrix}, \tag{2}$$

with "elements" a_{ij}. This matrix has m "rows" and n "columns."

The system (1) often is written in more compact summation notation:

$$\sum_{j=1}^{n} a_{ij}x_j = b_i, \tag{3}$$

but even this notation appears clumsy in comparison with the concise statement that matrix algebra makes possible:

$$A\mathbf{x} = \mathbf{b}. \tag{4}$$

Since they too form rectangular arrays, the column vectors

$$\mathbf{x} = \begin{pmatrix} x_1 \\ x_2 \\ \cdot \\ \cdot \\ \cdot \\ x_n \end{pmatrix} \quad \text{and} \quad \mathbf{b} = \begin{pmatrix} b_1 \\ b_2 \\ \cdot \\ \cdot \\ \cdot \\ b_m \end{pmatrix}$$

also can be considered matrices, whose "column count" is 1; the row count of $\mathbf{x}$ is n; the row count of $\mathbf{b}$ is m.

In general a "matrix" is any rectangular array of elements, arranged in rows and columns. If there are m elements in each row and n elements in each column of the array, the matrix is said to be an "m by n matrix" or a matrix of "size m by n," designated A_{mn}. An n by n matrix forms a "square matrix of order n," designated A_n. Those elements of an n-square matrix for which $i = j$ are called "main diagonal elements"; those elements for which $i + j = n + 1$ are called "secondary or counter diagonal elements." For any n-square matrix A, the "trace" or "spur" is defined as the sum of the main diagonal elements, that is:

$$\operatorname{tr} A = \sum_{i=1}^{n} a_{ii}. \tag{5}$$

A matrix of size m by n contains m times n elements; a matrix of order n contains n^2 elements. The individual element appearing in the ith row and jth column of the matrix A often is denoted either by the symbol $[A]_{i,j}$ or, by the subscripted letter a_{ij}. The elements of all matrices appearing in this text are complex numbers unless otherwise stated. Note that although a_{ij} is used exclusively to denote the elements of A, all the results usually hold even if a_{ij} denotes a submatrix.

The "ordinary sum" $A + B$ of two matrices $A = (a_{ij})$ and $B - (b_{kl})$ of sizes m by n and r by s, respectively, is defined for $m = r$ and $n = s$ (that is, when A and B are "compatible for addition") to be the matrix C of size m by s whose entry in the ith row and lth column is given by:

$$C_{il} = a_{ij} + b_{kl}; \qquad k = i, \quad j = l. \tag{6}$$

Let A, B, and C be compatible for addition; then:

Associative law: $\quad A + (B + C) = (A + B) + C$

Commutative law: $\quad A + B = B + A. \quad (7)$

The "ordinary product" AB of two matrices $A = (a_{ij})$ and $B = (b_{kl})$ of size m by n and size r by s, respectively, is defined for $n = r$ (that is, when A and B are "compatible for multiplication") to be the matrix C of size m by s whose entry in the ith row and lth column is given by

$$C_{il} = \sum_{\lambda=1}^{n} a_{i\lambda} b_{\lambda l}. \tag{8}$$

Let A, B, and C be compatible for multiplication; then:

Associative law: $\quad A(BC) = (AB)C$

Left distributive law: $\quad A(B + C) = AB + AC$

Right distributive law: $\quad (A + B)C = AC + BC.$

Thus:

$$\begin{aligned}(A + B)(A + B) &= A^2 + AB + BA + B^2,\\ (A + B)(A - B) &= A^2 - AB + BA - B^2,\\ (A - B)(A + B) &= A^2 + AB - BA - B^2,\\ &\text{and}\\ (A - B)(A - B) &= A^2 - AB - BA + B^2.\end{aligned} \tag{10}$$

Provided that A and B commute (that is, if $AB = BA$),

$$\begin{aligned}(A + B)(A + B) &= A^2 + 2AB + B^2,\\ (A + B)(A - B) &= A^2 - B^2,\\ (A - B)(A + B) &= A^2 - B^2,\\ &\text{and}\\ (A - B)(A - B) &= A^2 - 2AB + B^2,\end{aligned} \tag{11}$$

which are perhaps less symmetrical, and therefore less satisfying results.

Equations (10) become:

$$\begin{aligned}(A \pm B)(A \pm B) &= A^2 \pm (AB + BA) + B^2\\ &\text{and}\\ (A \pm B)(A \mp B) &= A^2 \mp (AB - BA) + B^2,\end{aligned} \tag{12}$$

where the expression $(AB + BA)$ is called the "anticommutator" of A and B, written $\{A, B\}$, and the expression $(AB\text{-}BA)$ is called the "commutator" of A and B, written $[A, B]$.

The "identity matrix of order n" is the matrix

$$I_n = \begin{pmatrix} 1 & 0 \cdots 0 & 0 \\ 0 & 1 \cdots 0 & 0 \\ \cdot & \cdot\;\cdot \quad \cdot & \cdot \\ \cdot & \cdot \quad \cdot \quad \cdot & \cdot \\ \cdot & \cdot \quad \cdot\;\cdot & \cdot \\ 0 & 0 \cdots 1 & 0 \\ 0 & 0 \cdots 0 & 1 \end{pmatrix} = \begin{pmatrix} 1 & & & \\ & 1 & & \mathbf{0} \\ & & \ddots & \\ \mathbf{0} & & & 1 \end{pmatrix} \tag{13}$$

which can be written as

$$I_n = (\delta_{ij}),$$

where δ_{ij} is the Kronecker delta symbol, defined by

$$\begin{cases} \delta_{ij} = 1, & \text{for } i = j, \\ \delta_{ij} = 0, & \text{otherwise.} \end{cases} \tag{14}$$

A "scalar matrix of order n" has the form αI_n, where α is a scalar number. Therefore

$$\alpha I = \begin{pmatrix} \alpha & & & \mathbf{0} \\ & \alpha & & \\ & & \ddots & \\ \mathbf{0} & & & \alpha \end{pmatrix}. \tag{15}$$

The matrix I is called the identity matrix because the equations

$$A_{mn} I_n = A_{mn} \qquad \text{and} \qquad I_m A_{mn} = A_{mn} \tag{16}$$

hold for an arbitrary matrix A_{mn}.

The set of all possible complex matrices of size m by n contains another neutral element besides the identity matrix, called a "zero matrix." This unique matrix 0 has the property that for any matrix A_{mn} in the set

$$A + 0 = A. \tag{17}$$

The zero matrix evidently is:

$$0 = \begin{pmatrix} 0 & 0 \cdots 0 \\ 0 & 0 \cdots 0 \\ \cdot & \cdot\ \cdot\quad \cdot \\ \cdot & \cdot\quad \cdot\quad \cdot \\ \cdot & \cdot\quad \cdot\ \cdot \\ 0 & 0 \cdots 0 \end{pmatrix}. \tag{18}$$

When this matrix is of size m by n, it is denoted by 0_{mn}; when it is of order n it is denoted by 0_n; when its size is understood from the context, either 0_{mn} or 0_n may be abbreviated 0.

In conclusion, each matrix A in the aforementioned set has a unique negative, designated by $-A$, such that

$$A + (-A) = 0. \tag{19}$$

If $A = (a_{ij})$, this negative is given by

$$-A = (-a_{ij}). \tag{20}$$

Should there exist a matrix B such that $BA = I_n$ or such that $AB = I_m$, then B is called a "left inverse" or a "right inverse" of A, respectively. If it happens that $AB = BA = I$, then A is said to be "nonsingular" (otherwise "singular"), the matrix B is called the "inverse" of A. The notation A^{-1} is standard for designating A inverse.

EXERCISES

11. If each off-diagonal element of a square matrix is zero, then the matrix is called a "diagonal matrix." Is diagonality preserved under ordinary matrix addition and multiplication?

12. In addition to having four "principal diagonals," one for each corner, an m by n matrix usually has many off-corner diagonals. Determine how many diagonals there are in all, classify them using the terminology "left or right main diagonal, left or right secondary diagonal, subdiagonal, intradiagonal, and superdiagonal," and finally, specify each type by an i, j relationship.

13. A "triangular matrix" is defined to be a square matrix having only zero elements on one side of its main diagonal. Nothing is said about whether the remaining elements are zero. Using the terms "upper or lower triangular" for $a_{ij} = 0$ below or above a principal diagonal, respectively, extend this definition to m by n matrices. Is either triangularity or countertriangularity preserved under ordinary matrix addition and multiplication?

14. (*a*) Compute the number of multiplications and additions needed to form AB when A and B are of size m by n and n by s, respectively.

(*b*) Interpret and verify $AI = IA = A$ for an m by n matrix A.

(*c*) Let A, B, and C be compatible for multiplication and addition as necessary. Show that $A(BC) = (AB)C$ and $A(B + C) = AB + AC$.

15. (*a*) Prove that the transpose of the ordinary product of two matrices equals the ordinary product of the transpose matrices taken in reverse order.

(*b*) Generalize the preceding result to n factors,

(*c*) Prove that the inverse of the ordinary matrix product of two nonsingular matrices equals the ordinary matrix product of the inverse matrices taken in reverse order.

(*d*) Let A and B be "nonsingular matrices," that is, let $\det A \neq 0, \det B \neq 0$. Does

$$[(AB)^{-1}]^T = (A^T)^{-1}(B^T)^{-1}?$$

16. A matrix A is called "idempotent" if $AA = A^2 = A$.

(*a*) Show that if A is idempotent, then $I - A$ is also idempotent.

(*b*) Find all 2-square idempotent matrices and show that all but one of these is singular, that is, has a zero determinant.

17. Find the matrices that commute with

$$A = \begin{pmatrix} 0 & 1 & 0 & 0 \\ 0 & 0 & 1 & 0 \\ 1 & -3 & 3 & 0 \\ 0 & 0 & 0 & 1 \end{pmatrix}$$

18. Construct 2-square matrices A and B such that $AB \neq BA$. Use them to manufacture a noncommuting pair of any order.

19. Prove that a square matrix all of whose elements are integers has an inverse with all elements integers if and only if

$$\det A = \pm 1.$$

The symbol $N(\ )$ is newly adopted in this text to designate a formal operator that reshapes a row or column vector by inserting the components of the vector in the same order as main diagonal elements in an otherwise zero matrix. In other words, $N(\mathbf{x})$ bends the linear array of row vector components x_i downward at approximately a 45° angle and then pads the result with $n(n - 1)$ zeros to create a square matrix. Thus

$$N(\mathbf{x}) = N(x_1, x_2, x_3) = \begin{pmatrix} x_1 & & \mathbf{0} \\ & x_2 & \\ \mathbf{0} & & x_3 \end{pmatrix}, \tag{21}$$

a form easily recognizable as a diagonal matrix. The capital letter "N," with its slanting line between two verticals, is suggestive of this nature of the "slant" operator.

The operation symbolized by $N(\ )$ cannot be iterated, because its application to a square matrix is not defined. For this reason, $N(N(\mathbf{x})) = N^2(\mathbf{x})$ is not a permissible expression. Note, however, that

$$[N(\mathbf{x})]^2 = N(\mathbf{x})N(\mathbf{x}) = \begin{pmatrix} {x_1}^2 & & & \mathbf{0} \\ & {x_2}^2 & & \\ & & \ddots & \\ \mathbf{0} & & & {x_n}^2 \end{pmatrix}, \tag{22}$$

so that henceforth $N^2(\mathbf{x})$ can be used safely to denote $[N(\mathbf{x})]^2$. With $x_i \neq 0$, define the reciprocal $N^{-1}(\mathbf{x})$ such that

$$N(\mathbf{x})N^{-1}(\mathbf{x}) = N^{-1}(\mathbf{x})N(\mathbf{x}) = I.$$

Clearly,

$$N^{-1}(\mathbf{x}) = \begin{pmatrix} \frac{1}{x_1} & & & & \mathbf{0} \\ & \frac{1}{x_2} & & & \\ & & \cdot & & \\ & & & \cdot & \\ & & & & \cdot \\ \mathbf{0} & & & & \frac{1}{x_n} \end{pmatrix} \tag{23}$$

Because $N(\mathbf{x})$ is a diagonal matrix,

$$N^T(\mathbf{x}) = N(\mathbf{x}). \tag{24}$$

Moreover, since all n-square diagonal matrices commute,

$$N(\mathbf{x})\, N(\mathbf{y}) = N(\mathbf{y})\, N(\mathbf{x}). \tag{25}$$

Notice that

$$(y_1, y_2, \cdots, y_n) \begin{pmatrix} x_1 & & & & \mathbf{0} \\ & x_2 & & & \\ & & \cdot & & \\ & & & \cdot & \\ & & & & \cdot \\ \mathbf{0} & & & & x_n \end{pmatrix} = (y_1 x_1, y_2 x_2, \cdots, y_n x_n) \tag{26}$$

and

$$(x_1, x_2, \cdots, x_n) \begin{pmatrix} y_1 & & & & \mathbf{0} \\ & y_2 & & & \\ & & \cdot & & \\ & & & \cdot & \\ & & & & \cdot \\ \mathbf{0} & & & & y_n \end{pmatrix} = (x_1 y_1, x_2 y_2, \cdots, x_n y_n). \tag{27}$$

Consequently, provided that the components commute,

$$\mathbf{y}N(\mathbf{x}) = \mathbf{x}N(\mathbf{y}).$$

EXERCISES

20. Show that provided the components commute,

$$N(\mathbf{x})\mathbf{y}^T = N(\mathbf{y})\mathbf{x}^T.$$

21. Find all matrices $X = N(x_1, x_2)$ that are solutions of the matrix equation $X^2 + 5X - 6I = 0$.

Let L denote the 1 by n matrix

$$L = (1, 1, \cdots, 1), \tag{28}$$

whose elements are all unity. The length of L, considering L as a vector in n-dimensional space, is

$$L = \sqrt{1^2 + 1^2 + \cdots + 1^2} = \sqrt{n}. \tag{29}$$

It can be seen immediately that

$$N(L) = N(L^T) = I, \tag{30}$$

where N denotes the slant operator and I denotes the familiar n by n identity matrix. To illustrate:

$$N(L) = N(1, 1, \cdots, 1) = \begin{pmatrix} 1 & & & & \mathbf{0} \\ & 1 & & & \\ & & \cdot & & \\ & & & \cdot & \\ \mathbf{0} & & & & 1 \end{pmatrix} = I_n. \tag{31}$$

Notice that the L matrix can be employed to return a "slant vector" or vector in diagonal matrix form to row or column form, as follows:

$$LN(\mathbf{x}) = LN(\mathbf{x}^T) = \mathbf{x}, \tag{32a}$$

$$N(\mathbf{x})L^T = N(\mathbf{x}^T)L^T = \mathbf{x}^T. \tag{32b}$$

For example,

$$(1, 1, 1)\begin{pmatrix} x_1 & & \mathbf{0} \\ & x_2 & \\ \mathbf{0} & & x_3 \end{pmatrix} = (x_1, x_2, x_3), \tag{33a}$$

$$\begin{pmatrix} x_1 & & \mathbf{0} \\ & x_2 & \\ \mathbf{0} & & x_3 \end{pmatrix}\begin{pmatrix} 1 \\ 1 \\ 1 \end{pmatrix} = \begin{pmatrix} x_1 \\ x_2 \\ x_3 \end{pmatrix}. \tag{33b}$$

It follows from equations (32) that

$$LN(\mathbf{x}) = \mathbf{x}N(L) = \mathbf{x}I = \mathbf{x}, \tag{34a}$$

$$N(\mathbf{x})L^T = N(L)\mathbf{x}^T = I\mathbf{x}^T = \mathbf{x}^T. \tag{34b}$$

Thus although L is a singular matrix, it forms a kind of inverse of the slant operator N, despite the fact that N is not itself a matrix.

EXERCISES

Express the system of simultaneous linear algebraic equations $A\mathbf{x}^T = \mathbf{b}^T$ in terms of A, $N(\mathbf{x})$, and L.

To illustrate further the usefulness of the L matrix, let $R = (r_1, r_2, \cdots, r_m)$ and $C = (c_1, c_2, \cdots, c_n)$ be two matrices defined as follows:

$$R^T = AL_n{}^T \quad \text{or} \quad R = L_n A^T, \tag{35a}$$

and

$$C^T = A^T L_m{}^T \quad \text{or} \quad C = L_m A, \tag{35b}$$

with an arbitrary m by n matrix A. Since

$$r_i = \sum_{\lambda=1}^{n} a_{i\lambda} l_\lambda = \sum_{\lambda=1}^{n} a_{i\lambda}, \tag{36a}$$

and

$$c_j = \sum_{\rho=1}^{m} l_\rho a_{\rho j} = \sum_{\rho=1}^{m} a_{\rho j}, \tag{36b}$$

the elements of matrices R and C are the row and column sums of A, respectively, that is A_i and A_j.

To express the fact that the row sums of a matrix B are unity, write either

$$B_i = \sum_{j=1}^{n} b_{ij} = 1 \tag{37a}$$

or

$$R_B = LB^T = L. \tag{37b}$$

EXERCISE 22

23. Write expressions similar to equations (30) stating that a matrix A has unit column sums. Then show that the column sums of $AN(\mathbf{x})$ are given by $\mathbf{x}$.

24. Prove that (a) row sums are invariant under postmultiplication by a matrix with unit row sums, and (b) column sums are invariant under premultiplication by a matrix with unit column sums.

25. Devise a rule for constructing $2n$-square matrices A and B having no zero elements such that $AB = 0$.

26. A "permutation matrix" is defined to be a matrix of zeros and ones having unit row and column sums. Find all matrices that commute with $N(1, 2, 3)$ and a 3-square cyclic permutation matrix.

Let J denote the "unit block" matrix defined as follows:

$$J_n = L_n{}^T L_n = \begin{pmatrix} 1 \\ 1 \\ \cdot \\ \cdot \\ \cdot \\ 1 \end{pmatrix} (1, 1, \cdots, 1) = \begin{pmatrix} 1 & 1 \cdots & 1 \\ 1 & 1 & 1 \\ \cdot & \cdot & \cdot \\ \cdot & \cdot & \cdot \\ \cdot & \cdot & \cdot \\ 1 & 1 \cdots & 1 \end{pmatrix}, \tag{38}$$

or, more generally, as:

$$J_{mn} = L_m{}^T L_n. \tag{39}$$

Notice that the product of L and L transpose in the order opposite to that in definition (38) is a scalar number:

$$L_n L_n{}^T = n. \tag{40}$$

With this result in mind, consider the pth power of J, J^p. From equations (38) and (40):

$$J^p = (L^T L)^p = L^T (LL^T)^{p-1} L = L^T (n)^{p-1} L = n^{p-1} L^T L,$$

so that

$$J_n^p = n^{p-1} J_n. \tag{41}$$

For example,

$$J_5{}^3 = \begin{pmatrix} 1 & 1 & 1 & 1 & 1 \\ 1 & 1 & 1 & 1 & 1 \\ 1 & 1 & 1 & 1 & 1 \\ 1 & 1 & 1 & 1 & 1 \\ 1 & 1 & 1 & 1 & 1 \end{pmatrix}^3 = 25 \begin{pmatrix} 1 & 1 & 1 & 1 & 1 \\ 1 & 1 & 1 & 1 & 1 \\ 1 & 1 & 1 & 1 & 1 \\ 1 & 1 & 1 & 1 & 1 \\ 1 & 1 & 1 & 1 & 1 \end{pmatrix} = 5^2 J_5$$

EXERCISES

27. Using the more general definition:

$$J(l) = L^T(l)L(l) = \begin{pmatrix} l_1 \\ l_2 \\ \cdot \\ \cdot \\ \cdot \\ l_n \end{pmatrix} (l_1, l_2, \cdots, l_n),$$

show that $J^p(l) = n_l^{\,p-1}J(l)$, where n_l is the number of 1's appearing in the binary number

$$l = l_1 2^{n-1} + l_2 2^{n-2} + \cdots + l_{n-1}2 + l_n.$$

28. A square matrix with nonnegative elements is said to be "doubly stochastic" if both its row and column sums are equal to unity. Assume that P is a doubly stochastic matrix. Show that P^n is also a doubly stochastic matrix using (*a*) the definition of ordinary matrix multiplication, (*b*) the L matrix, and (*c*) the J matrix.

29. Many examples of doubly stochastic matrices arise in Markov chain theory. Consider the following $(k + 1)$-square transition matrix

$$P = \begin{vmatrix} 0 & \frac{1}{k} & \frac{1}{k} & \cdots & \frac{1}{k} \\ \frac{1}{k} & 0 & \frac{1}{k} & \cdots & \frac{1}{k} \\ \frac{1}{k} & \frac{1}{k} & 0 & \cdots & \frac{1}{k} \\ \cdot & \cdot & \cdot & \cdot & \cdot \\ \cdot & \cdot & \cdot & \cdot & \cdot \\ \cdot & \cdot & \cdot & \cdot & \cdot \\ \frac{1}{k} & \frac{1}{k} & \frac{1}{k} & \cdots & 0 \end{vmatrix},$$

which describes the nomadic existence of a system that at each time step is forced to vacate its present state and, with equal probability, to occupy one of k adjacent states. Find a general expression for the nth order transition matrix $P^n = (p_{ij}^{(n)})$ and determine the asymptotic behavior of $p_{ij}^{(n)}$ by evaluating $\lim_{n\to\infty} P^n$.

30. The "rank" $\rho(A)$ of a matrix A is the order of the largest square submatrix of A whose determinant is not zero.

(a) Show that J_{mn} is a matrix of rank one.

(b) Determine the rank of $(n - 2)I + J$.

31. Prove that in the xy-plane the three points (x_1, y_1), (x_2, y_2), and (x_3, y_3) are collinear if and only if

$$\rho\begin{pmatrix} x_1 & y_1 & 1 \\ x_2 & y_2 & 1 \\ x_3 & y_3 & 1 \end{pmatrix} < 3.$$

32. Prove that $\rho(A) = \rho(A^T)$.

Another new operator is symbolized by ∇. This operator, called "upper triangulator," replaces an arbitrary m by n matrix A by ∇A the upper triangular part of A of size either m by m or n by n, whichever is smaller. For example, given

$$M_{2n} = \begin{pmatrix} m_{11} & m_{12} \cdots m_{1n} \\ m_{21} & m_{22} \cdots m_{2n} \end{pmatrix}, \tag{42}$$

it follows that

$$\nabla M = \begin{pmatrix} m_{11} & m_{12} \\ 0 & m_{22} \end{pmatrix}, \quad \text{and} \quad \nabla M^T = \begin{pmatrix} m_{11} & m_{21} \\ 0 & m_{22} \end{pmatrix}. \tag{43}$$

The related operator, called "lower triangulator," is defined as follows:

$$\llcorner M = (\nabla M^T)^T. \tag{44}$$

Thus

$$\llcorner M = \begin{pmatrix} m_{11} & 0 \\ m_{21} & m_{22} \end{pmatrix}. \tag{45}$$

In terms of these triangulators a useful "lance" operator is defined, as follows:

$$\boxbslash A = \llcorner(\nabla A) = \nabla(\llcorner A). \tag{46}$$

To illustrate, the "lance of M" is given by

$$\boxbslash M = \boxbslash M^T = \begin{pmatrix} m_{11} & 0 \\ 0 & m_{22} \end{pmatrix} = N(m_{11}, m_{22}). \tag{47}$$

Applied to a square matrix, say $A = (a_{ij})$, the lance operator strips it of off-diagonal elements to produce

$$\boxbslash A = \begin{pmatrix} a_{11} & & & 0 \\ & a_{22} & & \\ & & \ddots & \\ 0 & & & a_{nn} \end{pmatrix} \tag{48}$$

Since the off-diagonal elements $(I - \boxbslash)A$ are destroyed in this process, there can be no inverse lance operation. It is obvious that $I = \boxbslash J$. Also notice that

$$\boxbslash^2 M = \boxbslash(\boxbslash M) = \boxbslash M, \tag{49}$$

so

$$\boxbslash^r M = \boxbslash(\boxbslash^{r-1} M) = \boxbslash M. \tag{50}$$

As an n-dimensional row or column vector $\mathbf{x}$ is a specially thin matrix,

$$\boxbslash \mathbf{x} = x_1. \tag{51}$$

Further,

$$\boxbslash N(\mathbf{x}) = N(\mathbf{x}). \tag{52}$$

Moreover, since

$$\begin{pmatrix} x_1 \\ x_2 \\ \vdots \\ x_n \end{pmatrix} (y_1, y_2, \cdots, y_n) = \begin{pmatrix} x_1y_1 & x_1y_2 & \cdots & x_1y_n \\ x_2y_1 & x_2y_2 & \cdots & x_2y_n \\ \vdots & \vdots & & \vdots \\ x_ny_1 & x_ny_2 & \cdots & x_ny_n \end{pmatrix}, \tag{53}$$

and

$$\begin{pmatrix} y_1 \\ y_2 \\ \vdots \\ y_n \end{pmatrix} (x_1, x_2, \cdots, x_n) = \begin{pmatrix} y_1x_1 & y_1x_2 & \cdots & y_1x_n \\ y_2x_1 & y_2x_2 & \cdots & y_2x_n \\ \vdots & \vdots & & \vdots \\ y_nx_1 & y_nx_2 & \cdots & y_nx_n \end{pmatrix}, \tag{54}$$

it follows, provided the components commute, that

$$\boxbslash(\mathbf{x}^T\mathbf{y}) = \boxbslash(\mathbf{y}^T\mathbf{x}). \tag{55}$$

EXERCISES

33. Discuss:

(*a*) $(\boxbslash A)^T = \boxbslash^T A = \boxbslash A$,

(*b*) $\boxbslash A \boxbslash B = \boxbslash B \boxbslash A$,

(*c*) $(\boxbslash A)^{-1} \boxbslash A = \boxbslash^{-1} A \boxbslash A = I$,

(*d*) $A = \text{◹} A + \text{◺} A - \boxbslash A$.

34. Why is "grand sum" a good name for $L_m A L_n{}^T$?

35. The trace of a matrix A is the sum of its main diagonal elements, that is, $\mathrm{tr}A = \sum_{i=1}^{n} a_{ii}$. Prove that

$$\mathrm{tr}A = L \boxbslash (A_n) L^T.$$

36. Express the matrix

$$\begin{pmatrix} 1 & -1 \cdots & -1 \\ 0 & 1 \cdots & -1 \\ \cdot & \cdot\ \cdot & \cdot \\ \cdot & \cdot\quad\cdot & \cdot \\ \cdot & \cdot\quad\cdot\quad\cdot & \cdot \\ 0 & 0 \cdots & 1 \end{pmatrix}$$

in several ways using lance operators and triangulators.

37. The "Hadamard product" $C = A \text{Ⓗ} B$ of two m by n matrices $A = (a_{ij})$ and $B = (b_{ij})$ is defined as $C = (c_{ij})$, where $c_{ij} = a_{ij} b_{ij}$. Let M be an r-square matrix. Express $\text{◹} M$, $\text{◺} M$, and $\boxbslash M$ in terms of the identity matrix $I_r = H^0$ and nilpotent matrices:

$$H = \begin{pmatrix} 0 & 1 \cdots 0 \\ 0 & 0 \\ \cdot & \cdot \\ \cdot & \quad\cdot \\ \cdot & \quad\ \cdot\ 1 \\ 0 & \cdots\cdot 0 \end{pmatrix} \quad \text{and} \quad H^T = \begin{pmatrix} 0 & 0 \cdots\cdot & 0 \\ 1 & 0 & \cdot \\ \cdot & \cdot\ \cdot & \cdot \\ \cdot & \cdot\quad\cdot\quad\cdot & \cdot \\ \cdot & \cdot\quad\cdot\ \cdot & \cdot \\ 0 & \cdots\cdot 1 & 0 \end{pmatrix}$$

(*Note:* A matrix T is said to be "nilpotent" if $T^p = 0$ for some positive integer p.)

38. (*a*) What is the rank of an H matrix of order n?

(*b*) State a necessary and sufficient condition for the matrix

$$a_0 I + a_1 H + \cdots + a_{n-1} H^{n-1}$$

to be nonsingular.

(*c*) Find the inverse of the matrix $\sum_{k=0}^{\infty} H^k$.

39. A "lower Hessenberg matrix" is defined to be a square matrix A having only zero elements above the first superdiagonal (the elements $a_{i,i+1}$). Is the property of being of Hessenberg type preserved under ordinary matrix addition and multiplication or Hadamard multiplication?

If it is possible to gauge the importance of a scientific concept from the wealth of its synonyms, then the concept "eigenvalue," alias "characteristic root," alias "proper value," alias "latent root," alias "secular value," must be of fundamental importance throughout the world of mathematics, physics, and engineering.

The "eigenvalues" $\lambda(A)$ of an n-square matrix A are defined to be those scalar numbers λ for which the matrix equation

$$A\mathbf{x} = \lambda\mathbf{x} \tag{56}$$

has a nontrivial n by 1 solution vector $\mathbf{x} \neq 0$.

Equation (1) can be written

$$(A - \lambda I)\mathbf{x} = \mathbf{0}, \tag{57}$$

where I denotes the n-square identity and $\mathbf{0}$ denotes a zero column vector. From the theory of homogeneous systems of linear equations, it is assured that equation (57) will have a nonzero solution vector $\mathbf{x}$ if and only if the "secular determinant," that is,

$$\det(A - \lambda I), \tag{58}$$

vanishes. The roots of the resulting "characteristic equation"

$$\det(A - \lambda I) = 0 \tag{59}$$

are given by the n complex zeros $\lambda_1, \lambda_2, \cdots, \lambda_n$ of the "characteristic polynomial"

$$p(\lambda) = \det(A - \lambda I), \tag{60}$$

which is a polynomial of the nth degree in λ.

If the n numbers λ_i in this set $\lambda(A)$, including any multiple roots, are real, then they customarily are ordered as follows:

$$\lambda_1 \geq \lambda_2 \geq \cdots \geq \lambda_n. \tag{61}$$

A solution vector $\mathbf{x}$ corresponding to any particular eigenvalue $\lambda_i (i = 1, 2, \cdots, n)$ is referred to as a "column eigenvector" of A associated with λ_i. Similarly, the equation

$$\mathbf{x}A = \mathbf{x}\lambda = \lambda\mathbf{x} \tag{62}$$

gives rise to "row eigenvectors."

EXERCISES

40. (*a*) Explain what is meant by row equivalent matrices.

(*b*) Illustrate by reducing the matrix

$$A = \begin{pmatrix} 3 & 1 & 0 \\ 1 & 2 & 2 \\ 2 & 1 & 4 \end{pmatrix}$$

to so-called "Hermite normal (or row-echelon) form."

41. A matrix $A = (a_{ij})$ is called a "band matrix" if and only if $a_{ij} = 0$ for $|i - j| > k$ where k is a nonnegative integer. Find the eigenvalues of the band matrix

$$B = \begin{pmatrix} 3 & 1 & 1 & 0 \\ 1 & 3 & 1 & 1 \\ 1 & 1 & 3 & 1 \\ 0 & 1 & 1 & 3 \end{pmatrix}$$

exactly by calculating the coefficients of the characteristic polynomial and using a standard method to find its zeros.

42. A "tridiagonal matrix" is a band matrix with $k = 1$. Write the characteristic equation for the tridiagonal matrix

$$T = \begin{pmatrix} a & b & 0 \\ c & d & e \\ 0 & f & g \end{pmatrix}.$$

43. The Cayley-Hamilton theorem states: A square matrix A satisfies the equation $p(A) = 0$, where p is the characteristic polynomial of A. Verify it for the matrix

$$\begin{pmatrix} 1 & -1 \\ 3 & -2 \end{pmatrix}.$$

44. Show that tr A and det A appear as coefficients in $p(A)$.

45. (*a*) Prove that $\lambda(A) = \lambda(A^T)$, with A square.

(*b*) Prove that $\lambda(AB) = \lambda(BA)$, with A and B doubly compatible.

46. Prove that the column (row) eigenvectors belonging to a pair of different eigenvalues of a real symmetric matrix are orthogonal.

47. Assume that λ and μ are distinct eigenvalues of A. Prove that if $\mathbf{x}$ is a column eigenvector belonging to λ and $\mathbf{y}^T$ is a row eigenvector belonging to μ, then $\mathbf{x}$ and $\mathbf{y}$ are orthogonal.

48. The nonnegative square roots of the eigenvalues of A^*A are called the "singular values" of A. Prove that the singular values of A^* are the same as those of A. Here A^* denotes the complex conjugate transpose of A.

49. The matrix:

$$C = \begin{pmatrix} 0 & 1 & 0 & \cdots & 0 & 0 \\ 0 & 0 & 1 & \cdots & 0 & 0 \\ \cdot & \cdot & \cdot & \cdot\,\cdot & \cdot & \cdot \\ \cdot & \cdot & \cdot & \cdot\,\cdot & \cdot & \cdot \\ \cdot & \cdot & \cdot & \cdot & \cdot & \cdot \\ 0 & 0 & 0 & \cdots & 0 & 1 \\ -a_0 & -a_1 & -a_2 & \cdots & -a_{n-2} & -a_{n-1} \end{pmatrix}$$

is called a "companion matrix." Find its characteristic polynomial.

50. Among all the monic polynomials of the form

$$f(x) = x^n + a_{n-1}x^{n-1} + \cdots + a_1 x + a_0$$

that are nonconstant and have the matrix A as a zero there is one of lowest degree called the "minimum polynomial" of A. Find the "minimal equation" of

$$\begin{pmatrix} 0 & 0 & 1 \\ 0 & 1 & 0 \\ -1 & 0 & 1 \end{pmatrix}.$$

51. In what sense do the "Euclidean matrix norm" (a)

$$||A||_E = \left(\sum_{i=1}^{m} \sum_{j=1}^{n} a_{ij}\right)^{1/2}$$

and the "spectral matrix norm" (b)

$$||A||_S = [\max \lambda(AA^T)]^{1/2},$$

respectively, correspond to the "Euclidean vector norm"

$$||\mathbf{x}|| = (x_1{}^2 + x_2{}^2 + \cdots + x_n{}^2)^{1/2}.$$

Let $A = (a_{ij})$ be an m by n matrix with complex elements. The graph of A, which is drawn in the complex plane, consists of the displacement vectors a_{ij} with heads at the points a_{ij} and tails at the origin:

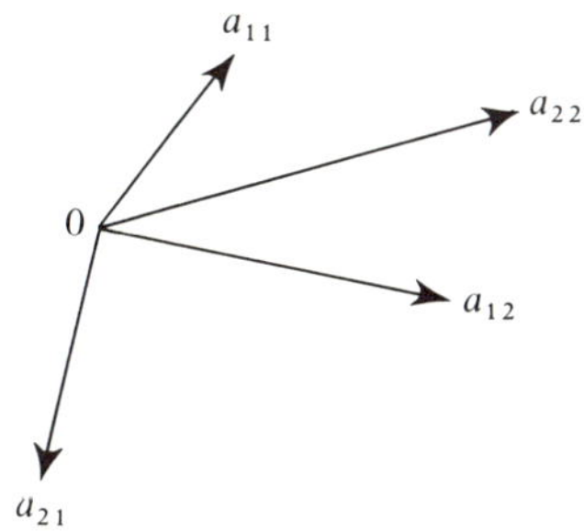

These m times n line segments, some possibly of zero length, all emanating from the origin, are termed "element vectors." For a particular matrix, these element vectors need not all be different. By the procedure just discussed, a graph or geometrical figure always can be associated with any given matrix—the "matrix graph."

Sometimes the matrix graph will seem quite degenerate, for example, the graph of the identity matrix is the vector $1 + 0 \cdot i$ taken n times, and the zero vector taken $n(n-1)$ times. At other times the matrix graph will be so complicated that it will be impractical to plot. The concept, nevertheless, often is illuminating. This is evidenced by the following geometrical interpretation of multiplication of a matrix A by a complex scalar $\alpha = re^{i\theta}$. In the formation of αA, the element vectors of the matrix graph undergo magnification or contraction in the amount given by the factor r and the graph as a whole rotates through an angle θ. Valuable insights sometimes are afforded by including other pieces of information in the picture, such as the determinant vector, the trace vector, and the eigenvalue vectors.

EXERCISES

52. Graph the matrix

$$\begin{pmatrix} 2 + i & -1 + 2i \\ -1 - i & 2 - 2i \end{pmatrix}.$$

Include and label: (*a*) the trace; (*b*) the row and column sums; (*c*) the grand sum. What meaning can be given to the Euclidean norm?

53. Prove that for complex scalars α and β:

(*a*) $\operatorname{tr} AB = \operatorname{tr} BA$.

(*b*) $\operatorname{tr}(\alpha A + \beta B) = \alpha \operatorname{tr} A + \beta \operatorname{tr} B$.

(*c*) $\operatorname{tr} A^* = \overline{\operatorname{tr} A}$.

(*d*) $||A||_E = \operatorname{tr} A^*A = \operatorname{tr} AA^*$.

54. (*a*) For sufficiently small values of ϵ, the numerical inversion of the n-square "Newman matrix"

$$M = aI + J = \epsilon^2 I + J$$

is influenced greatly by rounding errors. Using the definition

$$||M|| = \frac{\max |\lambda(M)|}{\min |\lambda(M)|},$$

show that M becomes progressively more ill-conditioned as $\epsilon \to 0$.

(*b*) Determine a matrix A such that

$$A^T A = M.$$

(*c*) Using matrix algebra produce M^{-1}. For what values of a does M fail to have an inverse?

55. The "general Newman matrix" is of the form

$$N = \alpha I + \beta J + \gamma K,$$

where α, β, and γ are scalars. Put $\alpha = 1$, $\beta = -a$, $\gamma = 0$.

(*a*) What is det N?
(*b*) What are the characteristic roots of N?
(*c*) For what values of a, if any, is N singular?
(*d*) What is N^{-1} when N is nonsingular?
(*e*) What is the minimal equation of N? (See page 70 for notation.)

56. (*a*) The "Hilbert matrix," defined by

$$H_{mn} = (h_{ij}) = ((i + j - 1)^{-1}) = \begin{pmatrix} 1 & \frac{1}{2} & \cdots & \frac{1}{n} \\ \frac{1}{2} & \frac{1}{3} & \cdots & \frac{1}{n+1} \\ \cdot & \cdot & \cdot & \cdot \\ \cdot & \cdot & \cdot & \cdot \\ \cdot & \cdot & \cdot & \cdot \\ \frac{1}{m} & \frac{1}{m+1} & \cdots & \frac{1}{m+n+1} \end{pmatrix},$$

provides another common test matrix for $m = n$, since it is so ill-conditioned. Calculate the inverse of a 3-square H rounding all numbers to four decimal places.

(*b*) The elements of the inverse Hilbert matrix* are given by:

*Marvin Marcus, *Basic Theorems in Matrix Theory*, NBS Applied Math Series, 57–1964 (for sale by Superintendent of Documents, Washington, D.C. 20402).

$$e_{ij} = (-1)^{i+j}\frac{(n+i-1)!(n+j-1)!}{(i+j-1)[(i-1)!(j-1)!]^2(n-i)!(n-j)!}.$$

Prove that these elements are all integers.

(*c*) Find the inverse of a 3-square H^{-1} by hand calculation again rounding all numbers to four decimal places. Comment on the relative precision of the two processes: $(H)^{-1} \to H^{-1}$ and $(H^{-1})^{-1} \to H$.

57. Show that if A is a symmetric matrix the condition number

$$||A||_s \cdot ||A^{-1}||_s$$

reduces to

$$\frac{\max|\lambda_i|}{\min|\lambda_i|}.$$

Let $A = (a_{ij})$ be an m by n matrix. Then, if the parenthetical frame ( ) is regarded as a container for the matrix elements a_{ij}, it has depth m and width n. Thus m and n are the "dimensions" of the container and the "dimensionality of A" is specified by the ordered pair (m, n).

To increase the dimensionality of a matrix new rows may be added either at the top or bottom or between rows and new columns may be added either at the right or left or between columns. It should be remembered, however, that in a matrix $A = (a_{ij})$, successive j-values get larger from left to right and that successive i-values get larger from top to bottom. This means that for any matrix the direction of increasing dimensionality is toward the lower right or "southeast corner" of the matrix.

A "dimensionality equation" is a rule that decides the dimension of an answer. Two important dimensionality equations are the following:

Ordinary Matrix Addition:

$$(m, n) + (m, n) = (m, n),$$

Ordinary Matrix Multiplication:

$$(m, r)(r, n) = (m, n).$$

For comparison, here are two additional dimensionality equations of commensurate importance, dealing with operations to be taken up later:

Direct Addition:

$$(m, n) \dotplus (r, s) = (m + r, n + s)$$

Direct Multiplication:

$$(m, n) \dot{\times} (r, s) = mr, ns).$$

EXERCISES

58. A matrix A is said to be "sparse" if most elements $a_{ij} = 0$. Otherwise, it is called "dense." Is either property enhanced under the operations of ordinary matrix arithmetic?

59. Let B denote the n-square matrix:

$$B = \begin{pmatrix} 1 & & & \mathbf{0} \\ 1 & -1 & & \\ 1 & -2 & 1 & \\ \cdot & & \cdot & \\ \cdot & \cdot & & \cdot \\ \cdot & & & \cdot \\ 1 & \cdots & & 1 \end{pmatrix}.$$

(*a*) Show that $B^2 = I$.

(*b*) Show that $BB^T = a_{ij}$, where a_{ij} is the binomial coefficient $\binom{i+j}{i}$, $0 \leq i, j \leq n - 1$.

60. (*a*) Let $\mathbf{x} = (x_1, x_2, \cdots, x_n)^T$ be a column vector. The "pth-power norm" of $\mathbf{x}$ is defined as

$$||\mathbf{x}||_p = \left(\sum_{i=1}^{n} |x_i|^p\right)^{1/p},$$

where p is a real number on the range $1 \leq p < \infty$. For $p = 2$,

$$||\mathbf{x}||_2 = \sqrt{x_1{}^2 + x_2{}^2 + \cdots + x_n{}^2}$$

gives the Euclidean length of $\mathbf{x}$. A closely associated norm, called the "maximum norm," is defined as:

$$||\mathbf{x}||_\infty = \max_{1 \leq i \leq n} |x_i|.$$

Generalize these definitions to m by n matrices.

(*b*) Let A be a real m by n matrix. The determination of an n by 1 vector $\mathbf{x}$ such that

$$||A\mathbf{x} - \mathbf{b}||_2$$

becomes minimized is known as the linear "least squares problem." Derive the so-called "normal equations":

$$A^T A\mathbf{x} = A^T\mathbf{b}.$$

61. The "absolute norm" of a matrix A is defined as $||A|| = \max_{i,j} |a_{ij}|$.

(*a*) Prove that if A is n-square and if $||A|| < 1/n$, then this is a sufficient condition that $A^r \to 0$ as $r \to \infty$.

(*b*) Prove that this is not a necessary condition by exhibiting a counter-example.

62. A "matrix norm" $||A||$ has the following properties:

I. $||A|| \geq 0$, $||A|| = 0$ if and only if $A = 0$.

II. $||\gamma A|| = |\gamma|\ ||A||$, with γ a complex scalar.

III. $||A + B|| \leq ||A|| + ||B||$,

IV. $||AB|| \leq ||A||\ ||B||$.

Properties I and III relate to a distance function in metric space. Property IV is peculiar to matrices. Show that (*a*) Euclidean, (*b*) Absolute, and (*c*) Spectral matrix norms have these properties.

The equation

$$A\mathbf{x} = \lambda\mathbf{x} \tag{63}$$

concerns a certain distinguished vector $\mathbf{x}$, called an 'eigenvector," which because of its strategic location is able to withstand the brunt of an n-dimensional linear transformation A (although its compatriots are buffeted about and sometimes suffer serious disorientation), and instead either expands or contracts in proportion to scalar multiplier λ, called an "eigenvalue."

This useful equation (63) asserts, for example, that if λ is an eigenvalue of A, then λ^2 is an eigenvalue of A^2, for by substitution:

$$A^2\mathbf{x} = A(A\mathbf{x}) = A\lambda\mathbf{x} = \lambda A\mathbf{x} = \lambda(A\mathbf{x}) = \lambda(\lambda\mathbf{x}) = \lambda^2\mathbf{x}. \tag{64}$$

Householder* has shown that if λ is any eigenvalue of a complex n-square matrix A, then $\bar{\lambda}\lambda$ lies in the field of values $F(A^*A)$ of A^*A. Let $\mathbf{x} \neq 0$ be an arbitrary vector assumed normalized such that $\mathbf{x}^*\mathbf{x} = 1$. Then, the field of values† of A may be represented by

$$F(A) = Q = \mathbf{x}^*A\mathbf{x}. \tag{65}$$

Form $Q\bar{Q} = \mathbf{x}^*A\mathbf{x}(\mathbf{x}^*A\mathbf{x})^* - \mathbf{x}^*(A\mathbf{x}\mathbf{x}^*A^*)\mathbf{x}$. Since $B = A\mathbf{x}\mathbf{x}^*A^*$ is hermitian, its eigenvalues are real. Moreover, B is normal; consequently, $QQ \leq \beta_1{}^2$ where $\beta_1{}^2$ is the largest eigenvalue of B. By Sylvester's law, however, B has rank zero or one. If it is one, then B has a single nonzero eigenvalue given by tr B. Let $\mathbf{b}$ denote $A\mathbf{x}$. Then tr B = tr $\mathbf{b}\mathbf{b}^* = \mathbf{b}^*\mathbf{b} = \mathbf{x}^*A^*A\mathbf{x}$. It follows that

$$Q\bar{Q} \leq \mathbf{x}^*A^*A\mathbf{x} \tag{66}$$

*Alston S. Householder, *Principles of Numerical Analysis*, 1953, p. 147.

†Marvin Marcus and Henryk Minc, *A Survey of Matrix Theory and Matrix Inadequalities*, Allyn and Bacon, Boston, 1964, pp. 103–109.

and therefore the field of values of A is in modulus less than or equal to $\rho_1{}^2$, where $\rho_1{}^2$ is the largest eigenvalue of A^*A.

EXERCISES

63. Show that

$$\lambda = \frac{\mathbf{x}^* A\mathbf{x}}{|x|^2}.$$

64. Fill in the details of the proof given in the last paragraph.

65. Let $A\mathbf{x} = \lambda\mathbf{x}$. Prove that if $f(z)$ is a polynomial in z with coefficients in the field F, then $\mathbf{x}$ is an eigenvector of $f(A)$ belonging to $f(\lambda)$.

66. If A is any real symmetric matrix with eigenvectors $\lambda_1 \leq \lambda_2 \leq \cdots \leq \lambda_n$, then the totality

$$\left\{\frac{\mathbf{x}^T A\mathbf{x}}{\mathbf{x}^T\mathbf{x}} : \mathbf{x} \text{ any real vector}\right\},$$

which is called the "field of values of A," satisfies the inequality

$$\lambda_1 \leq \frac{\mathbf{x}^T A\mathbf{x}}{\mathbf{x}^T\mathbf{x}} \leq \lambda_n.$$

Prove that (*a*)

$$\lambda_1 \leq \frac{\sum a_{ij}}{n} \leq \lambda_n$$

and (*b*)

$$\lambda_1 \leq a_{ii} \leq \lambda_n$$

for all the elements a_{ij} of A.

67. Let A be a matrix and $R(y)$ a rational function, that is, let $R(y)$ be the ratio of two polynomials in the undeterminate y. Assume λ is an eigenvalue of A to which $\mathbf{x}$ corresponds as eigenvector. Then $R(\lambda)$ is an eigenvalue of $R(A)$ and $\mathbf{x}$ is still the corresponding eigenvector.

The literature concerning bounds and inequalities on eigenvalues of complex finite matrices is large, now numbering perhaps several hundred papers. A particularly easy result to apply is the following theorem due to Geršgoren† (1931):

Let

$$P_i = \sum_{j(j \neq i)} |a_{ij}|, \quad Q_j = \sum_{i(i \neq j)} |a_{ij}|.$$

*Brauer's (1947) extension is intriguing: Every λ lies in at least one Cassini oval:

$$|z - a_{ii}|\,|z - a_{jj}| \leq P_i P_j (i \neq j)$$

and in at least one Cassini oval:

$$|z - a_{ii}|\,|z - a_{jj}| \leq Q_i Q_j (i \neq j)$$

Then λ lies in at least one of the circles:

$$|z - a_{ii}| \leq P_i,$$

and in at least one of the circles:

$$|z - a_{ii}| \leq Q_i.$$

To illustrate the use of the Geršgoren theorem, let

$$A = \begin{pmatrix} 2 & 1 & 0 & 0 \\ 1 & 4 & 1 & 1 \\ 0 & 1 & 6 & 4 \\ 0 & 1 & 4 & 8 \end{pmatrix}.$$

The Geršgoren circle theorem will be applied to find bounds for the largest and smallest eigenvalues of A.

Since $A^* = A$, A is hermitian and $\lambda(A)$ is a set of real numbers.

From Fig. 1 it is established that $\lambda_i(A) > 0$, $i = 1, 2, \cdots, n$, by observing that the Geršgoren circles are confined to the right open half plane.

As a consequence of being hermitian and of having eigenvalues strictly greater than zero, A is positive definite. The possible range of the eigenvalues is given by the intersection of the real axis with the union of the Geršgoren circles. Bounds for the largest and smallest eigenvalues are given by the endpoints for this interval, [1, 13].

As a check, the characteristic equation is

$$\lambda^4 - 20\lambda^3 + 121\lambda^2 - 280\lambda + 212;$$

its roots are:

$$\begin{aligned} \lambda_1 &= 1.5499, \\ \lambda_2 &= 2.8624, \\ \lambda_3 &= 4.1941, \\ \lambda_4 &= 11.3936. \end{aligned}$$

Now let

$$B = \begin{pmatrix} 1 & 1 & 0 \\ 0 & 1 & 0 \\ 1 & 0 & 1 \end{pmatrix}.$$

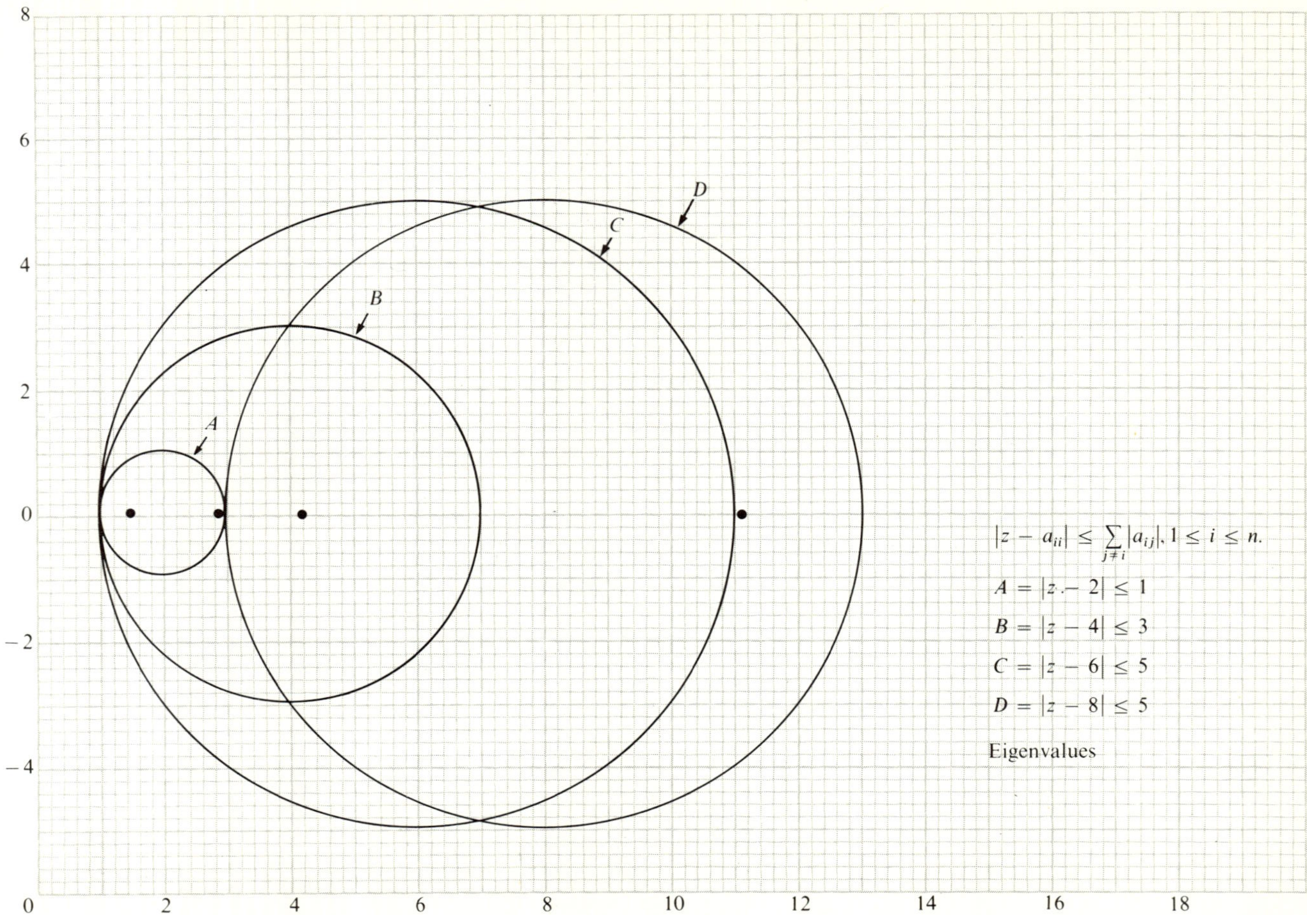

Figure 1. Geršgorin circles.

The characteristic equation for B is given by

$$\begin{vmatrix} 1-\lambda & 1 & 0 \\ 0 & 1-\lambda & 0 \\ 1 & 0 & 1-\lambda \end{vmatrix} = 0.$$

or $(1 - \lambda)^3 = 0$. Hence $\lambda_1 = \lambda_2 = \lambda_3 = 1$. Notice that:

1. B is not *symmetric* because $B \neq B^T$,
2. B is not *hermitian* because $B \neq B^*$,
3. B is not *normal* because $BB^* \neq B^*B$.

Let

$$F = \tfrac{1}{2}(B + B^*) = \frac{\begin{pmatrix} 1 & 1 & 0 \\ 0 & 1 & 0 \\ 1 & 0 & 1 \end{pmatrix} + \begin{pmatrix} 1 & 0 & 1 \\ 1 & 1 & 0 \\ 0 & 0 & 1 \end{pmatrix}}{2} = \tfrac{1}{2}\begin{pmatrix} 2 & 1 & 1 \\ 1 & 2 & 0 \\ 1 & 0 & 2 \end{pmatrix} = \begin{pmatrix} 1 & \frac{1}{2} & \frac{1}{2} \\ \frac{1}{2} & 1 & 0 \\ \frac{1}{2} & 0 & 1 \end{pmatrix}$$

Then:

1. F is *symmetric* since $F = F^T$,
2. is *hermitian* since $F = F^*$,
3. F is *normal* because it is hermitian.

Let

$$G = \frac{1}{2i}(B - B^*) = \frac{i}{2}(B^* - B) =$$

$$\frac{i}{2}\left\{\begin{pmatrix} 1 & 0 & 1 \\ 1 & 1 & 0 \\ 0 & 0 & 1 \end{pmatrix} - \begin{pmatrix} 1 & 1 & 0 \\ 0 & 1 & 0 \\ 1 & 0 & 1 \end{pmatrix}\right\} = \begin{pmatrix} 0 & -\frac{i}{2} & \frac{i}{2} \\ \frac{i}{2} & 0 & 0 \\ \frac{-i}{2} & 0 & 0 \end{pmatrix}$$

Then:

1. G is *skew-symmetric* since $G = -G^T$,
2. G is *hermitian* since $G = G^*$,
3. G is *normal* because it is hermitian.

Next write $B = F + iG$.

$$\begin{pmatrix} 1 & \frac{1}{2} & \frac{1}{2} \\ \frac{1}{2} & 1 & 0 \\ \frac{1}{2} & 0 & 1 \end{pmatrix} + i \begin{pmatrix} 0 & -\frac{i}{2} & \frac{i}{2} \\ \frac{i}{2} & 0 & 0 \\ -\frac{i}{2}. & 0 & 0 \end{pmatrix}$$

$$= \begin{pmatrix} 1 & \frac{1}{2} & \frac{1}{2} \\ \frac{1}{2} & 1 & 0 \\ \frac{1}{2} & 0 & 1 \end{pmatrix} + \begin{pmatrix} 0 & \frac{1}{2} & \frac{1}{2} \\ -\frac{1}{2} & 0 & 0 \\ \frac{1}{2} & 0 & 0 \end{pmatrix} = \begin{pmatrix} 1 & 1 & 0 \\ 0 & 1 & 0 \\ 1 & 0 & 1 \end{pmatrix}$$

The characteristic equation for F, however, is

$$\begin{vmatrix} 1-\lambda & \frac{1}{2} & \frac{1}{2} \\ \frac{1}{2} & 1-\lambda & 0 \\ \frac{1}{2} & 0 & 1-\lambda \end{vmatrix}$$

$$= (1-\lambda)^3 - [\tfrac{2}{4}(1-\lambda)] = \underline{\lambda^3 - 3\lambda^2 + \tfrac{5}{2}\lambda - \tfrac{1}{2} = 0}$$

and the characteristic equation for G is

$$\begin{vmatrix} -\lambda & -\frac{1}{2} & \frac{1}{2} \\ \frac{1}{2} & -\lambda & 0 \\ -\frac{1}{2} & 0 & -\lambda \end{vmatrix} = (-\lambda)^3 - [-\tfrac{1}{4}\lambda - \tfrac{1}{4}\lambda] = \underline{\lambda^3 - \tfrac{1}{2}\lambda = 0}.$$

Factoring out $(1 - \lambda)$ from the first expression leaves $-\lambda^2 + 2\lambda - \dfrac{1}{2} = 0$, so that

$$\lambda_1(F) = 1 - \frac{\sqrt{2}}{2}, \quad \lambda_2(F) = 1, \quad \text{and} \quad \lambda_3 = 1 + \frac{\sqrt{2}}{2}.$$

Factoring out λ from the second expression leaves $\lambda^2 - \dfrac{1}{2} = 0$, so that

$$\lambda_1(G) = -\frac{\sqrt{2}}{2}, \quad \lambda_2(G) = 0, \quad \text{and} \quad \lambda_3(G) = \frac{\sqrt{2}}{2}.$$

Now the real part of the field values Q is given by $\operatorname{Re} Q = (Q + \bar{Q})/2$ and the imaginary part of Q is given by $\operatorname{Im} Q = (Q - \bar{Q})/2i$. Recalling that these are bounded by

$$\lambda_1\left(\frac{B + B^*}{2}\right) \leq \operatorname{Re} Q \leq \lambda_n\left(\frac{B + B^*}{2}\right) \qquad 1 - \frac{\sqrt{2}}{2} \leq \operatorname{Re} Q \leq 1 + \frac{\sqrt{2}}{2}$$

$$\lambda_1\left(\frac{B - B^*}{2i}\right) \leq \operatorname{Im} Q \leq \lambda_n\left(\frac{B - B^*}{2}\right) \qquad -\frac{\sqrt{2}}{2} \leq \operatorname{Im} Q \leq \frac{\sqrt{2}}{2}.$$

The smallest real number belonging to the field of values of B is therefore $1 - \sqrt{2}/2$.

Let

$$U = \begin{pmatrix} 0 & 1 & 0 \\ 0 & 0 & 1 \\ 1 & 0 & 0 \end{pmatrix}.$$

The unitary transformation $U^{-1}BU$ brings B to triangular form with its eigenvalues along the main diagonal:

$$U^{-1}BU = \begin{pmatrix} 0 & 0 & 1 \\ 1 & 0 & 0 \\ 0 & 1 & 0 \end{pmatrix}\begin{pmatrix} 1 & 1 & 0 \\ 0 & 1 & 0 \\ 1 & 0 & 1 \end{pmatrix}\begin{pmatrix} 0 & 1 & 0 \\ 0 & 0 & 1 \\ 1 & 0 & 0 \end{pmatrix} = \begin{pmatrix} 1 & 0 & 1 \\ 1 & 1 & 0 \\ 0 & 1 & 0 \end{pmatrix}\begin{pmatrix} 0 & 1 & 0 \\ 0 & 0 & 1 \\ 1 & 0 & 0 \end{pmatrix}$$

$$= \begin{pmatrix} 1 & 1 & 0 \\ 0 & 1 & 1 \\ 0 & 0 & 1 \end{pmatrix} \cong \begin{pmatrix} \lambda_1 & c_{12} & c_{13} \\ 0 & \lambda_2 & c_{23} \\ 0 & 0 & \lambda_3 \end{pmatrix}$$

Following F. D. Murnaghan,* the equation of the bounding curve is

$$0 = p_1 p_2 p_3 - \tfrac{1}{4}(c_{23}\overline{c_{23}}p_1 + c_{13}\overline{c_{13}}p_2 + c_{12}\overline{c_{12}}p_3) + \tfrac{1}{8}(c_{12}c_{23}\overline{c_{13}}/t + \overline{c_{12}c_{23}}c_{13}t).$$

In the present case this specializes to

$$p_1 p_2 p_3 - \tfrac{1}{4}(p_1 + p_3) = 0.$$

The p_i are perpendiculars from the λ_i to the bounding curve. Hence p_1, p_2, and p_3 are all equal to $\mathbf{r}$, a radius vector from the point $x = \lambda_i = 1$. Taking this point as origin in (x_1, y) coordinates and noting $|\mathbf{r}| = \sqrt{x^2 + y^2}$,

$$p^3 - \tfrac{1}{4}(p + p) = p^3 - \frac{p}{2} = (x_1{}^2 + y^2)^{3/2} - \tfrac{1}{2}(x_1{}^2 + y^2)^{1/2} = 0.$$

* "On the Field of Values of a Square Matrix," *Proc. Nat. Acad. Sci.*, **18**, 246–248 (1932).

The equation for the boundary of the field of values in x, y coordinates ($x_1 = x - 1$) becomes

$$\{[(x-1)^2 + y^2] - \tfrac{1}{2}\} \sqrt{(x-1)^2 + y^2} = 0.$$

Letting $y = 0$ gives

$$\{(x-1)^2 - \tfrac{1}{2}\}(x-1) = 0$$
$$\underline{x^3 - 3x^2 + \tfrac{5}{2}x - \tfrac{1}{2} = 0}$$

Letting $x = 1$ gives

$$(y^2 - \tfrac{1}{2})y = 0$$
$$\underline{y^3 - \tfrac{1}{2}y = 0.}$$

EXERCISES

68. Prove that if for some positive integer k, $A^k = I$, then all characteristic roots of A are roots of unity. Show the converse is false by considering the matrix $\triangledown J_2$.

69. Use Geršgorin's inclusion theorem to give a lower bound for the eigenvalues of the matrix

$$A = \begin{pmatrix} 3 & 1 & 1 \\ 1 & 5 & 2 \\ 1 & 2 & 7 \end{pmatrix}.$$

C. Specialized Matrix Techniques

A matrix U is "unitary" if $U^\dagger U = I$, that is, if $U^\dagger = U^{-1}$. This property is helpful since the conjugation and transposition processes are much easier to carry out than the inversion process.

A matrix H is "hermitian" if $H^\dagger = H$. This property is also advantageous because: (1) the real part of H is symmetric, and (2) the imaginary part of H is antisymmetric. To illustrate, let

$$H = S + iA, \tag{1}$$

where S is the real part and A is the imaginary part. The conjugate of H is

$$H^* = S - iA. \tag{2}$$

The "hermitian conjugate" of H or "tranjugate" of H is

$$H^\dagger = \widetilde{H^*} = \widetilde{S - iA} = S + iA = H. \tag{3}$$

Observe the change in the meaning of the asterisk. Therefore

$$S = \tilde{S}, \qquad (S \text{ is symmetric}) \tag{4}$$

and

$$A = -\tilde{A}. \qquad (A \text{ is antisymmetric}) \tag{5}$$

Let

$$H = S + iA \tag{6}$$

be both unitary and hermitian or "hermitary."

Then

$$H^{\dagger} = H^{-1} = H \tag{7}$$

and it follows that

$$H^2 = I \qquad \text{(the identity matrix).} \tag{8}$$

Hence H is involutory. Moreover,

$$H^2 = (S + iA)(S + iA). \tag{9}$$

Then

$$(S^2 - A^2) + i(AS + SA) = I, \tag{10}$$

and from this it follows that

$$S^2 - A^2 = I, \tag{11}$$

and

$$AS + SA = 0 \qquad (S \text{ and } A \text{ anticommute}). \tag{12}$$

As an example take

$$H = \begin{pmatrix} \frac{1}{\sqrt{2}} & -\frac{i}{\sqrt{2}} \\ \frac{i}{\sqrt{2}} & -\frac{1}{\sqrt{2}} \end{pmatrix} = H^{\dagger} = H = S + iA \tag{13}$$

Then

$$S = \begin{pmatrix} \frac{1}{\sqrt{2}} & 0 \\ 0 & -\frac{1}{\sqrt{2}} \end{pmatrix}, \quad \text{and} \quad A = \begin{pmatrix} 0 & -\frac{1}{\sqrt{2}} \\ \frac{1}{\sqrt{2}} & 0 \end{pmatrix}. \tag{14}$$

Notice that

$$S = \frac{1}{\sqrt{2}}\sigma_z, \qquad \text{and} \quad A = \frac{-i}{\sqrt{2}}\sigma_y, \tag{15}$$

where

$$\sigma_z = \begin{pmatrix} 1 & 0 \\ 0 & -1 \end{pmatrix}, \qquad \text{and} \quad \sigma_y = \begin{pmatrix} 0 & -i \\ i & 0 \end{pmatrix} \tag{16}$$

are the familiar Pauli spin matrices.* Other hermitian-unitary or hermitary matrices may be formed from the basic set of Pauli spin matrices;

$$\begin{aligned} &I, && \frac{1}{\sqrt{2}}(\sigma_x + \sigma_y) = A, \\ &\sigma_x, && \frac{1}{\sqrt{2}}(\sigma_y + \sigma_z) = B, \\ &\sigma_y, && \frac{1}{\sqrt{2}}(\sigma_z + \sigma_x) = C, \\ &\sigma_z, && \frac{1}{\sqrt{3}}(\sigma_x + \sigma_y + \sigma_z) = D. \end{aligned} \tag{17}$$

As an exercise, it is interesting to diagonalize σ_x and σ_y by using two of the special matrices appearing in listing (17). Then for σ_y:

$$\begin{aligned} B\sigma_y B^{-1} &= \tfrac{1}{2}(\sigma_y + \sigma_z)\sigma_y(\sigma_y + \sigma_z) \\ &= \tfrac{1}{2}(\sigma_y^3 + \sigma_y^2\sigma_z + \sigma_z\sigma_y^2 + \sigma_z\sigma_y\sigma_z) \\ &= \tfrac{1}{2}(\sigma_y + 2\sigma_z + \sigma_z\sigma_y\sigma_z) \\ &= \tfrac{1}{2}(\sigma_y + 2\sigma_z - \sigma_y) \\ &= \sigma_z. \end{aligned} \tag{18}$$

Similarly for σ_x

$$C\sigma_x C^{-1} = \sigma_z. \tag{19}$$

70. (*a*) Prove that every n-square matrix is expressible as $S + T$, where S is symmetric and T is skew-symmetric.

*The third Pauli matrix is

$$\sigma_x = \begin{pmatrix} 0 & 1 \\ 1 & 0 \end{pmatrix}.$$

(*b*) Prove that the determinant of a skew-symmetric matrix of odd order is equal to zero.

(*c*) Prove that if T is a real skew-symmetric matrix, then $I + T$ is nonsingular.

71. A matrix A is called "skew-hermitian" if and only if $A = -A^*$. Prove that $AB - BA$ is skew-hermitian if A and B are skew-hermitian matrices.

72. A matrix A is called "normal" if and only if $A^*A = AA^*$.

(*a*) Prove that a necessary and sufficient condition for a matrix T that is triangular (◹$T = T$ or ◺$T = T$) to be normal is that it is diagonal (⧅$T = T$ or ◹$T =$ ◺T).

(*b*) Are unitary matrices normal?

73. (*a*) Establish the linear independence of eigenvectors belonging to distinct eigenvalues of A.

(*b*) Establish their orthogonality if A is "selfadjoint," another name for hermitian.

74. Prove that the matrix A is normal if and only if there exists a unitary matrix U such that $UA = A^*$. (*Note:* A^*, $A^\dagger$, $\bar{A}$, $\bar{A}^T$, and A^H are various ways of denoting the hermitian conjugate, which is defined simply as the complex conjugate of the transpose of A.)

Schur's lemma states: If A is any n-square matrix, then there exists a unitary matrix U such that:

$$UAU^* = \text{◺}T = \begin{pmatrix} t_{11} & & & 0 \\ t_{21} & t_{22} & & \\ \cdot & \cdot & \cdot & \\ \cdot & \cdot & & \cdot \\ \cdot & \cdot & & \cdot \\ t_{n1} & t_{n2} & \cdots & t_{nn} \end{pmatrix}. \tag{20}$$

To say that "A can be diagonalized" means that it is unitarily equivalent to a diagonal matrix:

$$UAU^* = \text{⧅}D = \begin{pmatrix} d_{11} & & & 0 \\ & d_{22} & & \\ & & \ddots & \\ 0 & & & d_{nn} \end{pmatrix}. \tag{21}$$

THEOREM. A matrix can be diagonalized if and only if it is normal.

Proof ("if" part). Let A be normal. Then $AA^* = A^*A$. Moreover,

$$UA^*AU^* = UAA^*U^*,$$

and since $U^*U = I$,

$$UA^*U^*UAU^* = UAU^*UA^*U^*, \tag{22}$$

which becomes

$$(UAU^*)^*UAU^* = UAU^*(UAU^*)^*.$$

Now choose U in such a way that UAU^* is a lower triangular matrix, $T = \triangle(t_{ij})$. Then equation (22) becomes

$$T^*T = TT^*$$

or

$$\sum_{k=1}^{n} t_{ik}\overline{t_{jk}} = \sum_{k=1}^{n} \overline{t_{ki}}t_{kj}. \tag{23}$$

Notice, however, that $t_{ij} = 0$ for $i < j$ and $\overline{t_{ji}} = 0$ for $i > j$.

Let $i = j = x$. Then

$$\sum_{k=1}^{n} t_{xk}\overline{t_{xk}} = \sum_{k=1}^{n} \overline{t_{kx}}t_{kx}, \tag{24}$$

which means that

$$\sum_{k=1}^{n} |t_{xk}|^2 = \sum_{k=1}^{n} |t_{kx}|^2, \tag{25}$$

But $t_{xk} = 0$ if $x < k$ and $t_{kx} = 0$ if $k < x$. Hence $t_{ij} = 0$ if $i \neq j$, which implies that T is diagonal and therefore A can be diagonalized.

Proof ("only if" part). The matrix A can be diagonalized if it is unitarily equivalent to a diagonal matrix, that is, $UAU^* = D$, where $D = \boxbslash D$. As a consequence $(UAU^*)^* = D^*$ is also diagonal and so is

$$DD^* = UAU^*(UAU^*)^*.$$

This yields a chain of equalities:

$$DD^* = UAU^*(UA^*U^*) = UAA^*U = D^*D \tag{26}$$

(since diagonal matrices commute) and thus

$$D^*D = (UAU^*)^*UAU^* = (UA^*U^*)UAU^* = UA^*AU^* = UAA^*U^*, \tag{27}$$

which implies that A is normal.

Let U operate on $[\sigma_y + I]$:

$$U^{-1}[\sigma_y + I]U = U^{-1}\sigma_y U + U^{-1}IU = \begin{pmatrix} 1 & 0 \\ 0 & -1 \end{pmatrix} + \begin{pmatrix} 1 & 0 \\ 0 & 1 \end{pmatrix} = \begin{pmatrix} 2 & 0 \\ 0 & 0 \end{pmatrix}. \tag{28}$$

Consider the 5 by 5 matrix:

$$M = \begin{pmatrix} 0 & -i & 0 & 0 & 0 \\ i & 0 & 0 & 0 & 0 \\ 0 & 0 & 3 & 0 & 0 \\ 0 & 0 & 0 & 1 & -i \\ 0 & 0 & 0 & i & 1 \end{pmatrix} = \begin{pmatrix} \sigma_y & \vdots & 0 & \vdots & 0 \\ \cdots & & \cdots & & \cdots \\ 0 & \vdots & 3 & \vdots & 0 \\ \cdots & & \cdots & & \cdots \\ 0 & \vdots & 0 & \vdots & \sigma_y + I \end{pmatrix}. \tag{29}$$

A suitable U for diagonalizing M is

$$U = \begin{pmatrix} U & \vdots & 0 & \vdots & 0 \\ \cdots & & \cdots & & \cdots \\ 0 & \vdots & I & \vdots & 0 \\ \cdots & & \cdots & & \cdots \\ 0 & \vdots & 0 & \vdots & U \end{pmatrix} = \begin{pmatrix} \frac{1}{\sqrt{2}} & \frac{1}{\sqrt{2}} & 0 & 0 & 0 \\ \frac{i}{\sqrt{2}} & \frac{-i}{\sqrt{2}} & 0 & 0 & 0 \\ 0 & 0 & 1 & 0 & 0 \\ 0 & 0 & 0 & \frac{1}{\sqrt{2}} & \frac{1}{\sqrt{2}} \\ 0 & 0 & 0 & \frac{i}{\sqrt{2}} & \frac{-i}{\sqrt{2}} \end{pmatrix}, \tag{30}$$

and therefore

$$U^{-1} = \begin{pmatrix} \frac{1}{\sqrt{2}} & \frac{-i}{\sqrt{2}} & 0 & 0 & 0 \\ \frac{1}{\sqrt{2}} & \frac{i}{\sqrt{2}} & 0 & 0 & 0 \\ 0 & 0 & 1 & 0 & 0 \\ 0 & 0 & 0 & \frac{1}{\sqrt{2}} & \frac{-i}{\sqrt{2}} \\ 0 & 0 & 0 & \frac{1}{\sqrt{2}} & \frac{i}{\sqrt{2}} \end{pmatrix}. \tag{31}$$

The diagonalized matrix, M', turns out to be

$$M' = \begin{pmatrix} 1 & 0 & 0 & 0 & 0 \\ 0 & -1 & 0 & 0 & 0 \\ 0 & 0 & 3 & 0 & 0 \\ 0 & 0 & 0 & 2 & 0 \\ 0 & 0 & 0 & 0 & 0 \end{pmatrix}. \tag{32}$$

Consider the product $M\mathbf{v}$ of matrix M by vector $\mathbf{v}$, which is to be transformed to a primed system, thus becoming $M'\mathbf{v}'$. Restate the product inserting the identity matrix as a buffer:

$$M\mathbf{v} = MUU^{-1}\mathbf{v}, \tag{33}$$

Then

$$U^{-1}M\mathbf{v} = U^{-1}MUU^{-1}\mathbf{v} = M'U^{-1}\mathbf{v}, \tag{34}$$

and finally, letting

$$U^{-1}\mathbf{v} = \mathbf{v}' \tag{35}$$

produces

$$U^{-1}(M\mathbf{v}) = M'\mathbf{v}'. \tag{36}$$

So the vector $M\mathbf{v}$ has been transformed to the vector $M'\mathbf{v}'$ by means of the matrix U. As a consequence, an arbitrary vector $\mathbf{r}$ in the unprimed system can be transformed to the primed system by the same matrix, U, as follows:

$$\mathbf{r}' \doteq U^{-1}\mathbf{r}. \tag{37}$$

Assume the vector to be transformed to the system in which the original matrix M is diagonal is given as

$$\mathbf{r} = \begin{pmatrix} 1 \\ a \\ i \\ b \\ -1 \end{pmatrix}. \tag{38}$$

Then

$$\mathbf{v}' = U^{-1}\mathbf{v} = \begin{pmatrix} U^{-1} & \vdots & \mathbf{0} \\ \cdots\cdots & I & \cdots\cdots \\ \mathbf{0} & \vdots & U^{-1} \end{pmatrix} \begin{pmatrix} 1 \\ a \\ \cdots \\ i \\ \cdots \\ b \\ -1 \end{pmatrix} = \begin{pmatrix} U^{-1}\begin{pmatrix} 1 \\ a \end{pmatrix} \\ I \quad (i) \\ U^{-1}\begin{pmatrix} b \\ -1 \end{pmatrix} \end{pmatrix} = \begin{pmatrix} \frac{1}{\sqrt{2}}(1 - ia) \\ \frac{1}{\sqrt{2}}(1 + ia) \\ i \\ \frac{1}{\sqrt{2}}(b + i) \\ \frac{1}{\sqrt{2}}(b - i) \end{pmatrix}. \tag{39}$$

Note here that

$$U = \begin{pmatrix} \frac{1}{\sqrt{2}} & \frac{1}{\sqrt{2}} \\ \frac{i}{\sqrt{2}} & \frac{-i}{\sqrt{2}} \end{pmatrix}. \tag{40}$$

The vector $\mathbf{v}$ is not claimed to be an eigenvector. The eigenvectors can be obtained from the columns of the U matrix:

$$\text{Eigenvalues} \rightarrow \quad +1, \quad -1, \quad 3, \quad +2, \quad 0.$$

$$\text{Eigenvectors} \rightarrow \begin{pmatrix} \frac{1}{\sqrt{2}} \\ \frac{i}{\sqrt{2}} \\ 0 \\ 0 \\ 0 \end{pmatrix}, \begin{pmatrix} \frac{1}{\sqrt{2}} \\ \frac{-i}{\sqrt{2}} \\ 0 \\ 0 \\ 0 \end{pmatrix}, \begin{pmatrix} 0 \\ 0 \\ 1 \\ 0 \\ 0 \end{pmatrix}, \begin{pmatrix} 0 \\ 0 \\ 0 \\ \frac{1}{\sqrt{2}} \\ \frac{i}{\sqrt{2}} \end{pmatrix}, \begin{pmatrix} 0 \\ 0 \\ 0 \\ \frac{1}{\sqrt{2}} \\ \frac{-i}{\sqrt{2}} \end{pmatrix}.$$

Note that the eigenvectors are orthogonal to each other, that is:

$$\tilde{\mathbf{v}}_{\lambda_i}{}^* \cdot \mathbf{v}_{\lambda_j} = \delta_{ij} = \begin{Bmatrix} 0, & i \neq j \\ 1, & i = j \end{Bmatrix}. \tag{41}$$

Consider another choice of Ʉ given by:

$$a = \frac{1}{\sqrt{2}}, \quad b = \frac{-i}{\sqrt{2}}, \tag{42}$$

which defines the 2 by 2 matrix U.

Then

$$Ʉ = \begin{pmatrix} \frac{1}{\sqrt{2}} & \frac{-i}{\sqrt{2}} & 0 & 0 & 0 \\ \frac{i}{\sqrt{2}} & \frac{-1}{\sqrt{2}} & 0 & 0 & 0 \\ 0 & 0 & 1 & 0 & 0 \\ 0 & 0 & 0 & \frac{1}{\sqrt{2}} & \frac{-i}{\sqrt{2}} \\ 0 & 0 & 0 & \frac{i}{\sqrt{2}} & \frac{-1}{\sqrt{2}} \end{pmatrix}. \tag{43}$$

This matrix is hermitary, and since the ith row is identical to the hermitian conjugate of the ith column, the eigenvectors can be taken as rows or as columns of U.

EXERCISES

75. (*a*) What are the eigenvalues and eigenvectors of the matrix

$$\begin{pmatrix} 2 & 0 & 0 \\ 0 & 5 & 11 \\ 0 & 4 & 9 \end{pmatrix}?$$

(*b*) Compute the eigenvalues and eigenvectors of the matrix

$J_2 \dotplus I_1$. (see page 105 for notation)

76. Prove that a normal matrix that has real eigenvalues is hermitian.

77. The exponential function exp A for a matrix argument is defined by the formal power series

$$e^A = I + \sum_{r=1}^{\infty} \frac{1}{r!} A^r.$$

Find $e^{-i\sigma_y}$.

78. An n by m matrix $X = A^+$ is called the "pseudo-inverse"* of an m by n matrix A if and only if:

(*a*) $AXA = A$.
(*b*) $XAX = X$.
(*c*) $(AX)^* = AX$.
(*d*) $(XA)^* = XA$.

Find the pseudo-inverse $\begin{pmatrix} x \\ y \end{pmatrix}$ of the matrix (a, b) where a and b are complex numbers.

The hermitian matrix,

$$H = \begin{pmatrix} 0 & -i & 0 \\ i & 0 & -i \\ 0 & i & 0 \end{pmatrix}, \tag{44}$$

is diagonalized by decomposing it into spin matrices, since their diagonalization already has been accomplished.

The matrix H is identified with the y component of the quantum mechanical matrix that represents angular momentum of a particle or a system of particles.†

The problem is to find a matrix U such that

$$UHU^{-1} = H_D, \tag{45}$$

where H_D is the diagonalized form of H. It follows from equation (45) that

$$UH = H_D U. \tag{46}$$

The characteristic equation for H is

$$|H - \lambda I| = \begin{vmatrix} -\lambda & -i & 0 \\ i & -\lambda & -i \\ 0 & i & -\lambda \end{vmatrix} = -\lambda^3 + 2\lambda = 0, \tag{47}$$

or

$$\lambda^3 - 2\lambda = \lambda(\lambda^2 - 2) = 0; \tag{48}$$

hence

$$\lambda_i = 0, \quad +\sqrt{2}, \quad -\sqrt{2}. \tag{49}$$

* R. Penrose, "A Generalized Inverse for Matrices," *Proc. Cambridge Philos. Soc.*, **51,** 406–413 (1955); A. Ben-Israel and S. J. Wersau, "An Elimination Method for Computing the Generalized Inverse for an Arbitrary Complex Matrix," *Journal of the ACM*, **10**, 532–537 (1963).
†Schiff, *Quantum Mechanics*, McGraw-Hill, 1949, p. 144.

Tentatively write U as

$$U = \begin{pmatrix} a & b & c \\ d & e & f \\ g & h & j \end{pmatrix}, \tag{50}$$

where the letter i is avoided because it is reserved to denote the imaginary unit. Then equation (38) becomes

$$\begin{pmatrix} a & b & c \\ d & e & f \\ g & h & j \end{pmatrix}\begin{pmatrix} 0 & -i & 0 \\ i & 0 & -i \\ 0 & i & 0 \end{pmatrix} = \begin{pmatrix} \sqrt{2} & 0 & 0 \\ 0 & 0 & 0 \\ 0 & 0 & -\sqrt{2} \end{pmatrix}\begin{pmatrix} a & b & c \\ d & e & f \\ g & h & j \end{pmatrix}, \tag{51}$$

where the diagonalized form for H has been inserted on the right side as specified in equation (46).

This produces the following system of equations:

$$\begin{cases} b = -i\sqrt{2}a \\ e = 0, \\ f = d, \\ h = i\sqrt{2}g, \\ \quad\text{and} \\ h = -i\sqrt{2}j. \end{cases} \tag{52}$$

From the last two equations in (52) it follows that:

$$j = -g. \tag{53}$$

Thus,

$$U = \begin{pmatrix} a & -i\sqrt{2}a & -a \\ d & 0 & d \\ g & i\sqrt{2}g & -g \end{pmatrix}. \tag{54}$$

Now, from experience with the matrix for angular momentum already referred to, it is known that the unitary matrix U is also hermitian, so

$$U = \begin{pmatrix} a & -i\sqrt{2}a & -a \\ i\sqrt{2}a & 0 & i\sqrt{2}a \\ -a & -i\sqrt{2}a & a \end{pmatrix}, \tag{55}$$

where a remains the only unknown. Solving for a by the unitary and hermitian properties of U:

$$UU^{-1} = UU^{\dagger} = UU = U^2 \tag{56}$$

gives

$$\begin{pmatrix} a & -i\sqrt{2}a & -a \\ i\sqrt{2}a & 0 & i\sqrt{2}a \\ -a & -i\sqrt{2}a & a \end{pmatrix}\begin{pmatrix} a & -i\sqrt{2}a & -a \\ i\sqrt{2}a & 0 & i\sqrt{2}a \\ -a & -i\sqrt{2}a & a \end{pmatrix} = \begin{pmatrix} 4a^2 & 0 & 0 \\ 0 & 4a^2 & 0 \\ 0 & 0 & 4a^2 \end{pmatrix}. \tag{57}$$

And since $U^2 = I$:

$$4a^2 = 1, \qquad \text{and} \quad a = \pm\tfrac{1}{2}. \tag{58}$$

Choosing the plus sign,

$$U = \begin{pmatrix} \frac{1}{2} & \frac{-i}{\sqrt{2}} & -\frac{1}{2} \\ \frac{i}{\sqrt{2}} & 0 & \frac{i}{\sqrt{2}} \\ -\frac{1}{2} & \frac{-i}{\sqrt{2}} & \frac{1}{2} \end{pmatrix}. \tag{59}$$

The matrix U that diagonalizes H is composed of the eigenvectors of H:

$$U = [V_1, V_2, V_3] = \left[\begin{pmatrix} \frac{1}{2} \\ \frac{i}{\sqrt{2}} \\ -\frac{1}{2} \end{pmatrix}\begin{pmatrix} \frac{-i}{\sqrt{2}} \\ 0 \\ \frac{-i}{\sqrt{2}} \end{pmatrix}\begin{pmatrix} -\frac{1}{2} \\ \frac{i}{\sqrt{2}} \\ \frac{1}{2} \end{pmatrix}\right]. \tag{60}$$

It can be checked readily that these vectors are orthogonal:

$$V_1{}^{\dagger}V_2 = V_1{}^{\dagger}V_3 = V_2{}^{\dagger}V_3 = 0. \tag{61}$$

The diagonalization of the original matrix is given by statement (45).

Explicitly carrying out this operation yields

$$\begin{pmatrix} \frac{1}{2} & \dfrac{-i}{\sqrt{2}} & -\frac{1}{2} \\ \dfrac{i}{2} & 0 & \dfrac{i}{\sqrt{2}} \\ -\frac{1}{2} & \dfrac{-i}{\sqrt{2}} & \frac{1}{2} \end{pmatrix} \begin{pmatrix} 0 & -i & 0 \\ i & 0 & -i \\ 0 & i & 0 \end{pmatrix} \begin{pmatrix} \frac{1}{2} & \dfrac{-i}{\sqrt{2}} & -\frac{1}{2} \\ \dfrac{i}{\sqrt{2}} & 0 & \dfrac{i}{\sqrt{2}} \\ -\frac{1}{2} & \dfrac{-i}{\sqrt{2}} & \frac{1}{2} \end{pmatrix}$$

$$= \begin{pmatrix} \frac{1}{2} & \dfrac{-i}{\sqrt{2}} & -\frac{1}{2} \\ \dfrac{i}{\sqrt{2}} & 0 & \dfrac{i}{\sqrt{2}} \\ -\frac{1}{2} & \dfrac{-i}{\sqrt{2}} & \frac{1}{2} \end{pmatrix} \begin{pmatrix} \dfrac{1}{\sqrt{2}} & 0 & \dfrac{1}{\sqrt{2}} \\ i & 0 & -i \\ \dfrac{-1}{\sqrt{2}} & 0 & \dfrac{-1}{\sqrt{2}} \end{pmatrix} \tag{62}$$

$$= \begin{pmatrix} \sqrt{2} & 0 & 0 \\ 0 & 0 & 0 \\ 0 & 0 & -\sqrt{2} \end{pmatrix} = H_D.$$

Since the eigenvectors of any system are directed along the principal axes of the system, in the diagonalized system the normalized eigenvectors are

$$V_1 = \begin{pmatrix} 1 \\ 0 \\ 0 \end{pmatrix}, \quad V_2 = \begin{pmatrix} 0 \\ 1 \\ 0 \end{pmatrix}, \quad \text{and} \quad V_3 = \begin{pmatrix} 0 \\ 0 \\ 1 \end{pmatrix}. \tag{63}$$

Note that the elements of these column vectors form the identity matrix:

$$I = |V_1, V_2, V_3| = \begin{pmatrix} 1 & 0 & 0 \\ 0 & 1 & 0 \\ 0 & 0 & 1 \end{pmatrix}. \tag{64}$$

To prove that the eigenvectors in the diagonal system are indeed correct, proceed as follows. The U matrix that diagonalizes H also transforms its eigenvectors, so

$$UV_\lambda = \begin{pmatrix} \frac{1}{2} & \frac{-i}{\sqrt{2}} & -\frac{1}{2} \\ \frac{i}{\sqrt{2}} & 0 & \frac{i}{\sqrt{2}} \\ -\frac{1}{2} & \frac{-i}{\sqrt{2}} & \frac{1}{2} \end{pmatrix} V_\lambda = \begin{pmatrix} V_1^\dagger \\ V_2^\dagger \\ V_3^\dagger \end{pmatrix} V_\lambda = \begin{pmatrix} V_1^\dagger V_\lambda \\ V_2^\dagger V_\lambda \\ V_3^\dagger V_\lambda \end{pmatrix}. \tag{65}$$

Therefore, for the first eigenvalue,

$$UV_1 = \begin{pmatrix} \frac{1}{2} & \frac{-i}{\sqrt{2}} & -\frac{1}{2} \\ \frac{i}{\sqrt{2}} & 0 & \frac{i}{\sqrt{2}} \\ -\frac{1}{2} & \frac{-i}{\sqrt{2}} & \frac{1}{2} \end{pmatrix} \begin{pmatrix} \frac{1}{2} \\ \frac{i}{\sqrt{2}} \\ -\frac{1}{2} \end{pmatrix} = \begin{pmatrix} V_1^\dagger \\ V_2^\dagger \\ V_3^\dagger \end{pmatrix} V_1 = \begin{pmatrix} V_1^\dagger V_1 \\ V_2^\dagger V_1 \\ V_3^\dagger V_1 \end{pmatrix} = \begin{pmatrix} 1 \\ 0 \\ 0 \end{pmatrix} \tag{66}$$

Similar results would be obtained for the second and third eigenvalues.

EXERCISE 79

79. (*a*) If H is hermitian, then $\lambda(H)$ is a set of real numbers.

(*b*) If $\lambda_i(H) \geq 0$, $i = 1, 2, \cdots, n$; then H is called "positive definite." Prove that all n roots of the determinantal equation:

$$|A - \lambda B| = 0,$$

with A real symmetric and B positive definite symmetric, are real.

(*c*) Prove that AA^T is positive definite provided that A is nonsingular.

(*d*) What about A^*A?

D. Matrix Rotations and Reflections

As described earlier, the main diagonal of a square matrix $A = (a_{ij})$ consists of the line of elements containing $a_{11}, a_{22}, \cdots, a_{nn}$. This line also may be called the "principal diagonal." Lines of elements running parallel to the main diagonals of A form other diagonals—the two lines adjacent to the main diagonal are called "co-diagonals." The "counter diagonals" of A are the lines of elements that run at right angles to the diagonals of A. The "secondary diagonal" of A is the main counterdiagonal—the line of elements consisting of $a_{1,n}, a_{2,n-1}, \cdots, a_{n,1}$. Consider the following definition:

DEFINITION. A matrix C with all elements not lying on the secondary diagonal equal to zero is called a "counterdiagonal matrix":

$$C = \begin{pmatrix} \mathbf{0} & & & c_{1,n} \\ & & & c_{2,n-1} \\ & & \cdot & \\ & \cdot & & \\ c_{n,1} & & & \mathbf{0} \end{pmatrix}. \tag{1}$$

Now take that particular counter-diagonal matrix defined by $c_{i,n+1-i} = 1$:

$$K = \begin{pmatrix} 0 \cdots 01 \\ 0 \cdots 10 \\ \cdot \quad \cdot \; \cdot \\ \cdot \quad \cdot \quad \cdot \\ \cdot \; \cdot \quad \cdot \\ 1 \cdots 00 \end{pmatrix}. \tag{2}$$

As this "counterscalar matrix" has n 1's along the secondary diagonal and 0's elsewhere, it is called the "counteridentity of order n" (see Fig. 2). Since

$$K = K^T = \tilde{K}, \tag{3}$$

where the T and tilde "~" are two equally popular ways of expressing the transpose of a matrix, the counteridentity is a "symmetric matrix."

$$K_2 = \begin{pmatrix} 0 & 1 \\ 1 & 0 \end{pmatrix} = \sigma_x \quad \text{Pauli} \qquad K_3 = \begin{pmatrix} 0 & 0 & 1 \\ 0 & 1 & 0 \\ 1 & 0 & 0 \end{pmatrix}$$

$$K_4 = \begin{pmatrix} 0 & 0 & 0 & 1 \\ 0 & 0 & 1 & 0 \\ 0 & 1 & 0 & 0 \\ 1 & 0 & 0 & 0 \end{pmatrix} = \begin{bmatrix} 0 & \sigma_x \\ \sigma_x & 0 \end{bmatrix} = \alpha_x \quad \text{Dirac}$$

Properties: Involutory Unitary Hermitian

$$K = K^{-1} \to K^{-1} = K^\dagger \to K^\dagger = K$$

Figure 2. Counteridentity matrix K.

In addition, since

$$K^2 = I, \qquad K = K^{-1}, \tag{4}$$

where the "2" signifies the square ($K^2 = KK$) and the "−1" signifies the inverse ($KK^{-1} = I$), the counteridentity is said to be an "involutory matrix." And since

$$KK^T = I, \qquad K^T = K^{-1} \tag{5}$$

K also is an orthogonal matrix. In the same way, since

$$K = K^* = \bar{K}^T, \tag{6a}$$

where the bar "-" over the K denotes the complex conjugate, K is a hermitian matrix. Observe that many books write this same property as

$$K = \tilde{K}^*, \tag{6b}$$

where the asterisk or star now replaces the bar to signify complex conjugate and the tilde replaces the T to signify transpose. Moreover, since

$$K^* = K^{-1}, \qquad \text{or} \quad \bar{K}^T = K^{-1}, \qquad \text{or} \quad \tilde{K}^* = K^{-1}, \tag{7}$$

K is a unitary matrix. Finally, although this does not exhaust its many peculiar properties, K is called a permutation matrix, because it has just a single one in each row and in each column and all the rest of the elements are zero.

The determinant of K may be found to satisfy:

$$\det K^2 = |K^2| = |K|\,|K| = |I| = 1; \tag{8}$$

hence

$$\det K = |K| = \pm 1. \tag{9}$$

It can be determined easily that the 1 by 1 K matrix has the determinant $|K| = +1$, the 2 by 2 K matrix has the determinant

$$K_2 = \begin{pmatrix} 0 & 1 \\ 1 & 0 \end{pmatrix} = 0 - 1 = -1, \tag{10}$$

the 3 by 3 K matrix has the determinant

$$|K_3| = \begin{pmatrix} 0 & 0 & 1 \\ 0 & 1 & 0 \\ 1 & 0 & 0 \end{pmatrix} = +1 \begin{pmatrix} 0 & 1 \\ 1 & 0 \end{pmatrix} = +1\,|K_2| = -1, \tag{11}$$

the 4 by 4 K matrix has the determinant

$$|K_4| = \begin{vmatrix} 0 & 0 & 0 & 1 \\ 0 & 0 & 1 & 0 \\ 0 & 1 & 0 & 0 \\ 1 & 0 & 0 & 0 \end{vmatrix} = -1 \begin{vmatrix} 0 & 0 & 1 \\ 0 & 1 & 0 \\ 1 & 0 & 0 \end{vmatrix} = -1\,|K_3| = +1, \quad (12)$$

and the 5 by 5 K matrix has the determinant

$$|K_5| = \begin{vmatrix} 0 & 0 & 0 & 0 & 1 \\ 0 & 0 & 0 & 1 & 0 \\ 0 & 0 & 1 & 0 & 0 \\ 0 & 1 & 0 & 0 & 0 \\ 1 & 0 & 0 & 0 & 0 \end{vmatrix} = +1 \begin{vmatrix} 0 & 0 & 0 & 1 \\ 0 & 0 & 1 & 0 \\ 0 & 1 & 0 & 0 \\ 1 & 0 & 0 & 0 \end{vmatrix} = +1\,|K_4| = +1. \quad (13)$$

A tabulation of the data appearing thus far in evaluations (10), (11), (12), and (13),

n	1	2	3	4	5
det	+1	−1	−1	+1	+1

leads to the following conjecture:

n	⋯	$4k$	$4k+1$	$4k+2$	$4k+3$	$4k+4$
det	⋯	+1	+1	−1	−1	+1

The proof is by induction on k. By the checkerboard rule of signs, the element in position i, j is given the sign:

$$s = (-1)^{i+j}.$$

For this reason the one in the first column and last row of K_n will have the sign

$$s = (-1)^{1+n}.$$

Thus when $n = 4k + 1, s = (-1)^{4k+2} = +1$; when $n = 4k + 2, s = -1$; when $n = 4k + 3, s = 1$, when $n = 4k + 4, s = -1$; and when $n = 4k + 5$, $s = 1$. Therefore, if $|K_{4k}| = 1$, then

$$\begin{aligned} |K_{4k+1}| &= 1\,|K_{4k}| = 1, \\ |K_{4k+2}| &= -1\,|K_{4k+1}| = -1 \\ |K_{4k+3}| &= 1\,|K_{4k+2}| = -1, \\ |K_{4k+4}| &= -1\,|K_{4k+3}| = 1, \end{aligned}$$

and

$$|K_{4k+5}| = 1\,|K_{4k+4}| = 1.$$

Since $|K_4| = 1$, it follows that $|K_{4k}| = 1$ for $k = 1$; the induction is complete, and the conjecture is verified. Consequently, the determinant of K_n may be written

$$|K_n| = (-1)^{[n/2]}, \tag{14}$$

where $[n/2]$ denotes the "largest integer less than or equal to $n/2$," for example,

n	1	2	3	4	5	6	7	8	$\cdots$
$[n/2]$	0	1	1	2	2	3	3	4	$\cdots$

EXERCISES

80. Prove result (14) an easier way. Evaluate tr K_n. Does tr AK equal tr KA?

81. Find a square root of K_2, that is, a matrix X such that

$$X^2 = \begin{pmatrix} 0 & 1 \\ 1 & 0 \end{pmatrix}.$$

82. The sum of the eigenvalues of a matrix A is given by its trace:

$$\operatorname{tr} A = \sum_{i=1}^{n} \lambda_i = \lambda_1 + \lambda_2 + \cdots + \lambda_n;$$

the product of the eigenvalues of a matrix A is given by its determinant:

$$|A| = \prod_{i=1}^{n} \lambda_i = \lambda_1 \lambda_2 \cdots \lambda_n;$$

the eigenvalues of a hermitian matrix are real:

$$\lambda_i = \bar{\lambda}_i;$$

and the eigenvalues of a unitary matrix are of unit magnitude:

$$|\lambda_i|^2 = \lambda_i \bar{\lambda}_i = 1.$$

Are these clues sufficient to determine the eigenvalues of the counteridentity matrix of order n? If so, what are they?

83. Does K_n correspond to a proper rotation? What is its effect on a vector $\mathbf{x} = (x_1, x_2, \cdots x_n)$ and $\mathbf{x}$ transpose?

84. Discover a counterdiagonal 3-square involutory matrix.

85. Define three additional new operators ◸, ⊠, and ◿, using the K matrix.

The equation

$$f(\lambda) = \det (A - \lambda I) = |A - \lambda I| = 0, \tag{15}$$

with A and the identity I both n-square matrices, is known as the "characteristic equation of A." Thus it is reasonable to call the equation

$$g(\mu) = \det (A - \mu K) = |A - \mu K| = 0, \tag{16}$$

with A and the counteridentity K both n-square matrices, the "countercharacteristic equation of A." A brief inspection of the general 2 by 2 case is informative.

Let

$$A = \begin{pmatrix} a & b \\ c & d \end{pmatrix},$$

then

$$f(\lambda) = \begin{vmatrix} a - \lambda & b \\ c & d - \lambda \end{vmatrix}$$

$$= (a - \lambda)(d - \lambda) - bc = \lambda^2 - (a + d)\lambda + \underline{ad - bc} = 0 \tag{17}$$

and

$$g(\mu) = \begin{vmatrix} a & b - \mu \\ c - \mu & d \end{vmatrix} = \underline{ad - bc} + (b + c)\mu - \mu^2 = 0 \tag{18}$$

show no apparent connection between the roots of equation (17):

$$\lambda_{\pm} = \frac{a + d \pm \sqrt{a^2 - 2ad + d^2 + 4bc}}{2} \tag{19}$$

and the roots of equation (18):

$$\mu_{\pm} = \frac{b + c \mp \sqrt{b^2 - 2bc + c^2 + 4ad}}{2}, \tag{20}$$

except that in magnitude the products of the eigenvalues and countereigenvalues:

$$|(\lambda_{+})(\lambda_{-})| = |(\mu_{+})(\mu_{-})| = |ad - bc| = |\det A|, \tag{21}$$

are equal to each other and to the absolute value of the determinant, a fact that remains true for any order n.

EXERCISES

86. Prove that if λ is an eigenvalue of A, it is a countereigenvalue of KA.

87. Prove that if μ is a countereigenvalue of A, it is an eigenvalue of KA.

88. Find a formula for determining the eigenvalues of the general "double diagonal matrix" defined by

$$X = \begin{pmatrix} x_{11} & 0 & \cdots & 0 & x_{1n} \\ 0 & x_{22} & \cdots & x_{2,n-1} & 0 \\ \cdot & \cdot & & \cdot & \cdot \\ \cdot & \cdot & & \cdot & \cdot \\ \cdot & \cdot & & \cdot & \cdot \\ 0 & x_{n-1,2} & \cdots & x_{n-1,n-1} & 0 \\ x_{n1} & 0 & \cdots & 0 & x_{nn} \end{pmatrix}.$$

Specialize this formula so that it applies strictly to counterdiagonal matrices.

89. What are the eigenvalues of the following matrices:

$$(a)\ X = \begin{pmatrix} 1 & 0 & 2 \\ 0 & 3 & 0 \\ 4 & 0 & 5 \end{pmatrix}.$$

$$(b)\ X = \begin{pmatrix} 0 & 0 & 0 & 0 & 1 \\ 0 & 0 & 0 & 2 & 0 \\ 0 & 0 & 4 & 0 & 0 \\ 0 & 8 & 0 & 0 & 0 \\ 16 & 0 & 0 & 0 & 0 \end{pmatrix}.$$

$$(c)\ X = \begin{pmatrix} 1 & 0 & 0 & -2 \\ 0 & 3 & -4 & 0 \\ 0 & 5 & -6 & 0 \\ 7 & 0 & 0 & -8 \end{pmatrix}.$$

90. Prove that the rank of a square matrix is not less than the number of its nonzero countereigenvalues.

If A is an arbitrary m by n matrix,

$$A = \begin{pmatrix} a_{11} & a_{12} \cdots a_{1n} \\ a_{21} & a_{22} \cdots a_{2n} \\ \cdot & \cdot \quad \cdot \quad \cdot \\ \cdot & \cdot \quad \cdot \quad \cdot \\ \cdot & \cdot \quad \cdot \quad \cdot \\ a_{m1} & a_{m2} \cdots a_{mn} \end{pmatrix}, \tag{22}$$

and K is the m-square or n-square counteridentity,

$$\underset{m \text{ or } n}{K} = \begin{pmatrix} 0 \cdots 0 & 1 \\ 0 \cdots 1 & 0 \\ \cdot \quad \cdot \quad \cdot & \cdot \\ \cdot \quad \cdot \quad \cdot & \cdot \\ \cdot \quad \cdot \quad \cdot & \cdot \\ 1 \cdots 0 & 0 \end{pmatrix}, \tag{23}$$

whichever order is appropriate, then the set of eight closely related matrices,

$$\{A\} = \{A, A^TK, KAK, KA^T, A^T, AK, KA^TK, KA\}, \tag{24}$$

forms the "octic set" of A. For displaying the gross character of the octic set, the following mnemonic based on the uppercase letter F is a handy device:

$$\mathsf{F} \quad \begin{matrix} \text{1st row of } A \\ i\text{th row of } A \\ \text{1st column of } A \end{matrix} \tag{25}$$

Consider the 3 by 4 matrix:

$$A = \begin{pmatrix} a & b & c & d \\ e & f & g & h \\ j & k & l & m \end{pmatrix}.$$

Using the mnemonic (25), the members of the octic set of A, called the "postures of A," may be calculated as follows:

$$A = \begin{pmatrix} a & b & c & d \\ e & f & g & h \\ j & k & l & m \end{pmatrix} = \mathsf{F} \tag{26a}$$

$$A^TK = \begin{pmatrix} a & b & c & d \\ e & f & g & h \\ j & k & l & m \end{pmatrix}^T \begin{pmatrix} 0 & 0 & 1 \\ 0 & 1 & 0 \\ 1 & 0 & 0 \end{pmatrix}$$

$$= \begin{pmatrix} a & e & j \\ b & f & k \\ c & g & l \\ d & h & m \end{pmatrix} \begin{pmatrix} 0 & 0 & 1 \\ 0 & 1 & 0 \\ 1 & 0 & 0 \end{pmatrix} = \begin{pmatrix} j & e & a \\ k & f & b \\ l & g & c \\ m & h & d \end{pmatrix} = \text{ㄱ} . \qquad (26b)$$

$$KAK = \begin{pmatrix} 0 & 0 & 1 \\ 0 & 1 & 0 \\ 1 & 0 & 0 \end{pmatrix} \begin{pmatrix} a & b & c & d \\ e & f & g & h \\ j & k & l & m \end{pmatrix} \begin{pmatrix} 0 & 0 & 0 & 1 \\ 0 & 0 & 1 & 0 \\ 0 & 1 & 0 & 0 \\ 1 & 0 & 0 & 0 \end{pmatrix}$$

$$= \begin{pmatrix} j & k & l & m \\ e & f & g & h \\ a & b & c & d \end{pmatrix} \begin{pmatrix} 0 & 0 & 0 & 1 \\ 0 & 0 & 1 & 0 \\ 0 & 1 & 0 & 0 \\ 1 & 0 & 0 & 0 \end{pmatrix} = \begin{pmatrix} m & l & k & j \\ h & g & f & e \\ d & c & b & a \end{pmatrix} = \text{Ⅎ} . \qquad (26c)$$

$$KA^T = \begin{pmatrix} 0 & 0 & 0 & 1 \\ 0 & 0 & 1 & 0 \\ 0 & 1 & 0 & 0 \\ 1 & 0 & 0 & 0 \end{pmatrix} \begin{pmatrix} a & b & c & d \\ e & f & g & h \\ j & k & l & m \end{pmatrix}^T$$

$$= \begin{pmatrix} 0 & 0 & 0 & 1 \\ 0 & 0 & 1 & 0 \\ 0 & 1 & 0 & 0 \\ 1 & 0 & 0 & 0 \end{pmatrix} \begin{pmatrix} a & e & j \\ b & f & k \\ c & g & l \\ d & h & m \end{pmatrix} = \begin{pmatrix} d & h & m \\ c & g & l \\ b & f & k \\ a & e & j \end{pmatrix} = \text{ш} . \qquad (26d)$$

$$A^T = \begin{pmatrix} a & b & c & d \\ e & f & g & h \\ j & k & l & m \end{pmatrix}^T = \begin{pmatrix} a & e & j \\ b & f & k \\ c & g & l \\ d & h & m \end{pmatrix} = \text{ㅠ} . \qquad (26e)$$

$$AK = \begin{pmatrix} a & b & c & d \\ e & f & g & h \\ j & k & l & m \end{pmatrix}\begin{pmatrix} 0 & 0 & 0 & 1 \\ 0 & 0 & 1 & 0 \\ 0 & 1 & 0 & 0 \\ 1 & 0 & 0 & 0 \end{pmatrix} = \begin{pmatrix} d & c & b & a \\ h & g & f & e \\ m & l & k & j \end{pmatrix} = \text{ꟻ}. \quad (26f)$$

$$KA^TK = \begin{pmatrix} 0 & 0 & 0 & 1 \\ 0 & 0 & 1 & 0 \\ 0 & 1 & 0 & 0 \\ 1 & 0 & 0 & 0 \end{pmatrix}\begin{pmatrix} a & b & c & d \\ e & f & g & h \\ j & k & l & m \end{pmatrix}^T\begin{pmatrix} 0 & 0 & 1 \\ 0 & 1 & 0 \\ 1 & 0 & 0 \end{pmatrix}$$

$$= \begin{pmatrix} 0 & 0 & 0 & 1 \\ 0 & 0 & 1 & 0 \\ 0 & 1 & 0 & 0 \\ 1 & 0 & 0 & 0 \end{pmatrix}\begin{pmatrix} a & e & j \\ b & f & k \\ c & g & l \\ d & h & m \end{pmatrix}\begin{pmatrix} 0 & 0 & 1 \\ 0 & 1 & 0 \\ 1 & 0 & 0 \end{pmatrix}$$

$$= \begin{pmatrix} d & h & m \\ c & g & l \\ b & f & k \\ a & e & j \end{pmatrix}\begin{pmatrix} 0 & 0 & 1 \\ 0 & 1 & 0 \\ 1 & 0 & 0 \end{pmatrix} = \begin{pmatrix} m & h & d \\ l & g & c \\ k & f & b \\ j & e & a \end{pmatrix} = \text{╜}. \quad (26g)$$

$$KA = \begin{pmatrix} 0 & 0 & 1 \\ 0 & 1 & 0 \\ 1 & 0 & 0 \end{pmatrix}\begin{pmatrix} a & b & c & d \\ e & f & g & h \\ j & k & l & m \end{pmatrix} = \begin{pmatrix} j & k & l & m \\ e & f & g & h \\ a & b & c & d \end{pmatrix} = \text{ꓞ}. \quad (26h)$$

Thus, as an aid to memory:

$A =$ F	$A^T =$ ╓
$A^TK =$ ╖	$AK =$ ꟻ
$KAK =$ Ⅎ	$KA^TK =$ ╜
$KA^T =$ ╙	$KA =$ ꓞ

and, to repeat,

F — 1st row of A; ith row of A; 1st column of A.

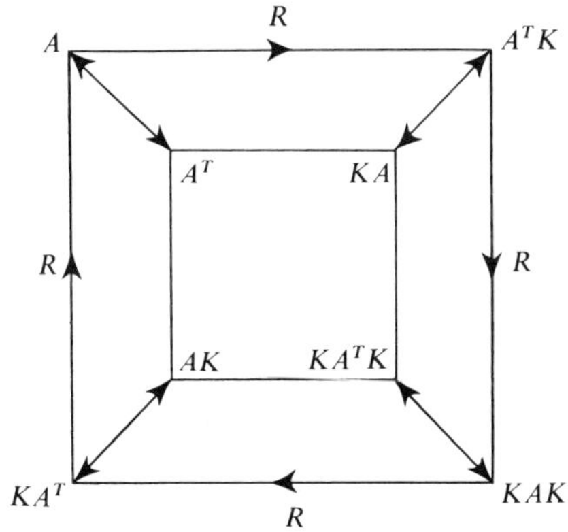

Figure 3. The octic set of A.

Consequently, the octic set of F (or A) has the following form:

$$\{F\} = \{\text{F}, ┳┓, Ⅎ, ┗┻, ┏┳, ꟻ, ┻┛, ⊧\}. \tag{27}$$

This suggests immediately that the octic set of a matrix is a representation of the dihedral group of degree four and order eight, which describes the possible motions of a rectangular lamina. This group D_4 is generated by a 90° rotation R and a flip T over the main diagonal, that is:

$$R^4 = e, \quad T^2 = e, \tag{28}$$

where e is the group identity.

Note that each member of the octic set of A has precisely the same elements as A. Also note that the octic set is closed under both transposition and ordinary matrix multiplication by K. In fact, if R denotes the operation $R(\ ) = (\ )^T K$; then

$$\begin{gathered} R^4T^0(A) = R^4(A) = R^3(A^TK) = R^2(KAK) = R(KA^T) = A, \\ \text{and} \\ R^4T(A) = R^4(A^T) = R^3(AK) = R^2(KA^TK) = R(KA) = A^T. \end{gathered} \tag{29}$$

This is illustrated in Fig. 3.

EXERCISES

91. Redraw Fig. 3 showing:

(*a*) The octic set of F.

(*b*) The octic set of I.

(*c*) The octic set of AA^T.

(*d*) The octic set of $\begin{pmatrix} a & b \\ c & d \end{pmatrix}$.

92. Express $(A + KAK)^2$ as the sum of four squares.

93. Prove that every matrix in an octic set has the same rank.

94. Decode the following matrix cryptogram:

$$M = \begin{pmatrix} l & m & s & d & - & a \\ u & s & k & n & - & i \\ w & y & e & - & o & n \\ p & i & - & t & s & i \\ i & o & n & g & h & e \\ a & t & i & n & s & s \end{pmatrix}.$$

Clue: The key to unscrambling the secret message is provided by the decomposition:

$$J \circledH M = (A + A^T K + KAK + KA^T) \circledH M,$$

where Ⓗ denotes the Hadamard product and A is an unknown matrix consisting of zeros and ones.

To find the determinants of the postures of A, first recall that

$$|AB| = |A|\,|B| = |B|\,|A|, \tag{30a}$$

and

$$|A^T| = |A|. \tag{30b}$$

Then:

$$|A| = |A|, \tag{31a}$$

$$|A^T K| = |A^T|\,|K| = (-1)^{[n/2]}|A|, \tag{31b}$$

$$|KAK| = |K|\,|A|\,|K| = |K^2|\,|A| = |A|, \tag{31c}$$

$$|KA^T| = |K|\,|A^T| = (-1)^{[n/2]}|A|, \tag{31d}$$

$$|A^T| = |A|, \tag{31e}$$

$$|AK| = |A|\,|K| = (-1)^{[n/2]}|A|, \tag{31f}$$

$$|KA^T K| = |K|\,|A^T|\,|K| = |K^2|\,|A| = |A|, \tag{31g}$$

$$|KA| = |K|\,|A| = (-1)^{[n/2]}|A|. \tag{31h}$$

Figure 4 shows the positions of these determinants for the octic set of A.

The matrices in the octic set of A have either λ_1 or μ_i for eigenvalues. As shown by Fig. 5, the correct determination is based upon information concerning the inclination of declination of the main diagonal elements.

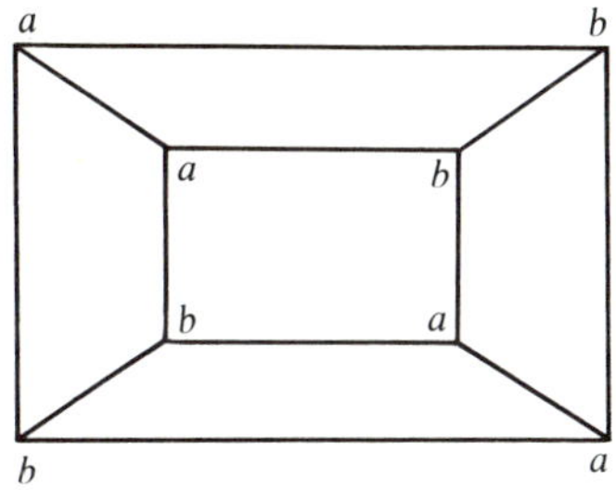

$$a = |A|$$
$$b = (-1)^{[n/2]}|A|$$

Figure 4. Distribution of determinants for the octic set of A.

Thus

$$\lambda_A = \lambda_{A^T} = \lambda_{KAK} = \lambda_{KA^TK} = \mu_{AK} = \mu_{KA^T} = \mu_{KA} = \mu_{A^TK}, \tag{32}$$

and

$$\mu_A = \mu_{A^T} = \mu_{KAK} = \mu_{KA^TK} = \lambda_{AK} = \lambda_{KA^T} = \lambda_{KA} = \lambda_{A^TK}. \tag{33}$$

Figure 6 shows the positions of these eigenvalues and countereigenvalues in the octic set of A.

The matrix A^C, called "A cispose," is defined by the equation

$$A^C - KA^TK = \begin{pmatrix} a_{mn} \cdots a_{2n} & a_{1n} \\ \cdot \quad \cdot & \cdot \\ \cdot \quad \cdot & \cdot \\ \cdot \quad \cdot & \cdot \\ a_{mn} \cdots a_{22} & a_{12} \\ a_{m1} \cdots a_{21} & a_{11} \end{pmatrix}.$$

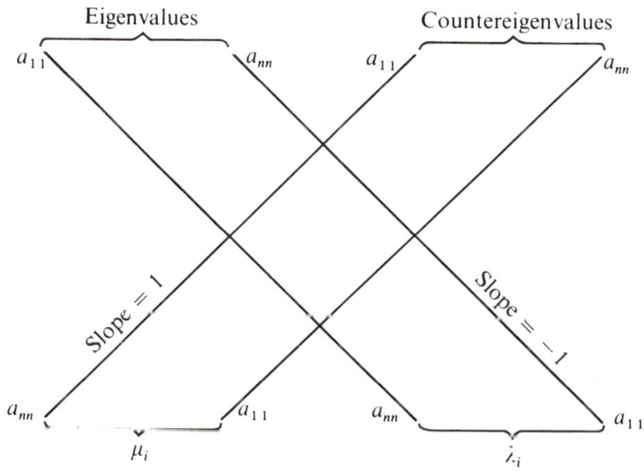

Figure 5. Eigenvalue determination key.

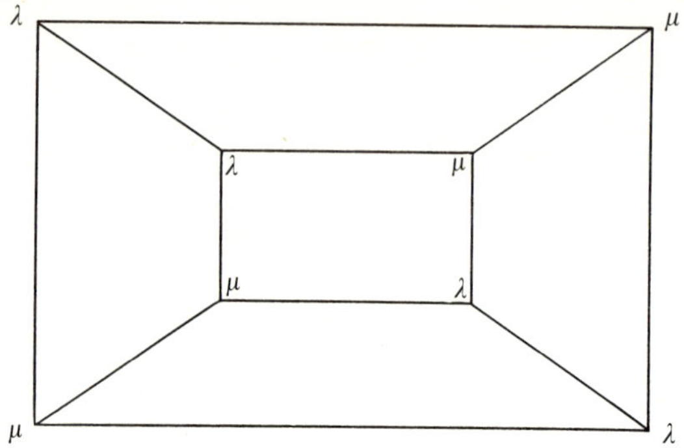

Figure 6. Distribution of eigenvalues and countereigenvalues for the octic set of A.

Just as $A = (a_{ij})$ implies that $A^T = (a_{ji})$, $A = (a_{ij})$ implies that $A^C = (a_{n-1-j,m-1-i})$. Loosely speaking, A^C is the "transpose of A" across the counterdiagonal or secondary diagonal. Therefore, occasionally A^C will be called the "countertranspose" of A.

THEOREM. The cispose of the identity is the identity.

Proof. $I^C = KI^TK = KIK = K^2 = I$.

THEOREM. The cispose of the counteridentity is the counteridentity.

Proof. $K^C = KK^TK = KKK = K^2K = IK \equiv K$.

THEOREM. The cispose of a cisposed matrix is the matrix itself.

Proof. $$(A^C)^C = K(A^C)^TK = K(KA^TK)^TK = K(K^TAK^T)K$$
$$= K(KAK)K = K^2AK^2 = A.$$

THEOREM. The cispose of a sum is the sum of the cisposes of each summand.

Proof. $$(A + B)^C = K(A + B)^TK = K(A^T + B^T)K = KA^TK + KB^TK$$
$$= KA^TK + KB^TK = A^C + B^C.$$

THEOREM. The cispose of the product of a scalar number and a matrix is equal to the product of the scalar number and the cisposed matrix.

Proof. $$(\gamma A)^C = K(\gamma A)^TK = K(A^T\gamma^T)K = KA^T\gamma K = K\gamma A^TK = \gamma KA^TK$$
$$= \gamma A^C.$$

THEOREM. The cispose of a product is the product of the cisposes of the factors taken in reverse order.

Proof. $(AB \cdots Z)^C = K(AB \cdots Z)^T K = K(Z^T \cdots B^T A^T)K$
$$= KZ^T KK \cdots KKB^T KKA^T K = Z^C \cdots B^C A^C.$$

THEOREM. The complex conjugate cispose is equal to the cisposed complex conjugate matrix.

Proof. $\overline{A^C} = \overline{KA^T K} = K\overline{A^T}K = K\bar{A}^T K = \bar{A}^C.$

Note that the complex conjugate cispose is denoted $\bar{A}^\star = \bar{A}^C$. Clearly, since $A^C = KA^T K$, $A^\star = KA^* K$.

THEOREM. If P commutes with Q, then P^C commutes with Q^C.

Proof. By hypothesis, $PQ = QP$. This implies that

$$(PQ)^T = (QP)^T, \quad Q^T P^T = P^T Q^T, \quad KQ^T P^T K = KP^T Q^T K,$$

and

$$KQ^T KKP^T K = KP^T KKQ^T K.$$

Hence $Q^C P^C = P^C Q^C$.

THEOREM. The product of two counterdiagonal matrices is diagonal.

Proof. Let the counterdiagonal matrices be expressed in terms of diagonal matrices D_1 and D_2 as follows:

$$A = D_1 K \qquad \text{and} \quad B = KD_2.$$

Since $D_1 D_2 = D_2 D_1 = D$, a diagonal matrix too, it follows that

$$AB = D_1 KKD_2 = D_1 K^2 D_2 = D_1 I D_2 = D.$$

Note that although diagonal matrices always commute, this is not true of counterdiagonal matrices, as is shown by the following example:

$$\begin{pmatrix} 0 & 0 & 1 \\ 0 & 2 & 0 \\ 3 & 0 & 0 \end{pmatrix}\begin{pmatrix} 0 & 0 & 2 \\ 0 & 1 & 0 \\ 2 & 0 & 0 \end{pmatrix} = \begin{pmatrix} 2 & 0 & 0 \\ 0 & 2 & 0 \\ 0 & 0 & 6 \end{pmatrix},$$

$$\begin{pmatrix} 0 & 0 & 2 \\ 0 & 1 & 0 \\ 2 & 0 & 0 \end{pmatrix}\begin{pmatrix} 0 & 0 & 1 \\ 0 & 2 & 0 \\ 3 & 0 & 0 \end{pmatrix} = \begin{pmatrix} 6 & 0 & 0 \\ 0 & 2 & 0 \\ 0 & 0 & 2 \end{pmatrix}.$$

The following theorem provides the necessary clarification:

THEOREM. If A and B are counterdiagonal matrices, then $AB = (BA)^C$.

Proof. $(BA)^C = K(BA)^TK = KA^TB^TK = KA^TKKB^TK = A^CB^C = AB.$

A through-the-looking-glass correspondence connects the identity matrix I and the counteridentity matrix K. The correspondence is more readily visible if the usual conditions for the symmetric, hermitian, orthogonal, and unitary properties are rewritten, respectively, as follows:

$$IA = A^TI, \quad IH^* = HI, \quad IO^* = O^TI, \quad \text{and} \quad IU^* = U^{-1}I.$$

Then the corresponding conditions for the countersymmetric, counterhermitian, counterorthogonal, and counterunitary properties, respectively, are simply:

$$KA = A^TK, \quad KH^* = HK, \quad KO^* = O^TK, \quad \text{and} \quad KU^* = U^{-1}K.$$

Many standard theorems about symmetric, hermitian, orthogonal, and unitary matrices have true counterparts, whose proofs are similar to the standard proofs except that they are sometimes several steps longer because the K matrix has to take a trip from one side of an expression to the other. The following theorems and proofs are samples of standard theorem counterparts.

THEOREM. The sum of two counterhermitian matrices is counterhermitian.

Proof. Let $KH^* = HK$ and $KH'^* = H'K$. Then

$$K(H^* + H'^*) = KH^* + KH'^* = HK + H'K = (H + H')K.$$

THEOREM. A necessary and sufficient condition for the product of two counterhermitian matrices to be counterhermitian is that the two matrices commute.

Proof. Let $KH^* = HK$ and $KH'^* = H'K$. Then

$$K(HH')^* = KH'^*H^* = H'KH^* = (H'H)K.$$

Hence, if $HH' = H'H$, it will follow that $K(HH')^* = (HH')K$ and the product will be counterhermitian.

THEOREM. The transform of a counterhermitian matrix by a counterunitary matrix is still counterhermitian.

Proof. Let $KH^* = HK$ and $KU^* = U^{-1}K$. To show that

$$K(UHU^{-1})^* = (UHU^{-1})K,$$

proceed as follows:

$$K(UHU^{-1})^* = K(U^{-1})^*H^*U^* = (U^{-1}K)^*H^*U^* = (KU^*)^*H^*U^*$$
$$= U(KH^*)U^* = UH(KU^*) = (UHU^{-1})K.$$

THEOREM. The product of two counterorthogonal matrices is counterorthogonal.

Proof. Let $KO^T = O^{-1}K$ and $KQ^T = Q^{-1}K$. Then

$$K(OQ)^T = (KQ^T)O^T = Q^{-1}(KO^T) = (Q^{-1}O^{-1})K = (OQ)^{-1}K.$$

THEOREM. The product of two counterunitary matrices is counterunitary.

Proof. Let $KU^* = U^{-1}K$ and $KV^* = V^{-1}K$. Then

$$K(UV)^* = (KV^*)U^* = V^{-1}(KU^*) = (V^{-1}U^{-1})K = (UV)^{-1}K.$$

THEOREM. The transform of a counterunitary matrix by a counterunitary matrix is still counterunitary.

Proof. Let $KU^* = U^{-1}K$ and $KV^* = V^{-1}K$. To show that

$$K(VUV^{-1})^* = (VUV^{-1})^{-1}K,$$

proceed as follows:

$$K(VUV^{-1})^* = K(V^{-1})^*U^*V^* = (V^{-1}K)^*U^*V^* = (KV^*)^*U^*V^*$$
$$= V(KU^*)V^* = VU^{-1}(KV^*) = (VU^{-1}V^{-1})K$$
$$= (VUV^{-1})^{-1}K.$$

THEOREM. The eigenvalues of a counterhermitian matrix are real.

Proof. $\mathbf{x}H = \lambda\mathbf{x}$, $H^*\mathbf{x}^* = \bar{\lambda}\mathbf{x}^*$, $KH^*\mathbf{x}^* = \bar{\lambda}KX^*$, $HK\mathbf{x}^* = \bar{\lambda}KX^*$, $\mathbf{x}HK\mathbf{x}^* = \bar{\lambda}XK\mathbf{x}^*$, $\lambda\mathbf{x}K\mathbf{x}^* = \bar{\lambda}\mathbf{x}K\mathbf{x}^*$, $\lambda = \bar{\lambda}$.

THEOREM. The eigenvalues of a counterunitary matrix lie on the unit circle.

Proof. $\mathbf{x}U = \lambda\mathbf{x}$, $U^*X^* = \bar{\lambda}X^*$, $KU^*\mathbf{x}^* = \bar{\lambda}K\mathbf{x}^*$, $U^{-1}K\mathbf{x}^* = \bar{\lambda}K\mathbf{x}^*$, $\mathbf{x}UU^{-1}K\mathbf{x}^* = \lambda\bar{\lambda}\mathbf{x}K\mathbf{x}^*$, $\mathbf{x}IK\mathbf{x}^* = \lambda\bar{\lambda}\mathbf{x}K\mathbf{x}^*$, $\lambda\bar{\lambda} = \mathbf{x}K\mathbf{x}^*/XK\mathbf{x}^* = 1$.

EXERCISES

95. State and prove the standard theorems upon which the counterpart theorems in this section are based.

96. Write the matrix

$$\begin{pmatrix} 2 & 5 & 1 \\ 7 & 4 & 3 \\ 6 & 1 & 0 \end{pmatrix}$$

as the sum of a countersymmetric and a skew countersymmetric matrix.

E. Matrix Symmetry

Octic Symmetry

There are many different kinds of matrix symmetry. This section focuses attention on 32 of 128 basic types built on the octic set of A, exploring some of their more obvious features.

Setting A equal to each of the members of the octic set in turn gives rise to a set of octic symmetries, which may be visualized as the rotations and reflections of a transparent rectangular lamina with the matrix elements of A inscribed (see Fig. 7). By performing simple operations such as negation or conjugation, or by requiring that A be hermitian, unitary, orthogonal, or antisymmetric, additional symmetry types are generated.

TYPE 1

If $A = A$, the matrix is "plain."

Example.

$$A = \begin{pmatrix} a & b & c & d \\ e & f & g & h \\ j & k & l & m \\ n & p & q & r \end{pmatrix} = \begin{pmatrix} a & b & c & d \\ e & f & g & h \\ j & k & l & m \\ n & p & q & r \end{pmatrix} \Rightarrow A = \begin{pmatrix} a & b & c & d \\ e & f & g & h \\ j & k & l & m \\ n & p & q & r \end{pmatrix} \tag{1}$$

THEOREM. The sum of two plain matrices is plain.

THEOREM. The product of two plain matrices is plain.

TYPE 2

If $A = A^T K$, the matrix is "right rotosymmetric."

Example.

$$A = \begin{pmatrix} a & b & c & d \\ e & f & g & h \\ j & k & l & m \\ n & p & q & r \end{pmatrix} = \begin{pmatrix} n & j & e & a \\ p & k & f & b \\ q & l & g & c \\ r & m & h & d \end{pmatrix} \Rightarrow A = \begin{pmatrix} a & b & c & a \\ c & f & f & b \\ b & f & f & c \\ a & c & b & a \end{pmatrix} \tag{2}$$

Moreover, A is left rotosymmetric (Type 4) and centrosymmetric (Type 3).

THEOREM. The sum of two right rotosymmetric matrices is right rotosymmetric.

THEOREM. The product of two right rotosymmetric matrices is centrosymmetric (Type 3). Their product is symmetric (Type 5) if and only if the matrices commute.

TYPE 3

If $A = KAK$, the matrix is "centrosymmetric."

Example.

$$A = \begin{pmatrix} a & b & c & d \\ e & f & g & h \\ j & k & l & m \\ n & p & q & r \end{pmatrix} = \begin{pmatrix} r & q & p & n \\ m & l & k & j \\ h & g & f & e \\ d & c & b & a \end{pmatrix} \Rightarrow A = \begin{pmatrix} a & b & c & d \\ e & f & g & h \\ h & g & f & e \\ d & c & b & a \end{pmatrix} \tag{3}$$

THEOREM. The sum of two centrosymmetric matrices is centrosymmetric.

THEOREM. The product of two centrosymmetric matrices is centrosymmetric.

TYPE 4

If $A = KA^T$, the matrix is "left rotosymmetric."

Example.

$$A = \begin{pmatrix} a & b & c & d \\ e & f & g & h \\ j & k & l & m \\ n & p & q & r \end{pmatrix} = \begin{pmatrix} d & h & m & r \\ c & g & l & q \\ b & f & k & p \\ a & e & j & n \end{pmatrix} \Rightarrow A = \begin{pmatrix} a & b & c & a \\ c & f & f & b \\ b & f & f & c \\ a & c & b & a \end{pmatrix} \tag{4}$$

Moreover, A is right rotosymmetric (Type 2) and centrosymmetric (Type 3).

THEOREM. The sum of two left rotosymmetric matrices is left rotosymmetric.

THEOREM. The product of two left rotosymmetric matrices is centrosymmetric (Type 3). Their product is symmetric (Type 5) if and only if the matrices commute.

TYPE 5

If $A = A^T$, the matrix is "symmetric."

Example.

$$A = \begin{pmatrix} a & b & c & d \\ e & f & g & h \\ j & k & l & m \\ n & p & q & r \end{pmatrix} = \begin{pmatrix} a & e & j & n \\ b & f & k & p \\ c & g & l & q \\ d & h & m & r \end{pmatrix} \Rightarrow A = \begin{pmatrix} a & b & c & d \\ b & f & k & h \\ c & k & l & m \\ d & h & m & r \end{pmatrix}. \tag{5}$$

THEOREM. The sum of two symmetric matrices is symmetric.

THEOREM. The product of two symmetric matrices is symmetric if and only if the matrices commute.

TYPE 6

If $A = AK$, the matrix is "vertically symmetric."

Example.

$$A = \begin{pmatrix} a & b & c & d \\ e & f & g & h \\ j & k & l & m \\ n & p & q & r \end{pmatrix} = \begin{pmatrix} d & c & b & a \\ h & g & f & e \\ m & l & k & j \\ r & q & p & n \end{pmatrix} \Rightarrow A = \begin{pmatrix} a & b & b & a \\ e & f & f & e \\ j & k & k & j \\ n & p & p & n \end{pmatrix}. \tag{6}$$

THEOREM. The sum of two vertically symmetric matrices is vertically symmetric.

THEOREM. The product of two vertically symmetric matrices is vertically symmetric.

TYPE 7

If $A = KA^TK$, the matrix is "countersymmetric."

Example.

$$A = \begin{pmatrix} a & b & c & d \\ e & f & g & h \\ j & k & l & m \\ n & p & q & r \end{pmatrix} = \begin{pmatrix} r & m & h & d \\ q & l & g & c \\ p & k & f & b \\ n & j & e & a \end{pmatrix} \Rightarrow A = \begin{pmatrix} a & b & c & d \\ e & f & g & c \\ j & k & f & b \\ n & j & e & a \end{pmatrix}. \tag{7}$$

THEOREM. The sum of two countersymmetric matrices is countersymmetric.

THEOREM. The product of two countersymmetric matrices is countersymmetric if and only if the matrices commute.

TYPE 8

If $A = KA$, then the matrix is "horizontally symmetric."

Example.

$$A = \begin{pmatrix} a & b & c & d \\ e & f & g & h \\ j & k & l & m \\ n & p & q & r \end{pmatrix} = \begin{pmatrix} n & p & q & r \\ j & k & l & m \\ e & f & g & h \\ a & b & c & d \end{pmatrix} \Rightarrow A = \begin{pmatrix} a & b & c & d \\ e & f & g & h \\ e & f & g & h \\ a & b & c & d \end{pmatrix}. \tag{8}$$

THEOREM. The sum of two horizontally symmetric matrices is horizontally symmetric.

THEOREM. The product of two horizontally symmetric matrices is horizontally symmetric.

$A = A$	Plain
$A = A^TK$	Right rotosymmetric
$A = KAK$	Centrosymmetric
$A = KA^T$	Left rotosymmetric
$A = A^T$	Symmetric
$A = AK$	Vertically symmetric
$A = KA^TK$	Countersymmetric
$A = KA$	Horizontally symmetric

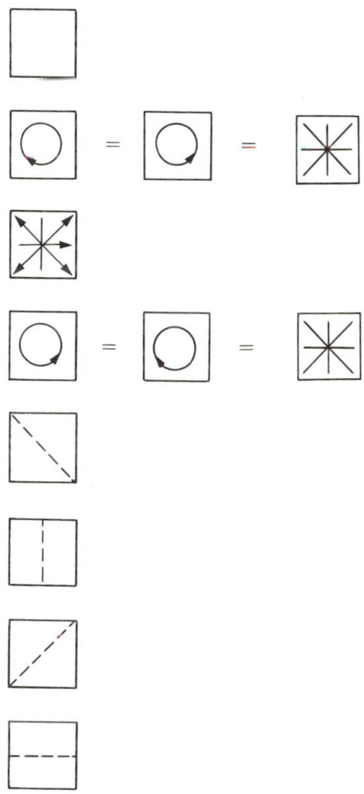

Figure 7. Basic octic symmetry of the octic set of $A = \{A\}$.

Skew Octic Symmetry

TYPE 9

If $A = -A$, the matrix is "zero."

Example.

$$A = \begin{pmatrix} a & b & c & d \\ e & f & g & h \\ j & k & l & m \\ n & p & q & r \end{pmatrix} = \begin{pmatrix} -a & -b & -c & -d \\ -e & -f & -g & -h \\ -j & -k & -l & -m \\ -n & -p & -q & -r \end{pmatrix} \Rightarrow$$

$$A = \begin{pmatrix} 0 & 0 & 0 & 0 \\ 0 & 0 & 0 & 0 \\ 0 & 0 & 0 & 0 \\ 0 & 0 & 0 & 0 \end{pmatrix} \tag{9}$$

THEOREM. The sum of two zero matrices is zero.

THEOREM. The product of two zero matrices is zero.

TYPE 10

If $A = -A^TK$, the matrix is "skew right rotosymmetric."

Example

$$A = \begin{pmatrix} a & b & c & d \\ e & f & g & h \\ j & k & l & m \\ n & p & q & r \end{pmatrix} \begin{pmatrix} -n & -j & -e & -a \\ -p & -k & -f & -b \\ -q & -l & -g & -c \\ -r & -m & -h & -d \end{pmatrix} \Rightarrow$$

$$A = \begin{pmatrix} a & b & c & -a \\ -c & f & -f & -b \\ -b & -f & f & -c \\ -a & c & b & a \end{pmatrix}. \tag{10}$$

Moreover, A is skew left rotosymmetric (Type 12) and skew centrosymmetric (Type 11).

THEOREM. The sum of two skew right rotosymmetric matrices is skew right rotosymmetric.

THEOREM. The product of two skew right rotosymmetric matrices is centrosymmetric.

TYPE 11

If $A = -KAK$, the matrix is "skew centrosymmetric."

Example.

$$A = \begin{pmatrix} a & b & c & d \\ e & f & g & h \\ j & k & l & m \\ n & p & q & r \end{pmatrix} = \begin{pmatrix} -r & -g & -p & -n \\ -m & -l & -k & -j \\ -h & -g & -f & -e \\ -d & -c & -b & - \end{pmatrix} \Rightarrow$$

$$A = \begin{pmatrix} a & b & c & d \\ e & f & g & h \\ -h & -g & -f & -e \\ -d & -c & -b & -a \end{pmatrix}. \qquad (11)$$

THEOREM. The sum of two skew centrosymmetric matrices is skew centrosymmetric.

THEOREM. The product of two skew centrosymmetric matrices is centrosymmetric.

TYPE 12

If $A = -KA^T$, the matrix is "skew left rotosymmetric."

Example

$$A = \begin{pmatrix} a & b & c & d \\ e & f & g & h \\ j & k & l & m \\ n & p & q & r \end{pmatrix} = \begin{pmatrix} -d & -h & -m & -r \\ -c & -g & -l & -q \\ -b & -f & -k & -p \\ -a & -e & -j & -n \end{pmatrix} \Rightarrow$$

$$A = \begin{pmatrix} a & b & c & -a \\ -c & f & -f & -b \\ -b & -f & f & -c \\ -a & c & b & a \end{pmatrix}. \qquad (12)$$

Moreover, A is skew right rotosymmetric (Type 10) and skew centrosymmetric (Type 11).

THEOREM. The sum of two skew left rotosymmetric matrices is skew left rotosymmetric.

THEOREM. The product of two skew left rotosymmetric matrices is centrosymmetric.

TYPE 13

If $A = -A^T$, the matrix is "skew symmetric."

Example.

$$A = \begin{pmatrix} a & b & c & d \\ e & f & g & h \\ j & k & l & m \\ n & p & q & r \end{pmatrix} = \begin{pmatrix} -a & -e & -j & -n \\ -b & -f & -k & -p \\ -c & -g & -l & -q \\ -d & -h & -m & -r \end{pmatrix} \Rightarrow$$

$$A = \begin{pmatrix} 0 & b & c & d \\ -b & 0 & g & h \\ -c & -g & 0 & m \\ -d & -h & -m & 0 \end{pmatrix}. \quad (13)$$

THEOREM. The sum of two skew symmetric matrices is skew symmetric.

THEOREM. The product of two skew symmetric matrices is symmetric if and only if the matrices commute.

TYPE 14

If $A = -AK$, the matrix is "skew vertically symmetric."

Example.

$$A = \begin{pmatrix} a & b & c & d \\ e & f & g & h \\ j & k & l & m \\ n & p & q & r \end{pmatrix} = \begin{pmatrix} -d & -c & -b & -a \\ -h & -g & -f & -e \\ -m & -l & -k & -j \\ -r & -q & -p & -n \end{pmatrix} \Rightarrow$$

$$A = \begin{pmatrix} a & b & -b & -a \\ e & f & -f & -e \\ j & k & -k & -j \\ n & p & -p & -n \end{pmatrix}. \quad (14)$$

THEOREM. The sum of two skew vertically symmetric matrices is skew vertically symmetric.

THEOREM. The product of two skew vertically symmetric matrices is skew vertically symmetric.

TYPE 15

If $A = -KA^TK$, the matrix is "skew countersymmetric."

Example.

$$A = \begin{pmatrix} a & b & c & d \\ e & f & g & h \\ j & k & l & m \\ n & p & q & r \end{pmatrix} = \begin{pmatrix} -r & -m & -h & -d \\ -q & -l & -g & -c \\ -p & -k & -f & -b \\ -n & -j & -e & -a \end{pmatrix} \Rightarrow$$

$$A = \begin{pmatrix} a & b & c & 0 \\ e & f & 0 & -c \\ j & 0 & -f & -b \\ 0 & -j & -e & -a \end{pmatrix}. \quad (15)$$

THEOREM. The sum of two skew countersymmetric matrices is skew countersymmetric.

THEOREM. The product of two skew countersymmetric matrices is countersymmetric if and only if the matrices commute.

TYPE 16

If $A = -KA$, the matrix is "skew horizontally symmetric."

Example.

$$A = \begin{pmatrix} a & b & c & d \\ e & f & g & h \\ j & k & l & m \\ n & p & q & r \end{pmatrix} = \begin{pmatrix} -n & -p & -q & -r \\ -j & -k & -l & -m \\ -e & -f & -g & -h \\ -a & -b & -c & -d \end{pmatrix} \Rightarrow$$

$$A = \begin{pmatrix} a & b & c & d \\ e & f & g & h \\ -e & -f & -g & -h \\ -a & -b & -c & -d \end{pmatrix}. \quad (16)$$

THEOREM. The sum of two skew horizontally symmetric matrices is skew horizontally symmetric.

THEOREM. The product of two skew horizontally symmetric matrices is skew horizontally symmetric.

Hermitian Octic Symmetry

TYPE 17

If $A = \bar{A}$, the matrix is "real."

Example.

$$A = \begin{pmatrix} a & b & c & d \\ e & f & g & h \\ j & k & l & m \\ n & p & q & r \end{pmatrix} = \begin{pmatrix} \bar{a} & \bar{b} & \bar{c} & \bar{d} \\ \bar{e} & \bar{f} & \bar{g} & \bar{h} \\ \bar{j} & \bar{k} & \bar{l} & \bar{m} \\ \bar{n} & \bar{p} & \bar{q} & \bar{r} \end{pmatrix} \Rightarrow$$

$$A = \begin{pmatrix} \mathrm{Re}(a) & \mathrm{Re}(b) & \mathrm{Re}(c) & \mathrm{Re}(d) \\ \mathrm{Re}(e) & \mathrm{Re}(f) & \mathrm{Re}(g) & \mathrm{Re}(h) \\ \mathrm{Re}(j) & \mathrm{Re}(k) & \mathrm{Re}(l) & \mathrm{Re}(m) \\ \mathrm{Re}(n) & \mathrm{Re}(p) & \mathrm{Re}(q) & \mathrm{Re}(r) \end{pmatrix}. \quad (17)$$

THEOREM. The sum of two real matrices is real.

THEOREM. The product of two real matrices is real.

TYPE 18

If $A = A^*K$, the matrix is "right rotohermitian."

Example.

$$A = \begin{pmatrix} a & b & c & d \\ e & f & g & h \\ j & k & l & m \\ n & p & q & r \end{pmatrix} = \begin{pmatrix} \bar{n} & j & \bar{e} & \bar{a} \\ \bar{p} & \bar{k} & \bar{f} & \bar{b} \\ q & \bar{l} & \bar{g} & \bar{c} \\ \bar{r} & \bar{m} & \bar{h} & \bar{d} \end{pmatrix} \Rightarrow A = \begin{pmatrix} a & b & c & \bar{a} \\ \bar{c} & f & \bar{f} & \bar{b} \\ \bar{b} & \bar{f} & f & \bar{c} \\ \bar{a} & c & b & a \end{pmatrix}. \quad (18)$$

Moreover, A is left rotohermitian (Type 20) and centrosymmetric (Type 11).

THEOREM. The sum of two right rotohermitian matrices is right rotohermitian.

THEOREM. The product of two right rotohermitian matrices is centrosymmetric.

TYPE 19

If $A = K\bar{A}K$, the matrix is "centrohermitian."

Example.

$$A = \begin{pmatrix} a & b & c & d \\ e & f & g & h \\ j & k & l & m \\ n & p & q & r \end{pmatrix} = \begin{pmatrix} \bar{r} & \bar{q} & \bar{p} & \bar{n} \\ \bar{m} & \bar{l} & \bar{k} & \bar{j} \\ \bar{h} & \bar{g} & \bar{f} & \bar{e} \\ \bar{d} & \bar{c} & \bar{b} & \bar{a} \end{pmatrix} \Rightarrow A = \begin{pmatrix} a & b & c & d \\ e & f & g & h \\ \bar{h} & \bar{g} & \bar{f} & \bar{e} \\ \bar{d} & \bar{c} & \bar{b} & \bar{a} \end{pmatrix}. \quad (19)$$

THEOREM. The sum of two centrohermitian matrices is centrohermitian.

THEOREM. The product of two centrohermitian matrices is centrohermitian.

TYPE 20

If $A = KA^*$, the matrix is "left rotohermitian."

Example.

$$A = \begin{pmatrix} a & b & c & d \\ e & f & g & h \\ j & k & l & m \\ n & p & q & r \end{pmatrix} = \begin{pmatrix} \bar{d} & \bar{h} & \bar{m} & \bar{r} \\ \bar{c} & \bar{g} & \bar{l} & \bar{q} \\ \bar{b} & \bar{f} & \bar{k} & \bar{p} \\ \bar{a} & \bar{e} & \bar{j} & \bar{n} \end{pmatrix} \Rightarrow A = \begin{pmatrix} a & b & c & \bar{a} \\ \bar{c} & f & \bar{f} & \bar{b} \\ \bar{b} & \bar{f} & f & \bar{c} \\ \bar{a} & c & b & a \end{pmatrix}. \quad (20)$$

Moreover, A is right rotohermitian (Type 18) and centrosymmetric (Type 11).

THEOREM. The sum of two left rotohermitian matrices is left rotohermitian.

THEOREM. The product of two left rotohermitian matrices is centrosymmetric.

TYPE 21

If $A = A^*$, the matrix is "hermitian."

Example.

$$A = \begin{pmatrix} a & b & c & d \\ e & f & g & h \\ j & k & l & m \\ n & p & q & r \end{pmatrix} = \begin{pmatrix} \bar{a} & \bar{e} & \bar{j} & \bar{n} \\ \bar{b} & \bar{f} & \bar{k} & \bar{p} \\ \bar{c} & \bar{g} & \bar{l} & \bar{q} \\ \bar{d} & \bar{h} & \bar{m} & \bar{r} \end{pmatrix} \Rightarrow$$

$$A = \begin{pmatrix} \mathrm{Re}(a) & b & c & d \\ \bar{b} & \mathrm{Re}(f) & g & h \\ \bar{c} & \bar{g} & \mathrm{Re}(l) & m \\ \bar{d} & \bar{h} & \bar{m} & \mathrm{Re}(r) \end{pmatrix}. \quad (21)$$

THEOREM. The sum of two hermitian matrices is hermitian.

THEOREM. The product of two hermitian matrices is hermitian if and only if they commute.

TYPE 22

If $A = \bar{A}K$, the matrix is "vertically hermitian."

Example.

$$A = \begin{pmatrix} a & b & c & d \\ e & f & g & h \\ j & k & l & m \\ n & p & q & r \end{pmatrix} = \begin{pmatrix} \bar{d} & \bar{c} & \bar{b} & \bar{a} \\ \bar{h} & \bar{g} & \bar{f} & \bar{e} \\ \bar{m} & \bar{l} & \bar{k} & \bar{j} \\ \bar{r} & \bar{q} & \bar{p} & \bar{n} \end{pmatrix} \Rightarrow A = \begin{pmatrix} a & b & \bar{b} & \bar{a} \\ e & f & \bar{f} & \bar{e} \\ j & k & \bar{k} & j \\ n & p & \bar{p} & \bar{n} \end{pmatrix}. \tag{22}$$

THEOREM. The sum of two vertically hermitian matrices is vertically hermitian.

THEOREM. The product of two vertically hermitian matrices is vertically hermitian provided the premultiplier is real.

TYPE 23

If $A = KA^*K$, the matrix is "counterhermitian."

Example.

$$A = \begin{pmatrix} a & b & c & d \\ e & f & g & b \\ j & k & l & m \\ n & p & q & r \end{pmatrix} = \begin{pmatrix} \bar{r} & \bar{m} & \bar{h} & \bar{d} \\ \bar{q} & \bar{l} & \bar{g} & \bar{c} \\ \bar{p} & \bar{k} & \bar{f} & \bar{b} \\ \bar{n} & \bar{j} & \bar{e} & \bar{a} \end{pmatrix} \Rightarrow$$

$$A = \begin{pmatrix} a & b & c & \mathrm{Re}(d) \\ e & f & \mathrm{Re}(g) & \bar{c} \\ j & \mathrm{Re}(k) & \bar{f} & \bar{b} \\ \mathrm{Re}(n) & j & \bar{e} & \bar{a} \end{pmatrix}. \tag{23}$$

THEOREM. The sum of two counterhermitian matrices is counterhermitian.

THEOREM. The product of two skew countersymmetric matrices is countersymmetric if and only if the matrices commute.

TYPE 24

If $A = K\bar{A}$, the matrix is "horizontally hermitian."

Example.

$$A = \begin{pmatrix} a & b & c & d \\ e & f & g & h \\ j & k & l & m \\ n & p & q & r \end{pmatrix} = \begin{pmatrix} \bar{n} & \bar{p} & \bar{q} & \bar{r} \\ \bar{j} & \bar{k} & \bar{l} & \bar{m} \\ \bar{e} & \bar{f} & \bar{g} & \bar{h} \\ \bar{a} & \bar{b} & \bar{c} & \bar{d} \end{pmatrix} \Rightarrow A = \begin{pmatrix} a & b & c & d \\ e & f & g & h \\ \bar{e} & \bar{f} & \bar{g} & \bar{h} \\ \bar{a} & \bar{b} & \bar{c} & \bar{d} \end{pmatrix}. \quad (24)$$

THEOREM. The sum of two horizontally hermitian matrices is horizontally hermitian.

THEOREM. The product of two horizontally hermitian matrices is horizontally hermitian provided that the postmultiplier is real.

Skew Hermitian Octic Symmetry

If $A = -\bar{A}$ the matrix is imaginary.

TYPE 25

If $A = KA^*K$, the matrix is "counterhermitian."

Example.

$$A = \begin{pmatrix} a & b & c & d \\ e & f & g & h \\ j & k & l & m \\ n & p & q & r \end{pmatrix} = \begin{pmatrix} -\bar{a} & -\bar{b} & -\bar{c} & -\bar{d} \\ -\bar{q} & -l & -\bar{g} & -\bar{c} \\ -\bar{j} & -\bar{k} & -\bar{l} & -\bar{m} \\ -\bar{n} & -\bar{p} & -\bar{q} & -\bar{r} \end{pmatrix} \Rightarrow$$

$$A = \begin{pmatrix} \mathrm{Im}(a) & \mathrm{Im}(b) & \mathrm{Im}(c) & \mathrm{Im}(d) \\ \mathrm{Im}(e) & \mathrm{Im}(f) & \mathrm{Im}(g) & \mathrm{Im}(h) \\ \mathrm{Im}(j) & \mathrm{Im}(k) & \mathrm{Im}(l) & \mathrm{Im}(m) \\ \mathrm{Im}(n) & \mathrm{Im}(p) & \mathrm{Im}(q) & \mathrm{Im}(r) \end{pmatrix}. \quad (25)$$

THEOREM. The sum of two imaginary matrices is imaginary.

THEOREM. The product of two imaginary matrices is real.

TYPE 26

If $A = -A^*K$, the matrix is "skew right rotohermitian."
Example.

$$A = \begin{pmatrix} a & b & c & d \\ e & f & g & h \\ j & k & l & m \\ n & p & q & r \end{pmatrix} = \begin{pmatrix} -\bar{n} & -\bar{j} & -\bar{e} & -\bar{a} \\ -\bar{p} & -\bar{k} & -\bar{f} & -\bar{b} \\ -\bar{q} & l & -\bar{g} & -\bar{c} \\ -\bar{r} & -\bar{m} & -\bar{h} & -\bar{d} \end{pmatrix} \Rightarrow$$

$$A = \begin{pmatrix} a & b & c & -\bar{a} \\ -\bar{c} & f & -\bar{f} & -\bar{b} \\ -\bar{b} & -\bar{f} & f & -\bar{c} \\ -\bar{a} & c & b & a \end{pmatrix}. \quad (26)$$

Moreover, A is skew left rotohermitian (Type 28) and centrosymmetric (Type 11).

THEOREM. The sum of two skew right rotohermitian matrices is skew right rotohermitian.

THEOREM. The product of two skew right rotohermitian matrices is centrosymmetric.

TYPE 27

If $A = -K\bar{A}K$, the matrix is "skew centrohermitian."
Example.

$$A = \begin{pmatrix} a & b & c & d \\ e & f & g & h \\ j & k & l & m \\ n & p & q & r \end{pmatrix} = \begin{pmatrix} -\bar{r} & -\bar{q} & -\bar{p} & -\bar{n} \\ -\bar{m} & -\bar{l} & -k & -\bar{j} \\ -\bar{h} & -\bar{g} & -\bar{f} & -\bar{e} \\ -\bar{d} & -\bar{c} & -\bar{b} & -\bar{a} \end{pmatrix} \Rightarrow$$

$$A = \begin{pmatrix} a & b & c & d \\ e & f & g & h \\ -\bar{h} & -\bar{g} & -\bar{f} & -\bar{e} \\ -\bar{d} & -\bar{c} & -\bar{b} & -\bar{a} \end{pmatrix}. \quad (27)$$

THEOREM. The sum of two skew centrohermitian matrices is skew centrohermitian.

THEOREM. The product of two skew centrohermitian matrices is centrohermitian if and only if the matrices commute.

TYPE 28

If $A = -KA^*$, the matrix is "skew left rotohermitian."

Example.

$$A = \begin{pmatrix} a & b & c & d \\ e & f & g & h \\ j & k & l & m \\ n & p & q & r \end{pmatrix} = \begin{pmatrix} -\bar{d} & -\bar{h} & -\bar{m} & -\bar{r} \\ -\bar{c} & -\bar{g} & -\bar{l} & -\bar{q} \\ -\bar{b} & -\bar{f} & -\bar{k} & -\bar{p} \\ -\bar{a} & -\bar{e} & -\bar{j} & -\bar{n} \end{pmatrix} \Rightarrow$$

$$A = \begin{pmatrix} a & b & c & -\bar{a} \\ -\bar{c} & f & -\bar{f} & -\bar{b} \\ -\bar{b} & -\bar{f} & f & -\bar{c} \\ -\bar{a} & c & b & a \end{pmatrix}. \quad (28)$$

Moreover, A is skew right rotohermitian (Type 26) and centrosymmetric (Type 11).

THEOREM. The sum of two skew left rotohermitian matrices is skew right rotohermitian.

THEOREM. The product of two skew left rotohermitian matrices is centrosymmetric.

TYPE 29

If $A = -A^*$, the matrix is "skew hermitian."

Example.

$$A = \begin{pmatrix} a & b & c & d \\ e & f & g & h \\ j & k & l & m \\ n & p & q & r \end{pmatrix} = \begin{pmatrix} -\bar{a} & -\bar{e} & -\bar{j} & -\bar{n} \\ -\bar{b} & -\bar{f} & -\bar{k} & -\bar{p} \\ -\bar{c} & -\bar{g} & -\bar{l} & -\bar{q} \\ -\bar{d} & -\bar{h} & -\bar{m} & -\bar{r} \end{pmatrix} \Rightarrow$$

$$A = \begin{pmatrix} \mathrm{Im}(a) & b & c & d \\ -\bar{b} & \mathrm{Im}(f) & g & h \\ -\bar{c} & -\bar{g} & \mathrm{Im}(l) & m \\ -\bar{d} & -\bar{h} & -\bar{m} & \mathrm{Im}(r) \end{pmatrix}. \quad (29)$$

THEOREM. The sum of two skew hermitian matrices is skew hermitian.

THEOREM. The product of two skew hermitian matrices is hermitian if and only if the matrices commute.

TYPE 30

If $A = -\bar{A}K$, the matrix is, "skew vertically hermitian."

Example.

$$A = \begin{pmatrix} a & b & c & d \\ e & f & g & h \\ j & k & l & m \\ n & p & q & r \end{pmatrix} = \begin{pmatrix} -\bar{d} & -\bar{c} & -\bar{b} & -\bar{a} \\ -\bar{h} & -\bar{g} & -\bar{f} & -\bar{e} \\ -\bar{m} & -\bar{l} & -\bar{k} & -j \\ -\bar{r} & -\bar{q} & -\bar{p} & -\bar{n} \end{pmatrix} \Rightarrow$$

$$A = \begin{pmatrix} a & b & -\bar{b} & -\bar{a} \\ e & f & -\bar{f} & -\bar{e} \\ j & k & -\bar{k} & -\bar{j} \\ n & p & -\bar{p} & -\bar{n} \end{pmatrix}. \quad (30)$$

THEOREM. The sum of two skew vertically hermitian matrices is skew vertically hermitian.

THEOREM. The product of two skew vertically hermitian matrices is vertically hermitian provided that the premultiplier is real.

TYPE 31

If $A = -KA^*K$, the matrix is "skew counter hermitian."

Example.

$$A = \begin{pmatrix} a & b & c & d \\ e & f & g & h \\ j & k & l & m \\ n & p & q & r \end{pmatrix} = \begin{pmatrix} -\bar{r} & -\bar{m} & -\bar{h} & -\bar{d} \\ -\bar{q} & -\bar{l} & -\bar{g} & \bar{c} \\ -\bar{p} & -\bar{k} & -\bar{f} & -\bar{b} \\ -\bar{n} & -\bar{j} & -\bar{e} & -\bar{a} \end{pmatrix} \Rightarrow$$

$$A = \begin{pmatrix} a & b & c & \mathrm{Im}(d) \\ e & f & \mathrm{Im}(g) & -\bar{c} \\ j & \mathrm{Im}(k) & -\bar{f} & -\bar{b} \\ \mathrm{Im}(n) & -\bar{j} & -\bar{e} & -\bar{a} \end{pmatrix}. \quad (31)$$

THEOREM. The sum of two skew counterhermitian matrices is skew counterhermitian.

THEOREM. The product of two skew counterhermitian matrices is counterhermitian if and only if the matrices commute.

TYPE 32

If $A = -K\bar{A}$, the matrix is "skew horizontally hermitian."

Example.

$$A = \begin{pmatrix} a & b & c & d \\ e & f & g & h \\ j & k & l & m \\ n & p & q & r \end{pmatrix} = \begin{pmatrix} -\bar{n} & -\bar{p} & -\bar{q} & -\bar{r} \\ -\bar{j} & -\bar{k} & -\bar{l} & -\bar{m} \\ -\bar{e} & -\bar{f} & -\bar{g} & -\bar{h} \\ -\bar{a} & -\bar{b} & -\bar{c} & -\bar{d} \end{pmatrix} \Rightarrow$$

$$A = \begin{pmatrix} a & b & c & d \\ e & f & g & h \\ -\bar{e} & -\bar{f} & -\bar{g} & -\bar{h} \\ -\bar{a} & -\bar{b} & -\bar{c} & -\bar{d} \end{pmatrix}. \quad (32)$$

THEOREM. The sum of two skew horizontally hermitian matrices is horizontally hermitian.

THEOREM. The product of two skew horizontally hermitian matrices is horizontally hermitian provided that the postmultiplier is real.

EXERCISES

97. A "right (left) circulant" is defined to be a matrix in which each row is the same as its predecessor if any, shifted one place to the right (left). Thus I, J, K, and L are circulants and

$$A = \begin{pmatrix} a_{11} & a_{12} & a_{13} \cdots a_{1,n-1} & a_{1n} \\ a_{1n} & a_{11} & a_{12} \cdots a_{1,n-2} & a_{1,n-1} \\ \cdot & \cdot & \cdot \quad \cdot \quad \cdot & \cdot \\ \cdot & \cdot & \cdot \quad \cdot \quad \cdot & \cdot \\ \cdot & \cdot & \cdot \quad \cdot \quad \cdot & \cdot \\ a_{13} & a_{14} & a_{15} \cdots a_{11} & a_{12} \\ a_{12} & a_{13} & a_{14} \cdots a_{1n} & a_{11} \end{pmatrix}$$

is an n-square right circulant.

(*a*) Describe the octic set of A.

(*b*) Discuss the symmetry types of various postures of A.

98. If A and B are m by n right circulants, then a matrix of the form $\llcorner\!\!\diagdown A + \diagup\!\!\lrcorner B$ is called a "striped matrix." Does ordinary matrix addition or multiplication preserve the left or right circulant or striped or counterstriped properties?

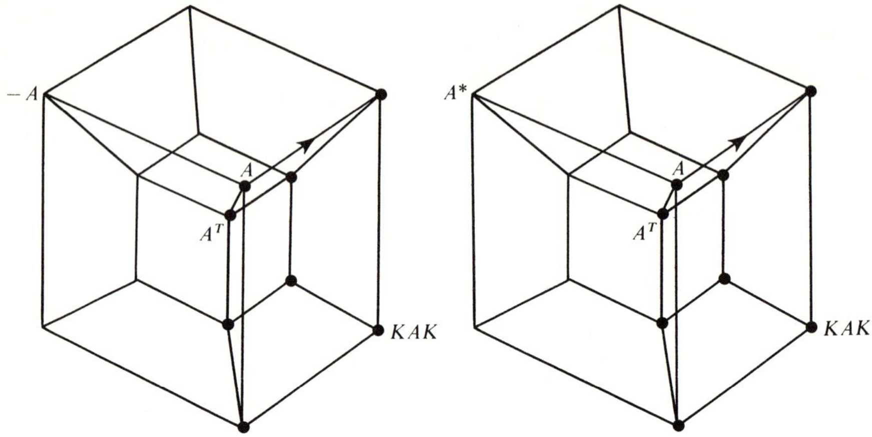

(*a*) The octic set of A extended by negation

(*b*) The octic set of A extended by tranjugation

Figure 8. Hypercubic aspects of matrix symmetry.

Defining Relation	Class
$A = \{\ \}$	Basic or regular octic symmetry
$A = -\{\ \}$	Skew
$A = \{\overline{\ \ }\}$	Hermitian
$A = -\{\overline{\ \ }\}$	Skew hermitian
$A = \{\ \}^{-1}$	Orthogonal
$A = -\{\ \}^{-1}$	Skew orthogonal
$A = \{\overline{\ \ }\}^{-1}$	Unitary
$A = -\{\overline{\ \ }\}^{-1}$	Skew unitary
$A = i\{\ \}$	Image
$A = -i\{\ \}$	Skew image
$A = i\{\overline{\ \ }\}$	Image hermitian
$A = -i\{\overline{\ \ }\}$	Skew image hermitian
$A = i\{\ \}^{-1}$	Image orthogonal
$A = -i\{\ \}^{-1}$	Skew image orthogonal
$A = i\{\overline{\ \ }\}^{-1}$	Image unitary
$A = -i\{\overline{\ \ }\}^{-1}$	Skew image unitary

Figure 9. Extended octic symmetries based on the octic set.

The illustrations shown in Fig. 8 provide views of several aspects of the underlying group structure of the many types of matrix symmetry listed in Fig. 9. Recall that a reflection of a matrix A through its center is simply a transformation in which the ranges of both subscripts i and j of the elements of A are reversed causing them to run backwards, so to speak, or decrement as follows:

$$i = m, m - 1, \cdots, 1;$$

$$j = n, n - 1, \cdots, 1.$$

Such a transformation has been shown to be brought about by a combined operation consisting of premultiplication and postmultiplication of A by the counteridentity matrix K. Hypothetically equating A with KAK, for example, then corresponds to a rigid rotation of a hyperube. Interesting symmetry types can be obtained by similar self-mappings of the set of vertices.

EXERCISES

99. Finish the sentence: "If for all pairs of subscripts (i, j) it is true that $a_{ij} = \bar{a}_{m+1-j,n+1-i}$, the matrix A is ."

100. Verify the following statements:

(*a*) The matrices AA^T and A^TKA are symmetric.

(*b*) The matrix $A + KAK$ is centrosymmetric.

(*c*) The matrix $AK - KA$ is skew centrosymmetric.

(*d*) A matrix does not have to be square in order to exhibit certain types of symmetry.

(*Suggestion:* List the types of symmetry open to an oblong matrix A.)

101. Prove the following representation theorem: Every counterhermitian matrix KH can be written in the form $A + iB$, where A is real and countersymmetric and B is real and skew countersymmetric.

102. Finish labelling the vertices of Fig. 8. (*Hint:* Consult the section on the Octic Set of a Matrix.)

103. Using the text as a model and using the defining relations as guides, develop and study the 12 remaining classes of octic symmetry. Are any trivial?

Let B be a matrix such that $B^3 = I$. It is informative to consider the following questions.

(*a*) Can the inverse of $C = xI + yB + zB^2$ be determined as a polynominal in B?

(*b*) For what sets of values of x, y, z will C be singular?

Since $B^3 = I$, the inverse as a polynominal in B can contain no terms B^3 or higher. Assume then that the inverse has the representation

$$C^{-1} = aI + bB + cB^2,$$

where the a, b and c are yet to be determined. Clearly,

$$(xI + yB + zB^2)(aI + bB + cB^2) = I$$

or

$$xaI + (ya + xb)B + (xc + yb + za)B^2 + (yc + zb)B^3 + zcB^4,$$

which reduces to

$$(xa + zb + yc)I + (ya + xb + zc)B + (xc + yb + za)B^2 = I.$$

Equating coefficients of like terms gives the system of simultaneous equations in the unknowns a, b, and c:

$$\left\{\begin{aligned} xa + zb + yc &= 1 \\ ya + xb + zc &= 0 \\ za + yb + xc &= 0 \end{aligned}\right\} \quad \text{or} \quad \underbrace{\begin{pmatrix} x & z & y \\ y & x & z \\ z & y & x \end{pmatrix}}_{\text{circulant } K} \underbrace{\begin{pmatrix} a \\ b \\ c \end{pmatrix}}_{\text{vector } v} = \underbrace{\begin{pmatrix} 1 \\ 0 \\ 0 \end{pmatrix}}_{\text{vector } \alpha}$$

Thus $Kv = \alpha$ and $v = K^{-1}\alpha$. Now if K is singular, it will be impossible to solve for a, b, and c, in which case no inverse will exist for the matrix C.

Hence C is singular if and only if $\det |K| = 0$. This happens when the expression

$$x^3 + y^3 + z^3 - 3xyz$$

vanishes.

We next establish that B is positive definite if and only if $B = I$. If $B = I$, then the characteristic equation is $(1 - \lambda)^n = 0$ and all the eigenvalues are $+1$. Since $I = I^*$, I is hermitian, and consequently B is positive definite.

On the other hand, if B is positive definite, it is hermitian, meaning $B = B^*$, and all its eigenvalues are real and greater than zero. Let λ_i be the eigenvalues of B. Then

$$\begin{aligned} B\mathbf{x} &= \lambda\mathbf{x} \\ B^2\mathbf{x} &= B\lambda\mathbf{x} = \lambda B\mathbf{x} = \lambda^2\mathbf{x} \\ B^3\mathbf{x} &= B\lambda^2\mathbf{x} = \lambda^2 B\mathbf{x} = \lambda^3\mathbf{x} \\ I\mathbf{x} &= \lambda^3\mathbf{x}. \end{aligned}$$

So the λ_i are cube roots of unity, but being real, are all equal to one.

It is possible to triangularize B by a unitary matrix U satisfying $U^*U = I$, such that U^*BU is a matrix with all elements zero below the main diagonal and with the eigenvalues of B on the main diagonal:

$$U^*BU = \begin{pmatrix} 1 & x_2 & \cdots & x_n \\ & 1 & \cdot & \cdot \\ & & \ddots & \vdots \\ \mathbf{0} & & & 1 \end{pmatrix} \cdot U^*B^*U = (U^*BU)^* = \begin{pmatrix} 1 & & & \mathbf{0} \\ \bar{x}_2 & 1 & & \\ \vdots & & \ddots & \\ \bar{x}_n & & \cdots & 1 \end{pmatrix}$$

Since $B = B^*$ implies $U^*BU = U^*B^*U$ or $U^*BU = I$, it follows that

$$(UU^*)B(UU^*) = UIU^*$$
$$B = I.$$

F. Direct Sums

The "direct sum" or, to use the preferred term, "concatenation" $\mathbf{z}$ of two row vectors $\mathbf{x} = (x_1, x_2, \cdots, x_n)$ and $\mathbf{y} = (y_1, y_2, \cdots, y_s)$, taken in that order, is defined by Marcus and Minc* to be a row vector also, which may be symbolized by

$$\mathbf{z} = \mathbf{x} \oplus \mathbf{y} = (x_1, x_2, \cdots, x_n, y_1, y_2, \cdots, y_s), \tag{1}$$

the vector obtained by abutting the components of vectors $\mathbf{x}$ and $\mathbf{y}$.

Similarly, the "direct sum" C of two matrices A and B, taken in that order, is defined to be a matrix also, which may be symbolized by

$$C = A \dotplus B = \begin{pmatrix} A & 0 \\ 0 & B \end{pmatrix}. \tag{2}$$

By this definition, as it is usually stated,

$$C = A_n \dotplus B_s = \left(\begin{array}{c:c} A_n & 0_{ns} \\ \hdashline 0_{sn} & B_s \end{array}\right). \tag{3}$$

For example, let

$$A = \begin{pmatrix} a_{11} & a_{12} \\ a_{21} & a_{22} \end{pmatrix} \quad \text{and} \quad B = \begin{pmatrix} b_{11} & b_{12} \\ b_{21} & b_{22} \end{pmatrix},$$

* Marvin Marcus and Henryk Minc, *A Survey of Matrix Theory and Matrix Inequalities*, Boston: Allyn and Bacon, 1964, p. 7.

then

$$A \dotplus B = \left(\begin{array}{cc:cc} a_{11} & a_{12} & 0 & 0 \\ a_{21} & a_{22} & 0 & 0 \\ \hdashline 0 & 0 & b_{11} & b_{12} \\ 0 & 0 & b_{21} & b_{22} \end{array}\right).$$

The customary requirement that A and B must be square is an unnecessary restriction. Therefore, in this book, the direct sum C of two matrices A and B of arbitrary sizes, say m by n and r by s, taken in that order, is defined to be a matrix also, as follows:

$$C_{(m+r),\,(n+s)} = A_{mn} \dotplus B_{rs} = \left(\begin{array}{c:c} A_{mn} & 0_{ms} \\ \hdashline 0_{rn} & B_{rs} \end{array}\right). \tag{4}$$

Clearly definition (4) holds for all constants m and r, particularly for $m = r = 1$. The result obtained by specializing definition (4) by setting $m = r = 1$ turns out to be another kind of direct sum for row vectors, which is at variance with definition (1). According to definition (4),

$$C_{2,\,(n+s)} = A_{1,n} \dotplus B_{1,s} = \left(\begin{array}{c:c} A_{1,n} & 0_{1,s} \\ \hdashline 0_{1,n} & B_{1,s} \end{array}\right), \tag{5}$$

and therefore

$$\mathbf{z} \neq \mathbf{x} \dotplus \mathbf{y} = \left(\begin{array}{c:c} x_1, x_2, \cdots, x_n, & 0,\ 0, \cdots,\ 0 \\ \hdashline 0,\ 0,\ \cdots, 0, & y_1, y_2, \cdots, y_s \end{array}\right), \tag{6}$$

which is obviously not a row vector.

Note, however, that concatenation (1) can be obtained from expression (6) by a further consolidation by a means of matrix multiplication:

$$\mathbf{z} = (1, 1) \left(\begin{array}{c:c} x_1, x_2, \cdots, x_n, & 0,\ 0, \cdots,\ 0 \\ \hdashline 0,\ 0,\ \cdots, 0, & y_1, y_2, \cdots, y_s \end{array}\right)$$

$$= (x_1, x_2, \cdots, x_n, y_1, y_2, \cdots, y_s) \tag{7}$$

so that:

$$\vec{\mathbf{x}} \oplus \vec{\mathbf{y}} = L_2\,(\mathbf{x} \dotplus \mathbf{y}), \tag{8}$$

where L_2 represents the two component L matrix.

Therefore, the gain in generality obtained by acceptance of definition (4), which permits A and B to have arbitrary sizes, causes no great loss other than making the words "direct sum" an outmoded usage when applied to definition (1), since the concatenation of row (or column) vectors already is implicit in this definition of the direct sum of two matrices.

A necessary and sufficient condition that $A \dotplus B = B \dotplus A$ is that $A = B$. Thus direct summation is not in general a commutative operation. It is, however, associative, that is,

$$A \dotplus (B \dotplus C) = (A \dotplus B) \dotplus C. \tag{9}$$

An important relationship connecting direct summation to ordinary matrix addition is

$$(A_{mn} \dotplus B_{rs}) + (C_{mn} \dotplus D_{rs}) = (A + C)_{mn} \dotplus (B + D)_{rs}, \tag{10}$$

which may be illustrated as follows:

$$\begin{pmatrix} A & \vdots & 0 \\ \cdots & \cdots & \cdots \\ 0 & \vdots & B \end{pmatrix} + \begin{pmatrix} C & \vdots & 0 \\ \cdots & \cdots & \cdots \\ 0 & \vdots & D \end{pmatrix} = \begin{pmatrix} A + C & \vdots & 0 \\ \cdots & \cdots & \cdots \\ 0 & \vdots & B + D \end{pmatrix}.$$

An equally important relationship connecting direct summation to ordinary matrix multiplication is

$$(A_{mn} \dotplus B_{rs})(C_{np} \dotplus D_{sq}) = (AB)_{mp} \dotplus (BD)_{rq}, \tag{11}$$

which may be illustrated as follows:

$$\begin{pmatrix} A & \vdots & 0 \\ \cdots & \cdots & \cdots \\ 0 & \vdots & B \end{pmatrix} \begin{pmatrix} C & \vdots & 0 \\ \cdots & \cdots & \cdots \\ 0 & \vdots & D \end{pmatrix} = \begin{pmatrix} AC & \vdots & 0 \\ \cdots & \cdots & \cdots \\ 0 & \vdots & BD \end{pmatrix}.$$

If A and C commute,

$$(A \dotplus B)(C \dotplus D) = (C \dotplus B)(A \dotplus D); \tag{12a}$$

similarly, if B and D commute,

$$(A \dotplus B)(C \dotplus D) = (A \dotplus D)(C \dotplus B); \tag{12b}$$

and if both commute,

$$(A \dotplus B)(C \dotplus D) = (C \dotplus D)(A \dotplus B).$$

Moreover, if A and B are square matrices, then

$$(A \dotplus B)^2 = (A \dotplus B)(A \dotplus B) = A^2 \dotplus B^2. \tag{13}$$

Finally, note that although the direct sum $A \dotplus B$ may seem to arise full-grown from A and B, its construction essentially is a matter of dealing successively with single rows or single columns.

EXERCISES

104. Explore the ramifications of a "main line direct sum," defined as

$$A \overset{m}{+} B = \begin{pmatrix} A_{mn} & \vdots & 0_m \\ \cdots\cdots & \vdots & \cdots\cdots \\ 0_r & \vdots & B_{rs} \end{pmatrix} \quad \text{or} \quad \begin{pmatrix} A_{mn} & \vdots & 0_m \\ \cdots\cdots & \vdots & \cdots\cdots \\ 0_r & \vdots & B_{rs} \end{pmatrix}.$$

Let $A = (a_{ij})$ be a general 2 by 3 matrix and $B = (b_{kl})$ be a general 4 by 5 matrix. Write out

$$A \overset{m}{+} B,\ A \overset{m}{+} B,\ A \overset{m}{+} B^T,\ A^T \overset{m}{+} B,\ A^T \overset{m}{+} B^T.$$

Are there any advantages or disadvantages in the main line direct sum not in definition (4)?

105. Let

$$A_1 = (\gamma_1\gamma_2\gamma_3\gamma_4), \quad B_{4,\,1} = \begin{pmatrix} \gamma_1 & \gamma_2 & \gamma_3 \\ \gamma_1 & \gamma_2 & \gamma_4 \\ \gamma_1 & \gamma_3 & \gamma_4 \\ \gamma_2 & \gamma_3 & \gamma_4 \end{pmatrix},$$

$$C_{6,\,1} = \begin{pmatrix} \gamma_1 & \gamma_2 \\ \gamma_1 & \gamma_3 \\ \gamma_1 & \gamma_4 \\ \gamma_2 & \gamma_3 \\ \gamma_2 & \gamma_4 \\ \gamma_3 & \gamma_4 \end{pmatrix}, \quad D_{4,\,1} = \begin{pmatrix} \gamma_1 \\ \gamma_2 \\ \gamma_3 \\ \gamma_4 \end{pmatrix}, \quad E_1 = I.$$

Evaluate GG^T where $G_{16,\,5} = A \dotplus B \dotplus C \dotplus D \dotplus E$.

106. To combine configuration and momentum spaces, with coordinates $\mathbf{q} = (q_1, q_2, q_3)$ and $\mathbf{p} = (p_1, p_2, p_3)$, respectively, into phase space with coordinates $(\mathbf{q}, \mathbf{p}) = (q_1, q_2, q_3, p_1, p_2, p_3)$, the concatenation proceeds as follows:

$$\begin{aligned} L_2(\mathbf{q} + \mathbf{p}) &= (1, 1)[(q_1, q_2, q_3) \dotplus (p_1, p_2, p_3)] \\ &= (1, 1) \begin{matrix} q_1 & q_2 & q_3 & 0 & 0 & 0 \\ 0 & 0 & 0 & p_1 & p_2 & p_3 \end{matrix} \\ &= (q_1 q_2 q_3 p_1 p_2 p_3). \end{aligned}$$

Formulate a concatenation for gamma space, with coordinates

$$(\mathbf{q}_1, \mathbf{q}_2, \cdots, \mathbf{q}_n, \mathbf{p}_1, \mathbf{p}_2, \cdots, \mathbf{p}_n),$$

wherein each point represents n particles.

Hint: Use the notation $\dot{\sum\limits_i^{N}} A_{(i)} = A_{(1)} \dotplus A_{(2)} \dotplus \cdots \dotplus A_{(N)}$.

107. Find the latent roots and corresponding characteristic vectors of the matrix $(I_2 - K_2) \dotplus 3$.

In set theory there is a distinction between the null set and the zero set. The former is defined to be the empty set, symbolized by $\Phi = \{\ \}$, a set devoid of members; the latter is defined to be the set with zero as a member, symbolized by $\Theta = \{0\}$, containing a single constituent.

As discussed earlier in the section on matrix algebra, ordinary matrix addition of the negative of an m by n matrix A to the matrix A itself produces a zero matrix of the same size as the addend matrices:

$$A_{mn} + (-A_{mn}) = 0_{mn}. \tag{14}$$

This almost trivial process can be viewed somewhat differently as an instance of ordinary matrix subtraction:

$$A_{mn} - A_{mn} = 0_{mn}. \tag{15}$$

Clearly, ordinary matrix subtraction of compatible matrices of positive dimensionality may yield a zero matrix of the same size as the subtrahend and minuend matrices, but obviously it never can yield a matrix devoid of elements. Nonetheless, it is possible and useful to talk about null matrices as a separate creation from zero matrices.

As the name might suggest to those familiar with set theory, a null matrix in this text is an empty matrix, that is, one that has no elements. A zero matrix 0, on the other hand, usually has elements and can be defined as a matrix with zero Euclidean norm. A "null matrix" Φ is then defined as a matrix with one or more zero dimensions.

A zero matrix arises as the solution of the equation

$$A + X = A; \tag{16}$$

a null matrix arises as the solution of the equation

$$A \dotplus X = A. \tag{17}$$

The only difference between equations (16) and (17) is in the kind of addition employed. Equation (16) calls for ordinary matrix addition, whereas equation (17) calls for direct matrix addition.

Recall that in a simple case $A \dotplus B$ looks like

$$A \dotplus B = \left(\begin{array}{cc:cc} a_{11} & a_{12} & 0 & 0 \\ a_{21} & a_{22} & 0 & 0 \\ \hdashline 0 & 0 & b_{11} & b_{12} \\ 0 & 0 & b_{21} & b_{22} \end{array}\right). \tag{18}$$

Substituting expression (17) does not change the appearance:

$$A \dotplus X \dotplus B = A \dotplus B \tag{19}$$

This result suggests that the unknown matrix X has degenerated by shrinking and somehow is concealed in the vicinity of the intersection point of the vertical and horizontal dotted lines.

Now examine the dimensionality equation:

$$(m, n) \dotplus (r, s) = (m + r, n + s). \tag{20}$$

Since it must be true that

$$(m, n) \dotplus (r, s) = (m, n) \tag{21}$$

in order for $A \dotplus X$ to be identical to the original matrix A, this implies that the dimensionality (r, s) of X must equal $(0, 0)$ and consequently that X is a square matrix of order zero. Such a matrix X, denoted by either Φ or Φ_0, is here called the "null matrix of order zero." Although there are infinitely many square zero matrices, there is only one square null matrix.

For Φ, direct addition becomes commutative:

$$A \dotplus \Phi_0 = \Phi_0 \dotplus A. \tag{22}$$

Moreover, since

$$\begin{aligned} A &= A \dotplus \Phi_0, \\ A &= (A \dotplus \Phi_0) \dotplus \Phi_0 = A \dotplus \Phi_0 \dotplus \Phi_0, \\ &\quad \cdot \quad \cdot \quad \cdot \\ A &= A \dotplus \Phi_0 \dotplus \cdots \dotplus \Phi_0, \\ A &= A \dotplus \dot{\Sigma} \Phi_0, \end{aligned} \tag{23}$$

it follows that

$$\dot{\Sigma} \Phi_0 \dotplus A = A \dotplus \dot{\Sigma} \Phi_0. \tag{24}$$

It is possible also to think of zero row or column counts being paired with finite column or row counts to form vacuous 0 by n or m by 0 matrices.

Another notation for the null matrix is

$$\Phi_0 = (\phi_{00}), \tag{25}$$

where the subscripted lowercase Greek letter ϕ_{00} does not represent a scalar number but rather represents a kind of "phantom element" that cannot be measured physically. In complete accord with this notation for the null square matrix, next define:

$$\Phi_{0n} = (\phi_{01}, \phi_{02}, \cdots, \phi_{0n}), \tag{26}$$

and

$$\Phi_{m0} = \begin{pmatrix} \phi_{10} \\ \phi_{20} \\ \cdot \\ \cdot \\ \cdot \\ \phi_{m0} \end{pmatrix}, \tag{27}$$

where ϕ_{0j} and ϕ_{i0} again are phantom elements, to be a null row and a null column matrix, respectively. That these new concepts have practical value will be shown in a discussion to follow of bordered matrices.

The null row matrix and null column matrix are "empty vectors"; but since the term "null vector" is entrenched so firmly in the literature to mean exactly the same thing as "zero vector" no attempt will be made here to dislodge it. (Either a useless extravagance or confusion generally results whenever names for things and things fail to correspond one-to-one.)

DEFINITION. Matrices that are empty or have phantom elements only are called "null matrices."

For simplicity, the symbol Φ will be used to denote either Φ_0, Φ_{0n}, or Φ_{m0}, provided that no doubt as to intended meaning can result. Also, occasionally, the symbol Φ_{mn} may be used with the understanding that either m or n or both are zero. Finally, it is worthwhile to identify Φ as a limiting case in such sequences as the following:

$$J_n, J_{n-1}, \cdots, J_1, J_0 = \Phi_0,$$
$$I_n, I_{n-1}, \cdots, I_1, I_0 = \Phi_0,$$
$$0_n, 0_{n-1}, \cdots, 0_1, 0_0 = \Phi_0,$$

and

$$L_n, L_{n-1}, \cdots, L_1, L_0 = \Phi_0.$$

EXERCISE 108

108. Find several sequences having (*a*) Φ_{0n} and (*b*) Φ_{m0} as a limiting case

An arbitrary matrix A can be bordered on either the left or right or both with additional columns to produce a wider matrix; it can be bordered on either the top or bottom or both with additional rows to produce a taller matrix. The resulting appendages here are termed "flanks" and "skirts," respectively. The direct sum $A \dotplus B$ serves as an example:

$$\left(\begin{array}{c:c} A & \begin{matrix}\text{right flank of } A \\ \text{top skirt of } B\end{matrix} \\ \hdashline \begin{matrix}\text{left flank of } B \\ \text{bottom skirt of } A\end{matrix} & B \end{array}\right)$$

By alternately appending rows and columns, matrices can be made to grow:

$$(\lambda_1),\quad (\lambda_1 \quad 0),\quad \begin{pmatrix} \lambda_1 & 0 \\ 0 & 0 \end{pmatrix},\quad \begin{pmatrix} \lambda_1 & 0 & 0 \\ 0 & 0 & 0 \end{pmatrix},\quad \begin{pmatrix} \lambda_1 & 0 & 0 \\ 0 & 0 & 0 \\ 0 & 0 & 0 \end{pmatrix};$$

$$(\lambda_2),\quad (\lambda_2 \quad 0),\quad \begin{pmatrix} \lambda_2 & 0 \\ 0 & 0 \end{pmatrix},\quad \begin{pmatrix} 0 & \lambda_2 & 0 \\ 0 & 0 & 0 \end{pmatrix},\quad \begin{pmatrix} 0 & 0 & 0 \\ 0 & \lambda_2 & 0 \\ 0 & 0 & 0 \end{pmatrix}:$$

$$(\lambda_3),\quad (0 \quad \lambda_3),\quad \begin{pmatrix} 0 & 0 \\ 0 & \lambda_3 \end{pmatrix},\quad \begin{pmatrix} 0 & 0 & 0 \\ 0 & 0 & \lambda_3 \end{pmatrix},\quad \begin{pmatrix} 0 & 0 & 0 \\ 0 & 0 & 0 \\ 0 & 0 & \lambda_3 \end{pmatrix}.$$

By ordinary addition of these terms,

$$\begin{aligned} &\begin{pmatrix} \lambda_1 & 0 & 0 \\ 0 & \lambda_2 & 0 \\ 0 & 0 & \lambda_3 \end{pmatrix} \\ &= (\lambda_1 \dotplus 0_1 \dotplus 0_1) \dotplus (0_1 \dotplus \lambda_2 \dotplus 0_1) \dotplus (0_1 \dotplus 0_1 \dotplus \lambda_3) \\ &= (\lambda_1 \dotplus 0_2) \dotplus (0_1 \dotplus \lambda_2 \dotplus 0_1) \dotplus (0_2 \dotplus \lambda_3). \end{aligned} \tag{28}$$

This expansion suggests that it might be possible to express $\begin{pmatrix} a & b & c \\ d & e & f \\ g & h & j \end{pmatrix}$ as the sum first of nine bordered matrices:

$$\begin{pmatrix} a & 0 & 0 \\ 0 & 0 & 0 \\ 0 & 0 & 0 \end{pmatrix} + \begin{pmatrix} 0 & b & 0 \\ 0 & 0 & 0 \\ 0 & 0 & 0 \end{pmatrix} + \begin{pmatrix} 0 & 0 & c \\ 0 & 0 & 0 \\ 0 & 0 & 0 \end{pmatrix} + \begin{pmatrix} 0 & 0 & 0 \\ d & 0 & 0 \\ 0 & 0 & 0 \end{pmatrix} + \begin{pmatrix} 0 & 0 & 0 \\ 0 & e & 0 \\ 0 & 0 & 0 \end{pmatrix}$$

$$+ \begin{pmatrix} 0 & 0 & 0 \\ 0 & 0 & f \\ 0 & 0 & 0 \end{pmatrix} + \begin{pmatrix} 0 & 0 & 0 \\ 0 & 0 & 0 \\ g & 0 & 0 \end{pmatrix} + \begin{pmatrix} 0 & 0 & 0 \\ 0 & 0 & 0 \\ 0 & h & 0 \end{pmatrix} + \begin{pmatrix} 0 & 0 & 0 \\ 0 & 0 & 0 \\ 0 & 0 & j \end{pmatrix},$$

and second, of direct sum terms. To see how a general matrix may be expressed in such fashion, consider the following degenerate formulations:

$$A_2 \dotplus \Phi_{01} = \begin{pmatrix} a_{11} & a_{12} & 0 \\ a_{21} & a_{22} & 0 \end{pmatrix}, \quad \Phi_{01} \dotplus A_2 = \begin{pmatrix} 0 & a_{11} & a_{12} \\ 0 & a_{21} & a_{22} \end{pmatrix},$$
$$A_2 \dotplus \Phi_{10} = \begin{pmatrix} a_{11} & a_{12} \\ a_{21} & a_{22} \\ 0 & 0 \end{pmatrix}, \text{ and } \Phi_{10} \dotplus A_2 = \begin{pmatrix} 0 & 0 \\ a_{11} & a_{12} \\ a_{21} & a_{22} \end{pmatrix}, \tag{29}$$

where Φ again denotes a null matrix. As further illustration of this principle,

$$(A_n \dotplus \Phi_{0n}) + (\Phi_{0n} \dotplus I_n) = (A_n \quad I_n), \tag{30a}$$

and

$$(I_n \dotplus \Phi_{0n}) + (\Phi_{0n} \dotplus A^{-1}) = (I_n \quad A^{-1}). \tag{30b}$$

Thus if A is an arbitrary m by n matrix,

$$A = (a_{ij}) = \sum_{i=1}^{m} \sum_{j=1}^{n} (0_{i-1,j-1} \dotplus a_{ij} \dotplus 0_{m-i,n-j}), \tag{31}$$

where $0 \equiv \Phi$ in null cases. Hence

$$J_{mn} = \sum_{i=1}^{m} \sum_{j=1}^{n} (0_{i-1,j-1} \dotplus 1 \dotplus 0_{m-i,n-j}), \tag{32a}$$

and

$$J = \sum_{i,j}^{n} (0_{i-1,j-1} \dotplus 1 \dotplus 0_{n-i,n-j}). \tag{32b}$$

A system of n normal random variables has the convariance matrix

$$C = \begin{pmatrix} 1 & 1 & 1 \cdots & 1 \\ 1 & 2 & 2 \cdots & 1 \\ 1 & 2 & 3 & 3 \\ \cdot & \cdot & \cdot\,\cdot & \cdot \\ \cdot & \cdot & \cdot & \cdot \\ \cdot & \cdot & \cdot & \cdot\,\cdot \\ 1 & 2 & 3 \cdots & n \end{pmatrix}. \tag{33}$$

To compute det C, subtract the $(n - i)$th row from the $(n - i + 1)$st row of C, letting $i = 1, 2, \cdots, n - 1$. This sequence of elementary operations yields

$$\nabla J = \begin{pmatrix} 1 & 1 & 1 \cdots & 1 \\ 0 & 1 & 1 \cdots & 1 \\ 0 & 0 & 1 \cdots & 1 \\ \cdot & \cdot & \cdot\,\cdot & \cdot \\ \cdot & \cdot & \cdot & \cdot \\ \cdot & \cdot & \cdot & \cdot\,\cdot \\ 0 & 0 & 0 \cdots & 1 \end{pmatrix}. \tag{34}$$

At this point, either apply the rule

$$\det \nabla A = \prod_{i=1}^{n} a_{ii} \tag{35}$$

or subtract the $(i + 1)$st row from the ith row of ∇J letting $i = 1, 2, \cdots, n - 1$. This sequence of elementary operations yields

$$I = \begin{pmatrix} 1 & 0 \cdots & 0 \\ 0 & 1 \cdots & 0 \\ \cdot & \cdot\,\cdot & \cdot \\ \cdot & \cdot & \cdot \\ \cdot & \cdot & \cdot\,\cdot \\ 0 & 0 \cdots & 1 \end{pmatrix}. \tag{36}$$

Consequently, $\det C = \det \nabla J = \det I = 1$.

EXERCISES

109. Compute the inverse matrix C^{-1} by flanking C with the unit matrix and applying elementary row operations to the resulting bordered matrix.

110. Let M be an m by n matrix. Notationally specify

$$(M \quad 0), \quad \begin{pmatrix} M \\ 0 \end{pmatrix}, \quad (0 \quad M), \quad \text{and} \quad \begin{pmatrix} 0 \\ M \end{pmatrix},$$

where 0 is always square.

111. Evaluate $(a + 0) + (0 + d) + [(b + \quad) + (0 + c)]K_2$.

112. Let A, B, C, and D all be n-square matrices. Express $\begin{pmatrix} A & B \\ C & D \end{pmatrix}$ two ways using $\Phi_{0,\,2n}$ and $\Phi_{2n,\,0}$.

113. To solve directly for the inverse of a nonsingular n-square matrix, border the matrix A, to be inverted on the right flank by the n-square identity matrix, obtaining $(A_n \; I_n)$, and then reduce this augmented matrix to the form $(I_n \; A^{-1})$ by performing successive row operations* specifically designed to change all the elements along the main diagonal of the given matrix to ones, and all the other elements of the given matrix to zeros. Using this method determine the inverses of the following matrices:

$$(a)\; A = \begin{pmatrix} a_{11} & a_{12} \\ a_{21} & a_{22} \end{pmatrix},$$

$$(b)\; B = \begin{pmatrix} b_{11} & b_{12} & b_{13} \\ b_{21} & b_{22} & b_{23} \\ b_{31} & b_{32} & b_{33} \end{pmatrix}.$$

Hint: Work with $A = \begin{pmatrix} a & b \\ c & d \end{pmatrix}$, $B = \begin{pmatrix} a & b & c \\ d & e & f \\ g & h & j \end{pmatrix}$.

The zero matrix—that m by n array whose elements are all zero—would seem to be a primitive concept. This conclusion is supported by the fact that although the product AB may be a zero matrix even though neither factor is zero, the well known divisors of zero are no more rudimentary than a zero

* When the elements of a matrix lie in a field, the following elementary operations are defined:

	Row	*Column*
Type I.	$I_{(i),\,(j)}$ interchange rows i and j	$I^{(i),\,(j)}$ interchange columns i and j
Type II.	$II_{(i)\,+\,c(j)}$ add c times row j to row i	$II^{(i)\,+\,c(j)}$ add c times column j to column i
Type III.	$III_{c(i)}$ multiply row i by c	$III^{c(i)}$ multiply column i by c.

matrix and often may be more complicated. To illustrate nonzero divisors of zero:

$$\begin{pmatrix} 0 & a \\ 0 & b \end{pmatrix}\begin{pmatrix} c & d \\ 0 & 0 \end{pmatrix} = \begin{pmatrix} 0 & 0 \\ 0 & 0 \end{pmatrix}.$$

It will be shown now that from one viewpoint the zero matrix has even stranger divisors among the null matrices. In a sense these divisors are conceptually more primitive.

To begin, consider the following progression of stick figures:

$$\Phi_{0l} \dotplus \Phi_{0k} = \Phi_{0k} \dotplus \Phi_{0l} = \Phi_{0,\,k+l} \tag{37a}$$

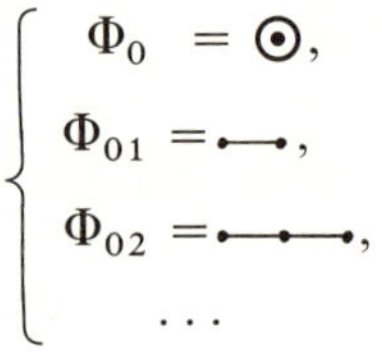

$$\Phi_{0} \dotplus \Phi_{k0} = \Phi_{k0} \dotplus \Phi_{0} = \Phi_{k+l,\,0} \tag{37b}$$

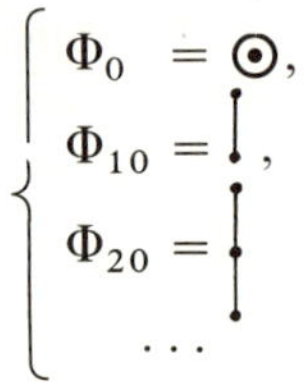

In terms of these figures, the following direct sums can be diagrammed:

$$\begin{aligned}
\Phi_{01} \dotplus \Phi_{10} &= 0 = 0_1, & \Phi_{10} \dotplus \Phi_{01} &= 0 = 0_1,\\
\Phi_{01} \dotplus \Phi_{20} &= \begin{matrix}0\\0\end{matrix} = 0_{21}, & \Phi_{10} \dotplus \Phi_{02} &= 00 = 0_{12},\\
\Phi_{02} \dotplus \Phi_{10} &= 00 = 0_{12}, & \Phi_{20} \dotplus \Phi_{01} &= \begin{matrix}0\\0\end{matrix} = 0_{21},\\
\Phi_{02} \dotplus \Phi_{20} &= \begin{matrix}00\\00\end{matrix} = 0_{2}, & \Phi_{20} \dotplus \Phi_{02} &= \begin{matrix}00\\00\end{matrix} = 0_{22},
\end{aligned} \tag{38}$$

and so forth. This leads to a general result:

$$\begin{aligned}
\Phi_{0n} \dotplus \Phi_{m0} &= 0_{mn},\\
\Phi_{m0} \dotplus \Phi_{0n} &= 0_{mn},
\end{aligned} \tag{39}$$

which follows as a limiting case of

$$0_{m-k,\,n-l} \dotplus 0_{k,\,l} = 0_{mn}, \tag{40}$$

upon identifying 0_0, 0_{0j}, 0_{i0} with the respective null matrices Φ_0, Φ_{0j}, Φ_{i0}.

Now making use of so-called "phantom elements" ϕ_{ij}, the ordinary matrix product of two null matrices may be written

$$\Phi_{03}\Phi_{30} = (\phi_{01}, \phi_{02}, \phi_{03}) \begin{pmatrix} \phi_{10} \\ \phi_{20} \\ \phi_{30} \end{pmatrix} = (\phi_{01}\phi_{10} + \phi_{02}\phi_{20} + \phi_{03}\phi_{30}) \tag{41}$$

From the dimensionality equation (0, 3)(3,0) = (0, 0), it follows that

$$\Phi_{03}\Phi_{30} = \Phi_0. \tag{42}$$

Using stick figures, this may be schematized:

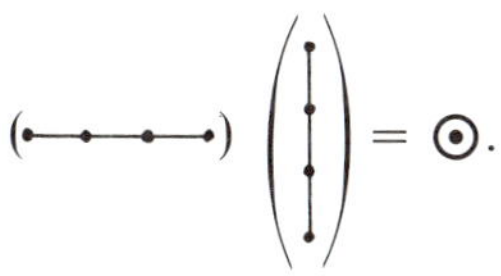

(43)

Next consider

$$\Phi_{m0}\Phi_{0n} = \begin{pmatrix} \phi_{10} \\ \phi_{20} \\ \cdot \\ \cdot \\ \cdot \\ \phi_{m0} \end{pmatrix} (\phi_{01}, \phi_{02}, \cdots, \phi_{0n}) = \begin{pmatrix} \phi_{10}\phi_{01} & \phi_{10}\phi_{02} & \cdots & \phi_{10}\phi_{0n} \\ \phi_{20}\phi_{01} & \phi_{20}\phi_{02} & \cdots & \phi_{20}\phi_{0n} \\ \cdot & \cdot & \cdot & \cdot \\ \cdot & \cdot & \cdot & \cdot \\ \cdot & \cdot & \cdot & \cdot \\ \phi_{m0}\phi_{01} & \phi_{m0}\phi_{02} & & \phi_{m0}\phi_{0n} \end{pmatrix}.$$

In the first product, as in a usual inner product, there is a decrease in dimensionality, or a contraction; in the second product, as in a usual outer product, there is an increase in dimensionality, or an expansion. In order to determine the meaning of the last product, call it Φ_{mn}, note that the original phantom elements ϕ_{i0} and ϕ_{0j} cannot be numbers, because even numbers like zero have size when viewed as 1 by 1 matrices. Moreover, since Φ_{mn} is the product of two matrices that must have no positive rank at all, it can have at most rank zero. Consequently, if a concrete representation is to be found for Φ_{mn}, it can be no more substantial than the numerical value 0_{mn}. That this is not totally unreasonable is suggested by the following stick figure diagram:

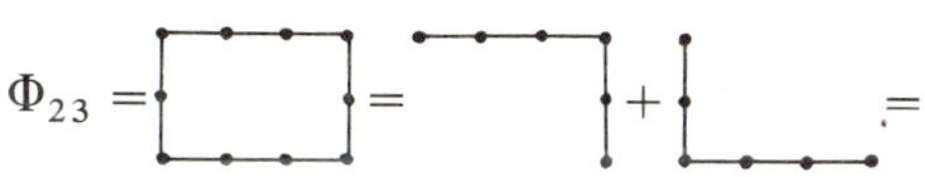

$$= \begin{pmatrix} 0 & 0 & 0 \\ 0 & 0 & 0 \end{pmatrix} + \begin{pmatrix} 0 & 0 & 0 \\ 0 & 0 & 0 \end{pmatrix} = 0_{23}. \tag{45}$$

Rather than depending upon this vague diagram, recall that

$$\Phi_0 \dotplus A = A \dotplus \Phi_0 = A. \tag{46}$$

Then the rule connecting ordinary matrix multiplication to direct addition,

$$(A \dotplus B)(C \dotplus D) = AC \dotplus BD, \tag{47}$$

which only holds provided both pairs A, C and B, D are compatible for ordinary multiplication, specializes to

$$\begin{aligned}(\Phi_{01} \dotplus \Phi_{10})(\Phi_{10} \dotplus \Phi_{01}) &= (\bar{0}])(\bar{0}]) = 0 \\ &= \phi_{01}\phi_{10} + \phi_{10}\phi_{01} = \Phi_0 + \Phi_{10}\Phi_{01} \\ &= \Phi_{10}\Phi_{01} = \Phi_1 = 0_1.\end{aligned} \tag{48}$$

Hence

$$\Phi_1 = 0_1. \tag{49}$$

In general,

$$\begin{aligned}(\Phi_{0n} + \Phi_{m0})(\Phi_{n0} + \Phi_{0r}) &= 0_{mn}0_{nr} = 0_{mr} \\ &= \Phi_{0n}\Phi_{n0} + \Phi_{m0}\Phi_{0r} = \Phi_0 + \Phi_{n0}\Phi_{0r} \\ &= \Phi_{m0}\Phi_{0r} = \Phi_{mr} = 0_{mr}.\end{aligned} \tag{50}$$

Hence

$$\Phi_{m0}\Phi_{0r} = \Phi_{mr} = 0_{mr}, \tag{51}$$

which indeed is a surprising factorization of zero.

Again, on the basis of the dimensionality equation $(0, m)(m, n) = (0, n)$, it seems reasonable to define

$$\Phi_{0m}A_{mn} = \Phi_{0n} \tag{52a}$$

and

$$A_{mn}\Phi_{n0} = \Phi_{m0}. \tag{52b}$$

EXERCISES

114. Prove that $\Phi_{00}{}^k = \overbrace{\Phi_{00}\Phi_{00}\cdots\Phi_{00}}^{k} = \Phi_{00}$.

115. If possible, evaluate the following expressions.

(*a*) $(A_n + \Phi_{0n})(A_n + \Phi_{n0})$.

(*b*) $(A_n + \Phi_{0n})(\Phi_{n0} + A_n)$.

(*c*) $(\Phi_{0n} + A_n)(A_n + \Phi_{n0})$.

(*d*) $(\Phi_{0n} + A_n)(\Phi_{n0} + A_n)$.

(*e*) $(A_n + \Phi_{n0})(A_n + \Phi_{0n})$.

(*f*) $(A_n + \Phi_{n0})(\Phi_{0n} + A_n)$.

(*g*) $(\Phi_{n0} + A_n)(A_n + \Phi_{0n})$.

(*h*) $(\Phi_{n0} + A_n)(\Phi_{0n} + A_n)$.

G. Direct Products

Let $A = (a_{ij})$ and $B = (b_{kl})$ be two matrices of dimensionality (m, n) and (r, s), respectively. Recall that for $m = s = 1$ and $n = r$, the ordinary matrix product AB reduces to a 1 by 1 matrix, as shown by the dimensionality equation:

$$(1, n)(n, 1) = (1, 1). \tag{1}$$

Again, but still under the condition that $n = r$, the ordinary matrix product AB yields a somewhat larger m by s matrix, in accordance with:

$$(m, n)(n, s) = (m, s). \tag{2}$$

Rather than imposing restrictions such as $n = r$, suppose that it is agreed upon to keep m, n, r, and s all different—although for generality permitting accidental agreements—and to aim at something really big in the way of products. It would be possible, for example, to devise meaningful prescriptions for forming products having size mn by rs or perhaps size mr by ns, as will be shown. In fact, it will be shown that, besides scalar multiplication and ordinary matrix multiplication, there are four different ways of defining large products using so-called "direct multiplication," three of them ($\cdot\times$, $\dot{\times}$, $\times\cdot$) applying basically the same notion; the fourth ($\underset{\cdot}{\times}$), a somewhat different notion.

In order to discuss direct multiplication more efficiently, it is convenient to introduce two new symbols, Λ and V, which are used in the next section on sequential notation. By supplying a key to interpreting these symbols, the following paragraphs should help the reader to acquire facility with this new notation.

Basically Λ denotes a "row" and V denotes a "column," and both ΛV and VΛ denote "matrices." Whenever written as a power, Λ^n means an "n-component row or subrow" and V^m means an "m-component column or subcolumn." At times a running variable is attached below the symbol and the "exponent" is centered above the symbol as a list limit. Then $\bigwedge_{j=1}^{n}$ indicates an "n-component row indexed by j" and $\bigvee_{i=1}^{n}$ indicates an "m-component column indexed by i."

Pairs such as $\Lambda^n V^m$ or $V^m \Lambda^n$ manufacture matrices from ordered sets of elements. In general, Λs and Vs can be combined in longer strings such as

$$\bigvee_{j=1}^{n} \bigwedge_{l=1}^{s} \bigvee_{k=1}^{r} \bigwedge_{i=1}^{m}, \tag{1}$$

which can be applied to a subscripted letter, like a $a_{ij,\,kl}$ to specify the structure of a partitioned matrix.

To reverse the process a string of Λs and Vs may be enclosed in parentheses and raised to the -1 power. When applied to the left of a matrix A, this specifies the undoing of the matrix into an ordered set of elements.

To understand the operations in Λ-V algebra, let us borrow some terminology from the television industry and label the Λs and Vs appearing in the following typical string:

$$\underset{\text{IV}}{\wedge}\ \underset{\text{III}}{\vee}\ \underset{\text{II}}{\wedge}\ \underset{\text{I}}{\vee} \tag{2}$$

as

$$\underset{\text{I}}{\vee} = \text{first vertical scanning operator,} \tag{3}$$

$$\underset{\text{II}}{\wedge} = \text{first horizontal scanning operator.}$$

$$\underset{\text{III}}{\vee} = \text{second vertical flyback operator,} \tag{4}$$

and

$$\underset{\text{IV}}{\wedge} = \text{second horizontal flyback operator.} \tag{5}$$

The counting path through the array of elements of a matrix produces rasters and subrasters as shown in Fig. 10.

In Fig. 10 the solid lines represent the motion of the scanning operators through the matrix rows or columns, represented by tick marks, and the dashed lines represent flyback operators. Notice that the counting paths in the two diagrams bear a transposed relationship to each other.

The Λs and Vs in a string are advanced sequentially and repeatedly as are

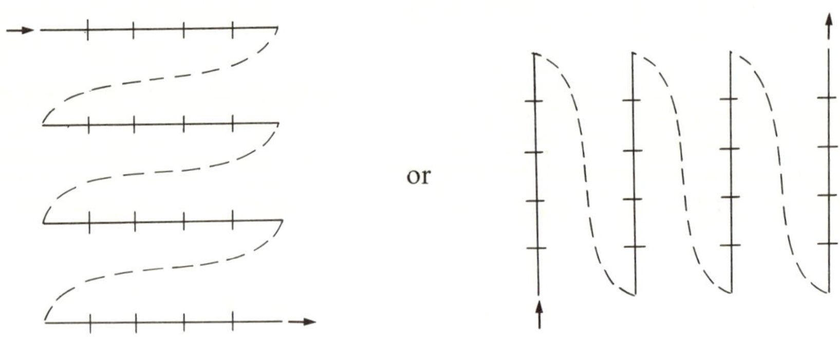

Figure 10. Wiggle worm diagrams showing counting path through matrix.

the digits displayed on an odometer, and really represent a generalized counting system with a variable base in the spirit of and including the decimal and binary counting systems. Consider a list limit of 2 for each Λ and V in the following:

$$\underset{\text{IV}}{\Lambda^2}\,\underset{\text{III}}{V^2}\,\underset{\text{II}}{\Lambda^2}\,\underset{\text{I}}{V^2}. \tag{6}$$

The Roman numeral subscripts will provide a record of list filling. Then

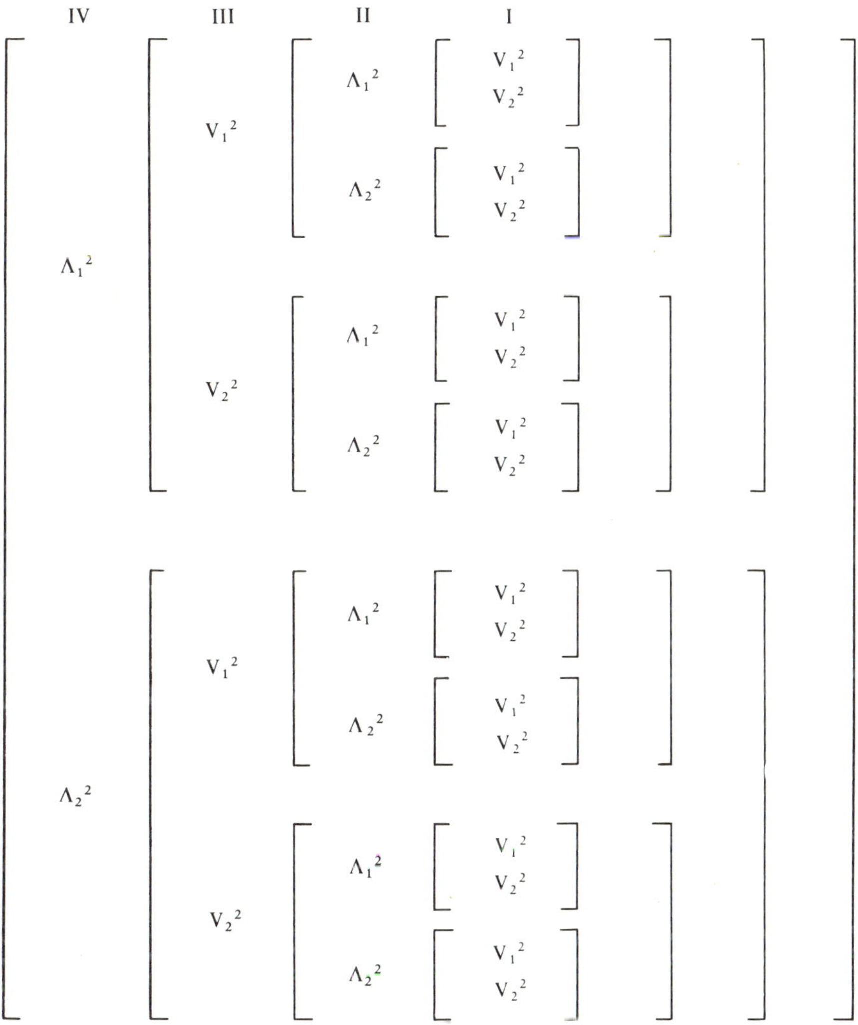

This can also be seen as:

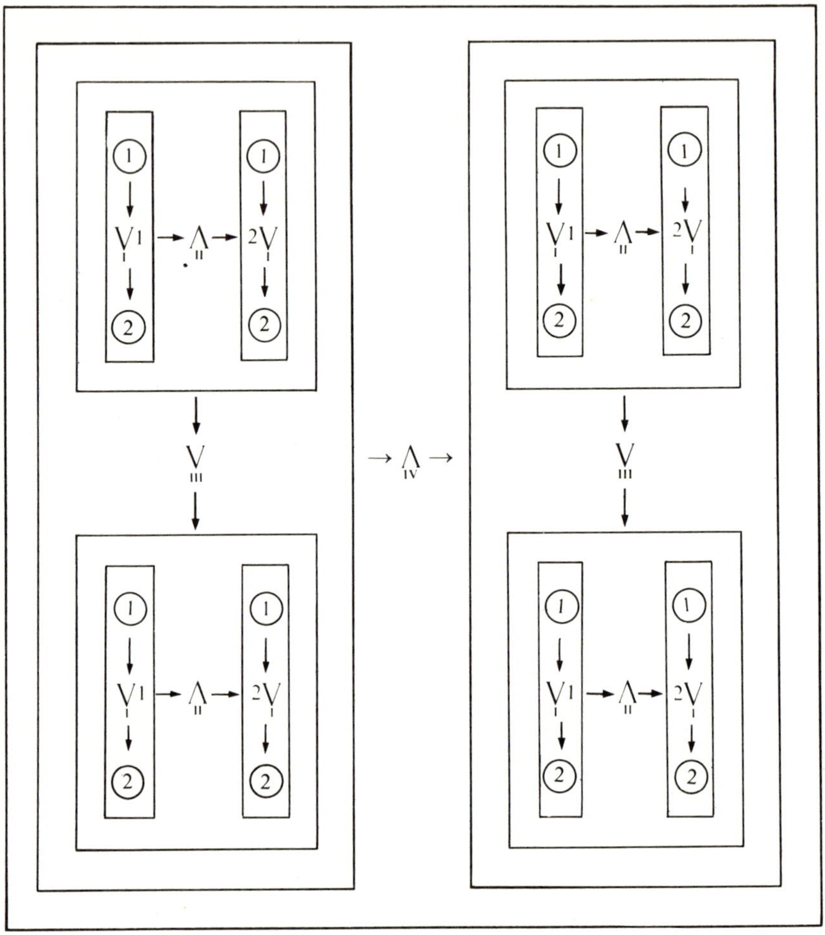

where arrows show the general route of motion.

The familiar summation sign $\sum$ (Greek capital letter sigma) provides a convenient notation for finite and countably infinite sums, as follows:

$$\sum_{j=1}^{n} a_j = a_1 + a_2 + \cdots + a_n, \tag{7a}$$

$$\sum_{j=1}^{\infty} a_j = a_1 + a_2 + a_3 + \cdots \tag{7b}$$

A similar symbol, the "horizontal list sign" Λ (Greek capital letter lambda) is a useful notation for finite and countably infinite sequences, as follows:

$$\bigwedge_{j=1}^{n} a_j = a_1, a_2, \cdots, a_n, \tag{8a}$$

$$\bigwedge_{j=1}^{\infty} a_j = a_1, a_2, a_3, \cdots. \tag{8b}$$

Because of its suggestive appearance, the Roman capital letter V may be used to denote the "vertical list sign:"

$$\bigvee_{i=1}^{n} a_j = \begin{matrix} a_1, \\ a_2, \\ \cdot \\ \cdot \\ \cdot\ , \\ a_m, \end{matrix} \tag{9a}$$

and

$$\bigvee_{i=1}^{\infty} a_i = \begin{matrix} a_1, \\ a_2, \\ a_3, \\ \cdot \\ \cdot \\ \cdot \end{matrix} \tag{9b}$$

In sequential notation row and column vectors are represented as follows:

$$\mathbf{x} = (x_1, x_2, \cdots, x_n) = \left(\bigwedge_{j=1}^{n} x_j\right), \tag{10a}$$

$$\mathbf{x}^T = \begin{pmatrix} x_1 \\ x_2 \\ \cdot \\ \cdot \\ \cdot \\ x_m \end{pmatrix} = \left(\bigvee_{i=1}^{m} x_i\right). \tag{10b}$$

Pairs such as $\bigwedge^{n} \bigvee^{m}$ manufacture matrices from ordered sets of elements. Thus, in sequential notation, matrices $A = (a_{ij})$ of size m by n become simply:

$$A = \left(\bigvee_{i=1}^{m} \bigwedge_{j=1}^{n} a_{ij}\right) = \left(\bigwedge_{j=1}^{n} \bigvee_{i=1}^{m} a_{ij}\right), \tag{11}$$

where i, j are the running variables and m, n are the list limits. For specification (11) the matrix A is built elementwise, first by rows and then by columns

Strings such as $\vee\wedge\vee\wedge$ can be applied to a subscripted letter like $A_{ij,kl}$ to specify the structure of a regularly partitioned matrix. If, for example,

$$M = \left(\begin{array}{cc:cc} m_{11,11} & m_{11,12} & m_{12,11} & m_{12,12} \\ m_{11,21} & m_{11,22} & m_{12,21} & m_{12,22} \\ \hdashline m_{21,11} & m_{21,12} & m_{22,11} & m_{22,12} \\ m_{21,21} & m_{21,22} & m_{22,21} & m_{22,22} \end{array}\right), \tag{12a}$$

then

$$M = \bigvee_{i=1}^{2} \bigwedge_{j=1}^{2} \bigvee_{k=1}^{2} \bigwedge_{l=1}^{2} m_{ij,kl}. \tag{12b}$$

Inverse strings such as $(\vee\wedge\ \vee\wedge)^{-1}$ can be applied to the left of a matrix to represent its undoing into an ordered set of elements. Thus to elongate or stretch the matrix A by rows:

$$\wedge\left(\bigvee_{i=1}^{m} \bigwedge_{j=1}^{n}\right)^{-1} A = \wedge\left(\bigvee_{i=1}^{m} \bigwedge_{j=1}^{n}\right)^{-1} \begin{pmatrix} a_{11} & a_{12} \cdots & a_{1n} \\ a_{21} & a_{22} \cdots & a_{2n} \\ \cdot & \cdot & \cdot \\ \cdot & \cdot & \cdot \\ \cdot & \cdot & \cdot \\ a_{m1} & a_{m2} & a_{mn} \end{pmatrix} \tag{13}$$

$$= (a_{11}, a_{12}, \cdots, a_{1m}, a_{21}, a_{22}, \cdots, a_{2m}, \cdots a_{m1}, a_{m2}, \cdots, a_{mn}).$$

Sequential notation affords a method for charting paths to follow in dealing with elements of a matrix. For example, consider the following $\wedge$-$\vee$ prescription for reading the elements of

$$A = \begin{vmatrix} a & b & c & d \\ e & f & g & h \\ j & k & l & m \\ n & p & q & r \end{vmatrix}$$

and for forming the elongations **x**, **y**, and **z**:

$$(\vee^4\wedge^4)^{-1} A = \begin{pmatrix} a \to b \to c \to d \\ \downarrow \\ e \to f \to g \to h \\ \downarrow \\ j \to k \to l \to m \\ \downarrow \\ n \to p \to q \to r \end{pmatrix}, \tag{14}$$

$$\mathbf{x} = \Lambda(\mathrm{V}^4\Lambda^4)^{-1}A = (a, b, c, d, e, f, g, h, j, k, l, m, n, p, q, r);$$

$$(\Lambda^2\mathrm{V}^2\Lambda^2\mathrm{V}^2)^{-1}A =$$ 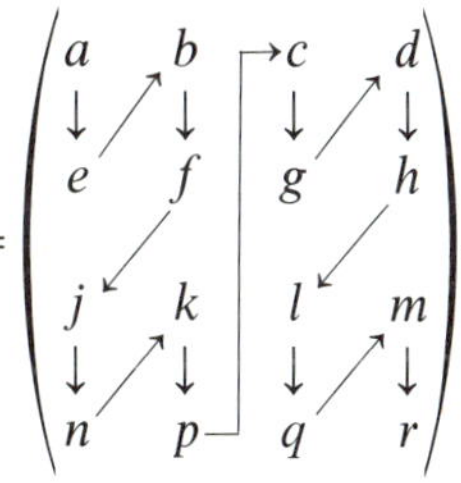 , (15)

$$\mathbf{y} = \Lambda(\Lambda^2\mathrm{V}^2\Lambda^2\mathrm{V}^2)^{-1}A = (a, e, b, f, j, n, k, p, c, g, d, h, l, q, m, r);$$

$$(\mathrm{V}^2\Lambda^4\mathrm{V}^2)^{-1}A =$$ 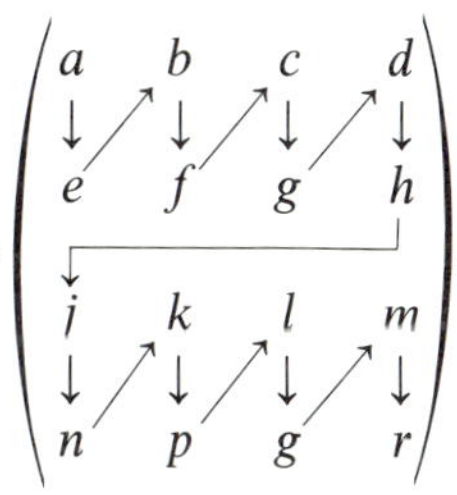 , (16)

$$\mathbf{z} = \Lambda(\mathrm{V}^2\Lambda^4\mathrm{V}^2)^{-1}A - (a, c, b, f, c, g, d, h, j, n, k, p, l, q, m, r).$$

EXERCISES

116. Let

$$A = \begin{vmatrix} a & b & c & d & e & f \\ g & h & j & k & l & m \\ n & p & q & r & s & t \\ u & v & w & x & y & z \end{vmatrix}.$$

(*a*) Using sequential notation write Λ-V prescriptions for the following regular lists of elements:

(1) *a b c d e f g h j k l m n p q r s t u v w x y z.*
(2) *a b c g h j n p q u v w d e f k l m r s t x y z.*
(3) *a b g h n p u v c d j k q r w x e f l m s t y z.*
(4) *a g n u b h p v c j q w d k r x e l s y f m t z.*

(*b*) List the following:

$$\Lambda(\mathrm{V}^2\Lambda^6\mathrm{V}^2)^{-1}A, \quad \mathrm{V}(\mathrm{V}^2\Lambda^3\mathrm{V}^2\Lambda^2)^{-1}A, \quad \text{and} \quad \Lambda(\mathrm{V}^2\Lambda^2\mathrm{V}^2\Lambda^3)^{-1}A.$$

(*c*) What further lists are possible?
(*d*) Chart all the resulting paths through the *A* matrix using arrow signs.

Table 1

Λ-V Operators

Summation sign

$$\sum_{j=1}^{n} a_j = a_1 + a_2 + a_3 + \cdots a_n$$

Product sign

$$\prod_{j=1}^{n} a_j = a_1\, a_2\, a_3 \cdots a_n$$

Horizontal list sign

$$\bigwedge_{j=1}^{n} a_j = a_1, a_2, a_3, \ldots, a_n$$

Vertical list sign

$$\bigvee_{j=1}^{n} a_j = \begin{matrix} a_1, \\ a_2, \\ a_3, \\ \cdot \\ \cdot \\ \cdot \\ a_n \end{matrix}$$

Basic Λ-V Operators

$$A = \begin{pmatrix} a & b & c \\ d & e & f \\ g & h & j \end{pmatrix}$$

$$\left(\bigvee_{i=1}^{3} \bigwedge_{j=1}^{3}\right)^{-1} A = \begin{pmatrix} a \rightarrow b \rightarrow c \\ \downarrow \\ d \rightarrow e \rightarrow f \\ \downarrow \\ g \rightarrow h \rightarrow j \end{pmatrix}$$

$$\left(\bigwedge_{j=1}^{3} \bigvee_{i=1}^{3}\right)^{-1} A = \begin{pmatrix} a & \rightarrow b & \rightarrow c \\ \downarrow & \downarrow & \downarrow \\ d & e & f \\ \downarrow & \downarrow & \downarrow \\ g & h & j \end{pmatrix}$$

Row elongation by rows:
$\Lambda A = \Lambda(\mathrm{V}\Lambda)^{-1} A = (a\ b\ c\ d\ e\ f\ g\ h\ j)$

Row elongation by columns:
$\Lambda A^T = \Lambda(\Lambda \mathrm{V})^{-1} A = (a\ d\ g\ b\ e\ h\ c\ f\ j)$

Column elongation by rows:

$$\mathrm{V}A = \mathrm{V}(\mathrm{V}\Lambda)^{-1} A = \begin{pmatrix} a \\ b \\ c \\ d \\ e \\ f \\ g \\ h \\ j \end{pmatrix}$$

Column elongation by columns:

$$\mathrm{V}A^T = \mathrm{V}(\Lambda \mathrm{V})^{-1} A = \begin{pmatrix} a \\ d \\ g \\ b \\ e \\ h \\ c \\ f \\ j \end{pmatrix}$$

Table 1 *continued*

Additional Λ-V Operators

$$\left(\bigvee_{j=1}^{3}\bigwedge_{i=j}^{3}\right)^{-1} A = \begin{pmatrix} a \rightarrow b \rightarrow c \\ & & \swarrow \\ d & e \rightarrow f \\ & & \downarrow \\ g & h & j \end{pmatrix}$$

$$\bigwedge_{i=j}^{-1} \begin{pmatrix} a_{11} & a_{12} & a_{13} \\ & \searrow & \\ a_{21} & a_{22} & a_{23} \\ & & \searrow \\ a_{31} & a_{32} & a_{33} \end{pmatrix}$$

$\Lambda(\mathrm{V}\Lambda)^{-1} A = (a\, b\, c\, e\, f\, j)$

can yield

$$(\mathrm{V}\Lambda)(\mathrm{V}\Lambda)^{-1} A = \begin{pmatrix} a & b & c \\ 0 & e & f \\ 0 & 0 & j \end{pmatrix}$$

$$\begin{pmatrix} a_{11} & 0 & 0 \\ 0 & a_{22} & 0 \\ 0 & 0 & a_{33} \end{pmatrix}$$

Smaller partitions:

$$\overset{2}{\Lambda}\,\overset{2}{\mathrm{V}}\,\overset{2}{\Lambda}\,\overset{2}{\mathrm{V}} \begin{pmatrix} a & b & c & d \\ e & f & g & h \\ j & k & l & m \\ n & p & q & r \end{pmatrix} =$$

$$= \left(\begin{array}{cc|cc} a & b & \nearrow c & d \\ \downarrow \nearrow & \downarrow & \downarrow \nearrow & \downarrow \\ e & f & g & h \\ \swarrow & & \swarrow & \\ j & k & l & m \\ \downarrow \nearrow & \downarrow & \downarrow \nearrow & \downarrow \\ n & p & q & r \end{array}\right)$$

Counting to infinity:

$$\begin{array}{ccccc} a_{11} & a_{12} & a_{13} & a_{14} & \cdots \\ \swarrow & \swarrow & \swarrow & & \\ a_{21} & a_{22} & a_{23} & a_{24} & \cdots \\ \swarrow & \swarrow & & & \\ a_{31} & a_{32} & a_{33} & & \\ \swarrow & & & & \\ a_{41} & a_{42} & & & \\ \vdots & & & & \end{array}$$

basic term $\bigwedge\bigvee_{i+j=n}$

117. Write prescriptions for the following paths.

(*a*)
$$\begin{pmatrix} a_{11} \rightarrow a_{12} \rightarrow a_{13} \rightarrow a_{14} \\ \hfill \downarrow \\ a_{21} \leftarrow a_{22} \leftarrow a_{23} \leftarrow a_{24} \\ \downarrow \hfill \\ a_{31} \rightarrow a_{32} \rightarrow a_{33} \rightarrow a_{34} \\ \hfill \downarrow \\ a_{41} \leftarrow a_{42} \leftarrow a_{43} \leftarrow a_{44} \end{pmatrix}$$

(*b*) 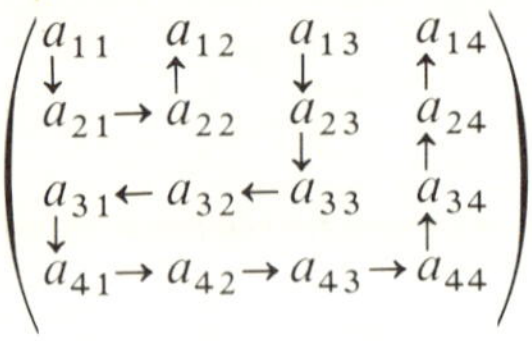

(*c*) 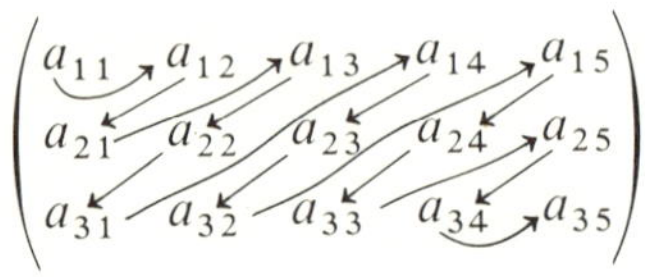

(*d*) 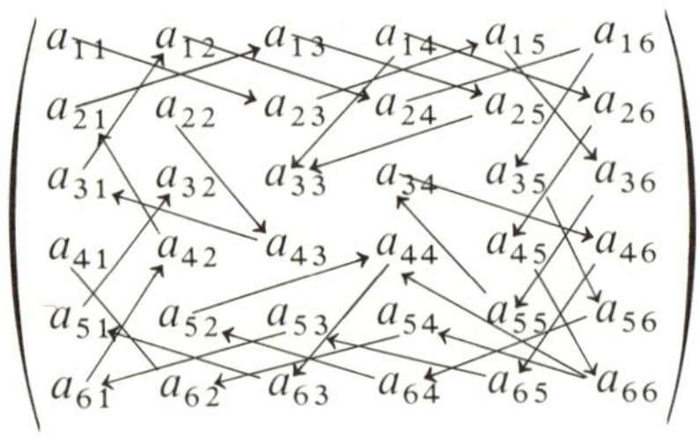

118. Let $\mathbf{x} = (1, 2, \cdots, 10)$. Complete the following specifications by supplying suitable indices and list limits:

(*a*) $$\mathrm{V}\Lambda\mathbf{x} = \begin{pmatrix} 1 & 2 & 3 & 4 \\ 0 & 5 & 6 & 7 \\ 0 & 0 & 8 & 9 \\ 0 & 0 & 0 & 10 \end{pmatrix}$$

(*b*) $$\Lambda\mathrm{V}\mathbf{x} = \begin{pmatrix} 1 & 0 & 0 & 0 \\ 2 & 3 & 0 & 0 \\ 4 & 5 & 6 & 0 \\ 7 & 8 & 9 & 10 \end{pmatrix}.$$

(*c*) $$N\mathbf{x} = \Lambda\mathrm{V}\mathbf{x} = \begin{pmatrix} 1 & 0 \cdots & 0 \\ 0 & 2 \cdots & 0 \\ \cdot & \cdot\;\cdot & \cdot \\ \cdot & \cdot\;\;\cdot & \cdot \\ \cdot & \cdot\;\;\;\cdot & \cdot \\ 0 & 0 \cdots & 10 \end{pmatrix}.$$

119. Devise indices, list limits, and, if necessary, side conditions to make the following definitions, based on the Kronecker delta, work.

(*a*) $I = \mathrm{V}\Lambda\, \delta_{ij}$.

(*b*) $J = \mathrm{V}\Lambda\, \delta_{ij}$.

(*c*) $K = \mathrm{V}\Lambda\, \delta_{ij}$.

(*d*) $L = \mathrm{V}\Lambda\, \delta_{ij}$.

120. Let $h_{ij} = 1$, if $j = i + 1$; and $h_{ij} = 0$ otherwise. Compute A^{-1} where

$$A = \mathop{\mathrm{V}}_{i=1}^{n} \mathop{\Lambda}_{j=1}^{n} (\delta_{ij} - h_{ij})$$

by skirting A with the unit matrix and performing elementary column operations.

If A and B are of size m by n and n by s, respectively, then the ordinary matrix product becomes

$$C = (c_{ij}) - AB = \left(\sum_{\lambda=1}^{n} a_{i\lambda} b_{\lambda j}\right) = \mathop{\Lambda}_{i=1}^{m} \mathop{\mathrm{V}}_{j=1}^{s} c_{ij}$$

$$- \mathop{\Lambda}_{i=1}^{m} \mathop{\mathrm{V}}_{j=1}^{s} \sum_{\lambda=1}^{n} (\mathrm{V}^{i}\Lambda^{\lambda})^{-1} A (\mathrm{V}^{\lambda}\Lambda^{i})^{-1} B,$$

$$C = \mathop{\Lambda}_{i=1}^{m} \mathop{\mathrm{V}}_{j=1}^{s} \sum_{\lambda=1}^{n} a_{i\lambda} b_{\lambda j} = \sum_{\lambda=1}^{n} \mathop{\Lambda}_{i=1}^{m} \mathop{\mathrm{V}}_{j=1}^{s} a_{i\lambda} b_{\lambda j},$$

$$C = \sum_{\lambda=1}^{n} \begin{pmatrix} a_{1\lambda} b_{\lambda 1} & a_{1\lambda} b_{\lambda 2} & \cdots & a_{1\lambda} b_{\lambda s} \\ a_{2\lambda} b_{\lambda 1} & a_{2\lambda} b_{\lambda 2} & \cdots & a_{2\lambda} b_{\lambda s} \\ \cdot & \cdot & \cdot & \cdot \\ \cdot & \cdot & \cdot & \cdot \\ \cdot & \cdot & \cdot & \cdot \\ a_{m\lambda} b_{\lambda 1} & a_{m\lambda} b_{\lambda 2} & \cdots & a_{m\lambda} b_{\lambda s} \end{pmatrix}.$$

Additional manipulations are possible using Λ and V. As a routine matter, of course, any m by n matrix can be subdivided into smaller parts, each part forming a submatrix. Consider, for example, the row matrix $A = (1\ 2\ 3\ 4\ 5)$. Reasoning schematically:

$$A = \underset{\Lambda^5 a_\lambda}{\boxed{1\ \ 2\ \ 3\ \ 4\ \ 5}} = \underset{\Lambda^{3+2} a_\lambda}{\boxed{1\ \ 2}\boxed{3\ \ 4\ \ 5}} = \underset{\Lambda^{3}\Lambda^{2} a_\lambda}{\boxed{1\ \ 2}\quad\boxed{3\ \ 4\ \ 5}}.$$

Similarly:

$$B = \underset{V^5\Lambda^5 b_{\lambda j}}{\left|\begin{array}{ccccc} a & b & c & d & e \\ f & g & h & j & k \\ l & m & n & o & p \\ q & r & s & t & u \\ v & w & x & y & z \end{array}\right|} = \underset{V^{3+2}\Lambda^{3+2} b_{\lambda j}}{\left|\begin{array}{cc|ccc} a & b & c & d & e \\ f & g & h & j & k \\ \hline l & m & n & o & p \\ q & r & s & t & u \\ v & w & x & y & z \end{array}\right|} = \underset{V^3 V^2 \Lambda^3 \Lambda^2 b_{\lambda j}.}{\begin{array}{cc} \boxed{\begin{array}{cc} a & b \\ f & g \end{array}} & \boxed{\begin{array}{ccc} c & d & e \\ h & j & k \end{array}} \\ \boxed{\begin{array}{cc} l & m \\ q & r \\ v & w \end{array}} & \boxed{\begin{array}{ccc} n & o & p \\ s & t & u \\ x & y & z \end{array}} \end{array}}.$$

Now A becomes

$$A = \Lambda^3\Lambda^2 a_\lambda = \bigwedge_{\lambda=3}^{5} a_\lambda \bigwedge_{\lambda=1}^{2} a_\lambda = (A_{11} \vdots A_{12})$$

and B becomes

$$B = V^3V^2\Lambda^3\Lambda^2 b_{\lambda j} = \bigvee_{\lambda=3}^{5} \bigwedge_{j=3}^{5} b_{\lambda j} \bigvee_{\lambda=3}^{5} \bigwedge_{j=1}^{2} b_{\lambda j} \bigvee_{\lambda=1}^{2} \bigwedge_{j=3}^{5} b_{\lambda j} \bigvee_{\lambda=1}^{2} \bigwedge_{j=1}^{2} b_{\lambda j}$$

$$= \begin{pmatrix} B_{11} & B_{12} \\ \cdots\cdots & \cdots\cdots \\ B_{21} & B_{22} \end{pmatrix}.$$

Consequently

$$C = AB = (A_{11}B_{11} + A_{12}B_{21}, A_{11}B_{12} + A_{12}B_{22})$$

or

$$C = \bigwedge_{\lambda=1}^{2} a_\lambda \bigvee_{\lambda=1}^{2} \bigwedge_{j=1}^{2} b_{\lambda j} + \bigwedge_{\lambda=3}^{5} a_\lambda \bigvee_{\lambda=3}^{5} \bigwedge_{j=1}^{2} b_{\lambda j}, \bigwedge_{\lambda=1}^{2} a_\lambda \bigvee_{\lambda=1}^{2} \bigwedge_{j=3}^{5} b_{\lambda j}$$

$$+ \bigwedge_{\lambda=3}^{5} a_\lambda \bigvee_{\lambda=3}^{5} \bigwedge_{j=3}^{5} b_{\lambda j} \quad ,$$

which is

$$C = \left[(12) \begin{pmatrix} a & b \\ f & g \end{pmatrix} + (3 \quad 4 \quad 5) \begin{pmatrix} l & m \\ q & r \\ v & w \end{pmatrix}, \right.$$

$$\left. (1 \quad 2) \begin{pmatrix} c & d & e \\ h & j & k \end{pmatrix} + (3 \quad 4 \quad 5) \begin{pmatrix} n & o & p \\ s & t & u \\ x & y & z \end{pmatrix} \right]$$

or

$$C = [(a + 2f, b + 2g) + (3l + 4q + 5v, 3m + 4r + 5w),\\ (c + 2h, d + 2j, e + 2k)\\ + (3n + 4s + 5x, 3o + 4t + 5y, 3p + 4u + 5z)].$$

Finally

$$C = (a + 2f + 3l + 4q + 5v, b + 2g + 3m + 4r + 5w,\\ c + 2h + 3n + 4s + 5x, d + 2j + 3o + 4t + 5y,\\ e + 2k + 3p + 4u + 5z)$$

and

$$C = AB = (1 \quad 2 \quad 3 \quad 4 \quad 5)\begin{pmatrix} a & b & c & d & e \\ f & g & h & j & k \\ l & m & n & o & p \\ q & r & s & t & u \\ v & w & x & y & z \end{pmatrix}$$

$$= (1 \quad 2 \quad 3 \quad 4 \quad 5)\left(\begin{array}{cc:ccc} a & b & c & d & e \\ f & g & h & j & k \\ \hdashline l & m & n & o & p \\ q & r & s & t & u \\ v & w & x & y & z \end{array}\right)$$

$$= [(1 \quad 2)(3 \quad 4 \quad 5)]\begin{pmatrix} \begin{pmatrix} a & b \\ f & g \end{pmatrix} & \begin{pmatrix} c & d & e \\ h & j & k \end{pmatrix} \\ \begin{pmatrix} l & m \\ q & r \\ v & w \end{pmatrix} & \begin{pmatrix} n & o & p \\ s & t & u \\ x & y & z \end{pmatrix} \end{pmatrix}$$

Let $A = (a_{ij})$ and $B = (b_{kl})$ be of size m by n and r by s, respectively. Consider all 24 possible direct products:

$$A \times B = (a_{ij}b_{kl}), \tag{17}$$

defined as follows:

(mr, ns):

$$A \overset{1}{\times} B = \bigvee_{i=1}^{m} \bigwedge_{j=1}^{n} \bigvee_{k=1}^{r} \bigwedge_{l=1}^{s} a_{ij} b_{kl}, \tag{18a}$$

(ms, nr):

$$A \overset{2}{\times} B = \bigvee_{i=1}^{m} \bigwedge_{j=1}^{n} \bigvee_{l=1}^{s} \bigwedge_{k=1}^{r} a_{ij} b_{kl}, \tag{18b}$$

(mn, rs):

$$A \overset{3}{\times} B = \bigvee_{i=1}^{m} \bigwedge_{k=1}^{r} \bigvee_{l=1}^{s} \bigwedge_{j=1}^{n} a_{ij} b_{kl}, \tag{18c}$$

(ms, rn):

$$A \overset{4}{\times} B = \bigvee_{i=1}^{m} \bigwedge_{l=1}^{s} \bigvee_{j=1}^{n} \bigwedge_{k=1}^{r} a_{ij} b_{kl}, \tag{18d}$$

(mn, sr):

$$A \overset{5}{\times} B = \bigvee_{i=1}^{m} \bigwedge_{l=1}^{s} \bigvee_{k=1}^{r} \bigwedge_{j=1}^{n} a_{ij} b_{kl}, \tag{18e}$$

(mr, sn):

$$A \overset{6}{\times} B = \bigvee_{j=1}^{n} \bigwedge_{i=1}^{m} \bigvee_{k=1}^{r} \bigwedge_{l=1}^{s} a_{ij} b_{kl}, \tag{18f}$$

(nr, ms):

$$A \overset{7}{\times} B = \bigvee_{j=1}^{n} \bigwedge_{i=1}^{m} \bigvee_{k=1}^{r} \bigwedge_{l=1}^{s} a_{ij} b_{kl}, \tag{18g}$$

(ns, mr):

$$A \overset{8}{\times} B = \bigvee_{j=1}^{n} \bigwedge_{i=1}^{m} \bigvee_{l=1}^{s} \bigwedge_{k=1}^{r} a_{ij} b_{kl}, \tag{18h}$$

(nm, rs):

$$A \overset{9}{\times} B = \bigvee_{j=1}^{n} \bigwedge_{k=1}^{r} \bigvee_{i=1}^{m} \bigwedge_{l=1}^{s} a_{ij} b_{kl}, \tag{18i}$$

(ns, rm):

$$A \overset{10}{\times} B = \bigvee_{j=1}^{n} \bigwedge_{k=1}^{r} \bigvee_{l=1}^{s} \bigwedge_{i=1}^{m} a_{ij} b_{kl}, \tag{18j}$$

(nm, sr):

$$A \overset{11}{\times} B = \bigvee_{j=1}^{n} \bigwedge_{l=1}^{s} \bigvee_{i=1}^{m} \bigwedge_{k=1}^{r} a_{ij}b_{kl}, \tag{18k}$$

(nr, sm):

$$A \overset{12}{\times} B = \bigvee_{j=1}^{n} \bigwedge_{l=1}^{s} \bigvee_{k=1}^{r} \bigwedge_{i=1}^{m} a_{ij}b_{kl}, \tag{18l}$$

(rn, ms):

$$A \overset{13}{\times} B = \bigvee_{k=1}^{r} \bigwedge_{i=1}^{m} \bigvee_{j=1}^{n} \bigwedge_{l=1}^{s} a_{ij}b_{kl}, \tag{18m}$$

(rs, mn):

$$A \overset{14}{\times} B = \bigvee_{k=1}^{r} \bigwedge_{i=1}^{m} \bigvee_{l=1}^{s} \bigwedge_{j=1}^{n} a_{ij}b_{kl}, \tag{18n}$$

(rm, ns):

$$A \overset{15}{\times} B = \bigvee_{k=1}^{r} \bigwedge_{j=1}^{n} \bigvee_{i=1}^{m} \bigwedge_{l=1}^{s} a_{ij}b_{kl}, \tag{18o}$$

(rs, nm):

$$A \overset{16}{\times} B = \bigvee_{k=1}^{r} \bigwedge_{j=1}^{n} \bigvee_{l=1}^{s} \bigwedge_{i=1}^{m} a_{ij}b_{kl}, \tag{18p}$$

(rm, sn):

$$A \overset{17}{\times} B = \bigvee_{k=1}^{r} \bigwedge_{l=1}^{s} \bigvee_{i=1}^{m} \bigwedge_{j=1}^{n} a_{ij}b_{kl}, \tag{18q}$$

(rn, sm):

$$A \overset{18}{\times} B = \bigvee_{k=1}^{r} \bigwedge_{l=1}^{s} \bigvee_{j=1}^{n} \bigwedge_{i=1}^{m} a_{ij}b_{kl}, \tag{18r}$$

(sn, mr):

$$A \overset{19}{\times} B = \bigvee_{l=1}^{s} \bigwedge_{i=1}^{m} \bigvee_{j=1}^{n} \bigwedge_{k=1}^{r} a_{ij}b_{kl}, \tag{18s}$$

(sr, mn):

$$A \overset{20}{\times} B = \bigvee_{l=1}^{s} \bigwedge_{i=1}^{m} \bigvee_{k=1}^{r} \bigwedge_{j=1}^{n} a_{ij}b_{ij}, \tag{18t}$$

(sm, nr):

$$A \overset{21}{\times} B = \bigvee_{l=1}^{s} \bigwedge_{j=1}^{n} \bigvee_{i=1}^{m} \bigwedge_{k=1}^{r} a_{ij} b_{kl}, \tag{18u}$$

(sr, nm):

$$A \overset{22}{\times} B = \bigvee_{l=1}^{s} \bigwedge_{j=1}^{n} \bigvee_{k=1}^{r} \bigwedge_{i=1}^{m} a_{ij} b_{kl}, \tag{18v}$$

(sm, rn):

$$A \overset{23}{\times} B = \bigvee_{l=1}^{s} \bigwedge_{k=1}^{r} \bigvee_{i=1}^{m} \bigwedge_{j=1}^{n} a_{ij} b_{kl}, \tag{18w}$$

(sn, rm):

$$A \overset{24}{\times} B = \bigvee_{l=1}^{s} \bigwedge_{k=1}^{r} \bigvee_{j=1}^{n} \bigwedge_{i=1}^{m} a_{ij} b_{kl}. \tag{18x}$$

Now, writing

$$\begin{pmatrix} a_{11} & a_{1n} \\ a_{m1} & a_{mn} \end{pmatrix}$$

for

$$\begin{pmatrix} a_{11} & a_{12} \cdots a_{1n} \\ a_{21} & a_{22} \cdots a_{2n} \\ \cdot & \cdot \quad\quad \cdot \\ \cdot & \cdot \quad\quad \cdot \\ \cdot & \cdot \quad\quad \cdot \\ a_{m1} & a_{m2} \cdots a_{mn} \end{pmatrix}$$

and so forth:

$$\begin{aligned} A \overset{1}{\times} B &= \left(\begin{array}{cc:cc} a_{11} b_{11} & a_{11} b_{1s} & a_{1n} b_{11} & a_{1n} b_{1s} \\ a_{11} b_{r1} & a_{11} b_{rs} & a_{1n} b_{r1} & a_{1n} b_{rs} \\ \hdashline a_{m1} b_{11} & a_{m1} b_{1s} & a_{mn} b_{11} & a_{mn} b_{1s} \\ a_{m1} b_{r1} & a_{m1} b_{rs} & a_{m1} b_{r1} & a_{mn} b_{rs} \end{array}\right) \\ &= \begin{pmatrix} a_{11} B & a_{1n} B \\ a_{m1} B & a_{mn} B \end{pmatrix} = A \times \cdot B, \end{aligned} \tag{19a}$$

$$
A \overset{2}{\times} B = \left(\begin{array}{cc:cc}
a_{11}b_{11} & a_{11}b_{r1} & a_{1n}b_{11} & a_{1n}b_{r1} \\
a_{11}b_{1s} & a_{11}b_{rs} & a_{1n}b_{1s} & a_{1n}b_{rs} \\
\hdashline
a_{m1}b_{11} & a_{m1}b_{r1} & a_{mn}b_{11} & a_{mn}b_{r1} \\
a_{m1}b_{1s} & a_{m1}b_{rs} & a_{mn}b_{rs} & a_{mm}b_{rs}
\end{array}\right)
$$

$$
= \begin{pmatrix} a_{11}B^T & a_{1n}B^T \\ a_{m1}B^T & a_{mn}B^T \end{pmatrix} = A \times\cdot B^T, \tag{19b}
$$

$$
A \overset{3}{\times} B = \left(\begin{array}{cc:cc}
a_{11}b_{11} & a_{11}b_{1s} & a_{11}b_{r1} & a_{11}b_{rs} \\
a_{1n}b_{11} & a_{1n}b_{1s} & a_{1n}b_{r1} & a_{1n}b_{rs} \\
\hdashline
a_{m1}b_{11} & a_{m1}b_{1s} & a_{m1}b_{r1} & a_{m1}b_{rs} \\
a_{mn}b_{11} & a_{mn}b_{1s} & a_{mn}b_{r1} & a_{mn}b_{rs}
\end{array}\right)
$$

$$
= \begin{pmatrix} a_{11} \\ a_{1n} \\ a_{m1} \\ a_{mn} \end{pmatrix} (b_{11}b_{1s}b_{r1}b_{rs}) = A \underset{\cdot}{\times} B, \tag{19c}
$$

$$
A \overset{4}{\times} B = \left(\begin{array}{cc:cc}
a_{11}b_{11} & a_{1n}b_{11} & a_{11}b_{r1} & a_{1n}b_{r1} \\
a_{11}b_{1s} & a_{1n}b_{1s} & a_{11}b_{rs} & a_{1n}b_{rs} \\
\hdashline
a_{m1}b_{11} & a_{mn}b_{11} & a_{m1}b_{r1} & a_{mn}b_{r1} \\
a_{m1}b_{1s} & a_{mn}b_{1s} & a_{m1}b_{rs} & a_{mn}b_{rs}
\end{array}\right) \tag{19d}
$$

$$
= \left(\begin{array}{c:c}
\begin{pmatrix} b_{11} \\ b_{1s} \end{pmatrix}(a_{11}a_{1n}) & \begin{pmatrix} b_{r1} \\ b_{rs} \end{pmatrix}(a_{11}a_{1n} \\
\hdashline
\begin{pmatrix} b_{11} \\ b_{1s} \end{pmatrix}(a_{m1}a_{mn}) & \begin{pmatrix} b_{r1} \\ b_{rs} \end{pmatrix}(a_{m1}a_{mn})
\end{array}\right)
$$

if elements commute

$$
= (A^T \dot{\times} B)^T,
$$

same or
otherwise

$$A \overset{5}{\times} B = \left(\begin{array}{cc|cc} a_{11}b_{11} & a_{11}b_{r1} & a_{11}b_{1s} & a_{11}b_{rs} \\ a_{1n}b_{11} & a_{1n}b_{r1} & a_{1n}b_{1s} & a_{1n}b_{rs} \\ \hline a_{m1}b_{11} & a_{m1}b_{r1} & a_{m1}b_{1s} & a_{m1}b_{rs} \\ a_{mn}b_{11} & a_{mn}b_{r1} & a_{mn}b_{1s} & a_{mn}b_{rs} \end{array}\right) \tag{19e}$$

$$= \begin{pmatrix} a_{11} \\ a_{1n} \\ a_{m1} \\ a_{mn} \end{pmatrix} (b_{11}b_{r1}b_{1s}b_{rs}) = A \underset{\cdot}{\times} B^T$$

$$A \overset{6}{\times} B = \left(\begin{array}{cc|cc} a_{11}b_{11} & a_{1n}b_{11} & a_{11}b_{1s} & a_{1n}b_{1s} \\ a_{11}b_{r1} & a_{1n}b_{r1} & a_{11}b_{rs} & a_{1n}b_{rs} \\ \hline a_{m1}b_{11} & a_{mn}b_{11} & a_{m1}b_{1s} & a_{mn}b_{1s} \\ a_{m1}b_{r1} & a_{mn}b_{r1} & a_{m1}b_{rs} & a_{mn}b_{rs} \end{array}\right) \tag{19f}$$

$$= \left(\begin{array}{c|c} \begin{pmatrix} b_{11} \\ b_{r1} \end{pmatrix} (a_{11}a_{1n}) & \begin{pmatrix} b_{1s} \\ b_{rs} \end{pmatrix} (a_{11}a_{1n}) \\ \hline \begin{pmatrix} b_{11} \\ b_{r1} \end{pmatrix} (a_{m1}a_{mn}) & \begin{pmatrix} b_{1s} \\ b_{rs} \end{pmatrix} (a_{m1}a_{mn}) \end{array}\right)$$

if elements commute

$$= (A^T \dot{\times} B^T)^T,$$

same or
otherwise

$$A \overset{7}{\times} B = \left(\begin{array}{cc|cc} a_{11}b_{11} & a_{11}b_{1s} & a_{m1}b_{11} & a_{m1}b_{1s} \\ a_{11}b_{r1} & a_{11}b_{rs} & a_{m1}b_{r1} & a_{m1}b_{rs} \\ \hline a_{1n}b_{11} & a_{1n}b_{1s} & a_{mn}b_{11} & a_{mn}b_{1s} \\ a_{1n}b_{r1} & a_{1n}b_{rs} & a_{mn}b_{r1} & a_{mn}b_{rs} \end{array}\right)$$

$$= \begin{pmatrix} a_{11}B & a_{m1}B \\ a_{1n}B & a_{mn}B \end{pmatrix} = A^T \times\cdot\, B,$$

$$A \overset{8}{\times} B = \begin{pmatrix} a_{11}b_{11} & a_{11}b_{r1} & \vdots & a_{m1}b_{11} & a_{m1}b_{r1} \\ a_{11}b_{1s} & a_{11}b_{rs} & \vdots & a_{m1}b_{1s} & a_{m1}b_{rs} \\ \cdots & \cdots & \cdots & \cdots & \cdots \\ a_{1n}b_{11} & a_{1n}b_{r1} & \vdots & a_{mn}b_{11} & a_{mn}b_{r1} \\ a_{1n}b_{1s} & a_{1n}b_{rs} & \vdots & a_{mn}b_{1s} & a_{mn}b_{rs} \end{pmatrix}$$

$$= \begin{pmatrix} a_{11}B^T & a_{m1}B^T \\ a_{1n}B^T & a_{mn}B^T \end{pmatrix} = A^T \times B^T, \qquad (19h)$$

$$A \overset{9}{\times\cdot} B = \begin{pmatrix} a_{11}b_{11} & a_{11}b_{1s} & \vdots & a_{11}b_{r1} & a_{11}b_{rs} \\ a_{m1}b_{11} & a_{m1}b_{1s} & \vdots & a_{m1}b_{r1} & a_{m1}b_{rs} \\ \cdots & \cdots & \cdots & \cdots & \cdots \\ a_{1n}b_{11} & a_{1n}b_{1s} & \vdots & a_{1n}b_{r1} & a_{1n}b_{rs} \\ a_{mn}b_{11} & a_{mn}b_{1s} & \vdots & a_{mn}b_{r1} & a_{mn}b_{rs} \end{pmatrix} \qquad (19i)$$

$$= \begin{pmatrix} a_{11} \\ a_{m1} \\ a_{1n} \\ a_{mn} \end{pmatrix} (b_{11}b_{1s}b_{r1}b_{rs}) = A^T \underset{\cdot}{\times} B,$$

$$A \overset{10}{\times} B = \begin{pmatrix} a_{m1}b_{r1} & a_{m1}b_{11} & \vdots & a_{11}b_{r1} & a_{m1}b_{r1} \\ a_{m1}b_{rs} & a_{m1}b_{1s} & \vdots & a_{11}b_{rs} & a_{m1}b_{rs} \\ \cdots & \cdots & \cdots & \cdots & \cdots \\ a_{1n}b_{11} & a_{mn}b_{11} & \vdots & a_{1n}b_{r1} & a_{mn}b_{r1} \\ a_{1n}b_{1s} & a_{mn}b_{1s} & \vdots & a_{1n}b_{rs} & a_{mn}b_{rs} \end{pmatrix} \qquad (19j)$$

$$= \begin{pmatrix} \begin{pmatrix} b_{11} \\ b_{1s} \end{pmatrix} (a_{11}a_{m1}) & \vdots & \begin{pmatrix} b_{r1} \\ b_{rs} \end{pmatrix} (a_{11}a_{m1}) \\ \cdots & \cdots & \cdots \\ \begin{pmatrix} b_{11} \\ b_{1s} \end{pmatrix} (a_{1n}a_{mn}) & \vdots & \begin{pmatrix} b_{r1} \\ b_{rs} \end{pmatrix} (a_{1n}a_{mn}) \end{pmatrix}$$

if elements commute

$$= (A \times B)^T,$$

same or otherwise

$$
\begin{aligned}
A \overset{11}{\times} B &= \left(\begin{array}{cc:cc}
a_{11}b_{11} & a_{11}b_{r1} & a_{11}b_{1s} & a_{11}b_{rs} \\
a_{m1}b_{11} & a_{m1}b_{r1} & a_{m1}b_{1s} & a_{m1}b_{rs} \\
\hdashline
a_{1n}b_{11} & a_{1n}b_{r1} & a_{1n}b_{1s} & a_{1n}b_{rs} \\
a_{mn}b_{11} & a_{mn}b_{r1} & a_{mn}b_{1s} & a_{mn}b_{rs}
\end{array}\right) \\
&= \begin{pmatrix} a_{11} \\ a_{m1} \\ a_{1n} \\ a_{mn} \end{pmatrix} (b_{11}\,b_{r1}\,b_{1s}\,b_{rs}) = A^T \underset{\cdot}{\times} B^T,
\end{aligned}
\tag{19k}
$$

$$
\begin{aligned}
A \overset{12}{\times} B &= \left(\begin{array}{cc:cc}
a_{11}b_{11} & a_{m1}b_{11} & a_{11}b_{1s} & a_{m1}b_{1s} \\
a_{11}b_{r1} & a_{m1}b_{r1} & a_{11}b_{rs} & a_{m1}b_{rs} \\
\hdashline
a_{1n}b_{11} & a_{mn}b_{11} & a_{1n}b_{1s} & a_{mn}b_{1s} \\
a_{1n}b_{r1} & a_{mn}b_{r1} & a_{1n}b_{rs} & a_{mn}b_{rs}
\end{array}\right) \\
&= \left(\begin{array}{c:c}
\begin{pmatrix} b_{11} \\ b_{r1} \end{pmatrix} (a_{11}\,a_{m1}) & \begin{pmatrix} b_{1s} \\ b_{rs} \end{pmatrix} (a_{11}\,a_{m1}) \\
\hdashline
\begin{pmatrix} b_{11} \\ b_{r1} \end{pmatrix} (a_{1n}\,a_{mn}) & \begin{pmatrix} b_{1s} \\ b_{rs} \end{pmatrix} (a_{1n}\,a_{mn})
\end{array}\right)
\end{aligned}
\tag{19l}
$$

if elements commute

$$= (A \dot{\times} B^T)^T,$$

same or otherwise

$$
\begin{aligned}
A \overset{13}{\times} B &= \left(\begin{array}{cc:cc}
a_{11}b_{11} & a_{11}b_{1s} & a_{m1}b_{11} & a_{m1}b_{1s} \\
a_{1n}b_{11} & a_{1n}b_{1s} & a_{mn}b_{11} & a_{mn}b_{1s} \\
\hdashline
a_{11}b_{r1} & a_{11}b_{rs} & a_{m1}b_{r1} & a_{m1}b_{rs} \\
a_{1m}b_{r1} & a_{1n}b_{rs} & a_{mn}b_{r1} & a_{mn}b_{rs}
\end{array}\right) \\
&= \left(\begin{array}{c:c}
\begin{pmatrix} a_{11} \\ a_{1n} \end{pmatrix} (b_{11}\,b_{1s}) & \begin{pmatrix} a_{m1} \\ a_{mn} \end{pmatrix} (b_{11}\,b_{1s}) \\
\hdashline
\begin{pmatrix} a_{11} \\ a_{1n} \end{pmatrix} (b_{r1}\,b_{rs}) & \begin{pmatrix} a_{m1} \\ a_{mn} \end{pmatrix} (b_{r1}\,b_{rs})
\end{array}\right) = A^T \dot{\times} B,
\end{aligned}
\tag{19m}
$$

$$A \overset{14}{\times} B = \begin{pmatrix} a_{11}b_{11} & a_{1n}b_{11} & a_{m1}b_{11} & a_{mn}b_{11} \\ a_{11}b_{1s} & a_{1n}b_{1s} & a_{m1}b_{1s} & a_{mn}b_{1s} \\ a_{11}b_{r1} & a_{1n}b_{r1} & a_{m1}b_{r1} & a_{mn}b_{r1} \\ a_{11}b_{rs} & a_{1n}b_{rs} & a_{m1}b_{rs} & a_{mn}b_{rs} \end{pmatrix} \tag{19n}$$

$$= \begin{pmatrix} b_{11} \\ b_{1s} \\ b_{r1} \\ b_{rs} \end{pmatrix} (a_{11}\, a_{1n} a_{m1}\, a_{mn}) = (A^T \mathbin{\dot{\times}} B^T)^T$$

same or otherwise

if elements commute

$$A \overset{15}{\times} B = \left(\begin{array}{cc:cc} a_{11}b_{11} & a_{11}b_{1s} & a_{1n}b_{11} & a_{1n}b_{1s} \\ a_{m1}b_{11} & a_{m1}b_{1s} & a_{mn}b_{11} & a_{mn}b_{1s} \\ \hdashline a_{11}b_{r1} & a_{11}b_{rs} & a_{1n}b_{r1} & a_{1n}b_{rs} \\ a_{m1}\, b_{r1} & a_{m1}b_{rr} & a_{mn}b_{r1} & a_{mn}b_{rs} \end{array}\right) \tag{19o}$$

$$= \left(\begin{array}{c:c} \begin{pmatrix} a_{11} \\ a_{m1} \end{pmatrix} (b_{11}\, b_{1s}) & \begin{pmatrix} a_{1n} \\ a_{mn} \end{pmatrix} (b_{11}\, b_{1s}) \\ \hdashline \begin{pmatrix} a_{11} \\ a_{m1} \end{pmatrix} (b_{r1}\, b_{rs}) & \begin{pmatrix} a_{1n} \\ a_{mn} \end{pmatrix} (b_{r1}\, b_{rs}) \end{array}\right) = A \mathbin{\dot{\times}} B$$

$$A \overset{16}{\times} B = \left(\begin{array}{cc:cc} a_{11}b_{11} & a_{m1}b_{11} & a_{1n}b_{11} & a_{mn}b_{11} \\ a_{11}b_{1s} & a_{m1}b_{1s} & a_{1n}b_{1s} & a_{mn}b_{1s} \\ \hdashline a_{11}b_{r1} & a_{m1}b_{r1} & a_{1n}b_{r1} & a_{mn}b_{r1} \\ a_{11}b_{rs} & a_{m1}b_{rs} & a_{1n}b_{rs} & a_{mn}b_{rs} \end{array}\right) \tag{19p}$$

$$= \begin{pmatrix} b_{11} \\ b_{1s} \\ b_{n1} \\ b_{rs} \end{pmatrix} (a_{11}\, a_{m1}\, a_{1n} a_{mn}) = (A \mathbin{\underset{\cdot}{\times}} B^T)^T$$

same or otherwise

if elements commute

$$A \overset{17}{\times} B = \left(\begin{array}{cc|cc} a_{11}b_{11} & a_{1n}b_{11} & a_{11}b_{1s} & a_{1n}b_{1s} \\ a_{m1}b_{11} & a_{mn}b_{11} & a_{m1}b_{1s} & a_{mn}b_{1s} \\ \hline a_{11}b_{r1} & a_{1n}b_{r1} & a_{11}b_{rs} & a_{1n}b_{rs} \\ a_{m1}b_{r1} & a_{mn}b_{r1} & a_{m1}b_{rs} & a_{mn}b_{rs} \end{array}\right) \tag{19q}$$

$$= \begin{pmatrix} Ab_{11} & Ab_{1s} \\ Ab_{r1} & Ab_{rs} \end{pmatrix} = A \cdot\!\times B,$$

$$A \overset{18}{\times} B = \left(\begin{array}{cc|cc} a_{11}b_{11} & a_{m1}b_{11} & a_{11}b_{1s} & a_{m1}b_{1s} \\ a_{1n}b_{11} & a_{mn}b_{11} & a_{1n}b_{1s} & a_{mn}b_{1s} \\ \hline a_{11}b_{r1} & a_{m1}b_{r1} & a_{11}b_{rs} & a_{m1}b_{rs} \\ a_{1n}b_{r1} & a_{mn}b_{r1} & a_{1n}b_{rs} & a_{mn}b_{rs} \end{array}\right) \tag{19r}$$

$$= \begin{pmatrix} A^Tb_{11} & A^Tb_{1s} \\ A^Tb_{r1} & A^Tb_{rs} \end{pmatrix} = A^T \cdot\!\times B,$$

$$A \overset{19}{\times} B = \left(\begin{array}{cc|cc} a_{11}b_{11} & a_{11}b_{r1} & a_{m1}b_{11} & a_{m1}b_{r1} \\ a_{1n}b_{11} & a_{1n}b_{r1} & a_{mn}b_{11} & a_{mn}b_{r1} \\ \hline a_{11}b_{1s} & a_{11}b_{rs} & a_{m1}b_{1s} & a_{m1}b_{rs} \\ a_{1n}b_{1s} & a_{1n}b_{rs} & a_{mn}b_{1s} & a_{mn}b_{rs} \end{array}\right) \tag{19s}$$

$$= \left(\begin{array}{c|c} \begin{pmatrix} a_{11} \\ a_{1n} \end{pmatrix} (b_{11} b_{r1}) & \begin{pmatrix} a_{m1} \\ a_{mn} \end{pmatrix} (b_{11} b_{r1}) \\ \hline \begin{pmatrix} a_{11} \\ a_{1n} \end{pmatrix} (b_{1s} b_{rss}) & \begin{pmatrix} a_{m1} \\ a_{mn} \end{pmatrix} (b_{1s} b_{rs}) \end{array}\right) = A^T \dot{\times} B^T,$$

$$A \overset{20}{\times} B = \left(\begin{array}{cc|cc} a_{11}b_{11} & a_{1n}b_{11} & a_{m1}b_{11} & a_{mn}b_{11} \\ a_{11}b_{r1} & a_{1n}b_{r1} & a_{m1}b_{r1} & a_{mn}b_{r1} \\ \hline a_{11}b_{1s} & a_{1n}b_{1s} & a_{m1}b_{1s} & a_{mn}b_{1s} \\ a_{11}b_{rs} & a_{1n}b_{rs} & a_{m1}b_{rs} & a_{mn}b_{rs} \end{array}\right) \tag{19t}$$

$$= \begin{pmatrix} b_{11} \\ b_{r1} \\ b_{1s} \\ b_{rs} \end{pmatrix} (a_{11}\, a_{1n}\, a_{m1}\, a_{mn}) = (A^T \underset{\cdot}{\times} B)^T$$

same or otherwise

if elements commute

$$A \overset{21}{\times} B = \left(\begin{array}{cc:cc} a_{11}b_{11} & a_{11}b_{r1} & a_{1n}b_{11} & a_{1n}b_{r1} \\ a_{m1}b_{11} & a_{m1}b_{r1} & a_{mn}b_{11} & a_{mn}b_{r1} \\ \hdashline a_{11}b_{1s} & a_{11}b_{rs} & a_{1n}b_{1s} & a_{1n}b_{rs} \\ a_{m1}b_{1s} & a_{m1}b_{rs} & a_{mn}b_{1s} & a_{mn}b_{rs} \end{array}\right)$$

$$= \left(\begin{array}{c:c} \begin{pmatrix} a_{11} \\ a_{m1} \end{pmatrix} (b_{11}\, b_{r1}) & \begin{pmatrix} a_{1n} \\ a_{mn} \end{pmatrix} (b_{11}\, b_{r1}) \\ \hdashline \begin{pmatrix} a_{11} \\ a_{m1} \end{pmatrix} (b_{1s}\, b_{rs}) & \begin{pmatrix} a_{1n} \\ a_{mn} \end{pmatrix} (b_{1s}\, b_{rs}) \end{array}\right) = A \dot{\times} B^T, \tag{19u}$$

$$A \overset{22}{\times} B = \left(\begin{array}{cc:cc} a_{11}b_{11} & a_{m1}b_{11} & a_{1n}b_{11} & a_{mn}b_{11} \\ a_{m1}b_{11} & a_{m1}b_{r1} & a_{1n}b_{r1} & a_{mn}b_{r1} \\ \hdashline a_{11}b_{1s} & a_{m1}b_{1s} & a_{1n}b_{1s} & a_{mn}b_{1s} \\ a_{11}b_{rs} & a_{m1}b_{rs} & a_{1n}b_{rs} & a_{mn}b_{rs} \end{array}\right) \tag{19v}$$

$$= \begin{pmatrix} b_{11} \\ b_{r1} \\ b_{1s} \\ b_{rs} \end{pmatrix} (a_{11}\, a_{m1}\, a_{1n}\, a_{mn}) = (A \underset{\cdot}{\times} B)^T$$

same or otherwise

if elements commute

$$A \overset{23}{\times} B = \left(\begin{array}{cc:cc} a_{11}b_{11} & a_{1n}b_{11} & a_{11}b_{r1} & a_{1n}b_{r1} \\ a_{m1}b_{11} & a_{mn}b_{11} & a_{m1}b_{r1} & a_{mn}b_{r1} \\ \hdashline a_{11}b_{1s} & a_{1n}b_{1s} & a_{11}b_{rs} & a_{1n}b_{rs} \\ a_{m1}b_{1s} & a_{mn}b_{1s} & a_{m1}b_{rs} & a_{mn}b_{rs} \end{array}\right) \tag{19w}$$

$$= \begin{pmatrix} Ab_{11} & Ab_{r1} \\ Ab_{1s} & Ab_{rs} \end{pmatrix} = A \cdot\!\times B^T$$

$$
A \overset{24}{\times} B = \left(\begin{array}{cc:cc} a_{11}b_{11} & a_{m1}b_{11} & a_{11}b_{r1} & a_{m1}b_{r1} \\ a_{1n}b_{11} & a_{mn}b_{11} & a_{1n}b_{r1} & a_{mn}b_{r1} \\ \hdashline a_{11}b_{1s} & a_{m1}b_{1s} & a_{11}b_{rs} & a_{m1}b_{rs} \\ a_{1m}b_{1s} & a_{mn}b_{1s} & a_{1n}b_{rs} & a_{mn}b_{rs} \end{array}\right)
$$

$$
= \begin{pmatrix} A^T b_{11} & A^T b_{r1} \\ A^T b_{1s} & A^T b_{rs} \end{pmatrix} = A^T \cdot\!\times B^T. \tag{19x}
$$

EXERCISES

121. What is a necessary and sufficient condition for the direct product of two matrices to be zero?

122. Let $\mathbf{a} = (a_1, a_2, a_3)$ and $\mathbf{b} = (b_1, b_2, b_3)$. Evaluate the cells of this cube.

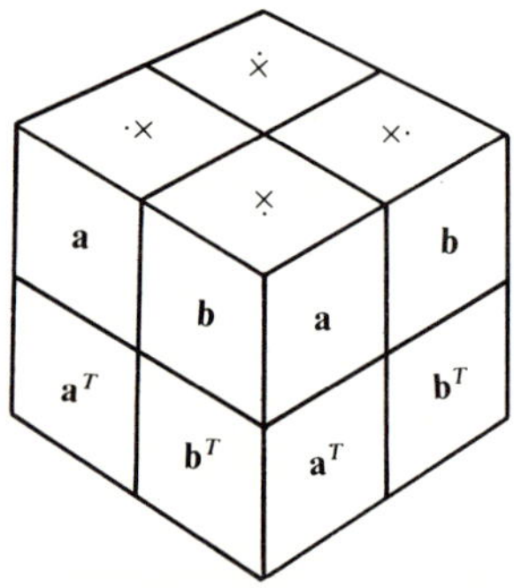

Is $\dot{\times}$ or $\underset{\cdot}{\times}$ the better intermediary between $\times\!\cdot$ and $\times\!\cdot$?

123. Show that:

(*a*) $(A + B) \times\!\cdot\ C = A \times\!\cdot\ C + B \times\!\cdot\ C$,

(*b*) $(AB) \times\!\cdot\ (CD) = (A \times\!\cdot\ C)(B \times\!\cdot\ D)$,

(*c*) $\operatorname{tr}(A \times\!\cdot\ B) = (\operatorname{tr} A)(\operatorname{tr} B)$.

124. Let A be of size m by n where $m > n$.

(*a*) Prove that AA^T is always singular and give an upper bound for its rank.

(*b*) Is the same theorem necessarily true for A^TA?

(*c*) What happens when ordinary matrix multiplication is replaced by direct multiplication?

(*d*) How about Hadamard multiplication?

Table of Direct Products

Table 2 provides considerable insight into the interrelationships of the 24 possible direct products found in the preceding section.

Table 2 Complete Set of 24 Direct Products

	Interrelated			Self-Related
	"Right" $\times$	"Upper" $\dot{\times}$	"Left" $\times$	"Lower" $\underset{\cdot}{\times}$
AB	$\overset{1}{\underset{(mr,ns)}{X}} = \bigvee_{i}^{m}\bigwedge_{j}^{n}\bigvee_{k}^{r}\bigwedge_{l}^{s}$ Standard	$\overset{15}{\underset{(mr,ns)}{X}} = \bigvee_{k}^{r}\bigwedge_{j}^{n}\bigvee_{i}^{m}\bigwedge_{l}^{s}$	$\overset{17}{\underset{(mr,ns)}{X}} = \bigvee_{k}^{r}\bigwedge_{l}^{s}\bigvee_{i}^{m}\bigwedge_{j}^{n}$	$\overset{3}{\underset{(mn,rs)}{X}} = \bigvee_{i}^{m}\bigwedge_{k}^{r}\bigvee_{j}^{n}\bigwedge_{l}^{s}$
AB^T	$\overset{2}{\underset{(ms,nr)}{X}} = \bigvee_{i}^{m}\bigwedge_{j}^{n}\bigvee_{l}^{s}\bigwedge_{k}^{r}$	$\overset{21}{\underset{(ms,nr)}{X}} = \bigvee_{l}^{s}\bigwedge_{j}^{n}\bigvee_{i}^{m}\bigwedge_{k}^{r}$	$\overset{23}{\underset{(ms,nr)}{X}} = \bigvee_{l}^{s}\bigwedge_{k}^{r}\bigvee_{i}^{m}\bigwedge_{j}^{n}$	$\overset{5}{\underset{(mn,rs)}{X}} = \bigvee_{i}^{m}\bigwedge_{l}^{s}\bigvee_{j}^{n}\bigwedge_{k}^{r}$
A^TB	$\overset{7}{\underset{(nr,ms)}{X}} = \bigvee_{j}^{n}\bigwedge_{i}^{m}\bigvee_{k}^{r}\bigwedge_{l}^{s}$	$\overset{13}{\underset{(nr,ms)}{X}} = \bigvee_{k}^{r}\bigwedge_{i}^{m}\bigvee_{j}^{n}\bigwedge_{l}^{s}$	$\overset{18}{\underset{(nr,ms)}{X}} = \bigvee_{k}^{r}\bigwedge_{l}^{s}\bigvee_{j}^{n}\bigwedge_{i}^{m}$	$\overset{9}{\underset{(mn,rs)}{X}} = \bigvee_{j}^{n}\bigwedge_{k}^{r}\bigvee_{i}^{m}\bigwedge_{l}^{s}$
A^TB^T	$\overset{8}{\underset{(ns,mr)}{X}} = \bigvee_{j}^{n}\bigwedge_{i}^{m}\bigvee_{l}^{s}\bigwedge_{k}^{r}$	$\overset{19}{\underset{(ns,mr)}{X}} = \bigvee_{l}^{s}\bigwedge_{i}^{m}\bigvee_{j}^{n}\bigwedge_{k}^{r}$	$\overset{24}{\underset{(ns,mr)}{X}} = \bigvee_{l}^{s}\bigwedge_{k}^{r}\bigvee_{j}^{n}\bigwedge_{i}^{m}$	$\overset{11}{\underset{(mn,rs)}{X}} = \bigvee_{j}^{n}\bigwedge_{l}^{s}\bigvee_{i}^{m}\bigwedge_{k}^{s}$
$(AB)^T$		$\overset{10}{\underset{(ns,mr)}{X}} = \bigvee_{j}^{n}\bigwedge_{k}^{r}\bigvee_{l}^{s}\bigwedge_{i}^{m}$		$\overset{22}{\underset{(rs,mn)}{X}} = \bigvee_{l}^{s}\bigwedge_{j}^{n}\bigvee_{k}^{r}\bigwedge_{i}^{m}$
$(AB^T)^T$		$\overset{12}{\underset{(nr,ms)}{X}} = \bigvee_{j}^{n}\bigwedge_{i}^{s}\bigvee_{k}^{r}\bigwedge_{i}^{m}$		$\overset{16}{\underset{(rs,mn)}{X}} = \bigvee_{k}^{r}\bigwedge_{j}^{n}\bigvee_{l}^{s}\bigwedge_{i}^{m}$
$(A^TB)^T$		$\overset{4}{\underset{(ms,nr)}{X}} = \bigvee_{i}^{m}\bigwedge_{k}^{r}\bigvee_{l}^{s}\bigwedge_{j}^{n}$		$\overset{20}{\underset{(rs,mn)}{X}} = \bigvee_{l}^{s}\bigwedge_{i}^{m}\bigvee_{k}^{r}\bigwedge_{j}^{n}$
$(A^TB^T)^T$		$\overset{6}{\underset{(mr,ns)}{X}} = \bigvee_{i}^{m}\bigwedge_{l}^{s}\bigvee_{k}^{r}\bigwedge_{j}^{n}$		$\overset{14}{\underset{(rs,mn)}{X}} = \bigvee_{k}^{r}\bigwedge_{i}^{n}\bigvee_{l}^{s}\bigwedge_{j}^{m}$

Note: Four indices have $4 \times 3 \times 2 \times 1 = 4! = 24$ combinations. Further information about the dimensionality of the direct products is contained in Fig. 11.

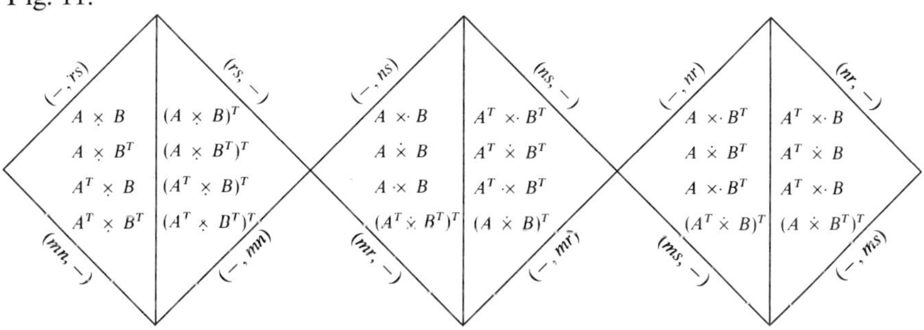

Figure 11. Direct product dimensionality.

125. What can be said about the various types of matrix symmetry under application of (*a*) direct matrix addition and (*b*) direct matrix multiplication?

126. The "Kronecker product," $K = (k_{rs})$, of two n^2-square matrices A and B is defined to be the n^2 by n^2 matrix $K = A \times B$, where

$$k_{rs} = a_{i_1 j_1} b_{i_2 j_2},$$

with

$$\begin{gathered} r = i_2 + n(i_1 - 1), \\ s = j_2 + n(j_1 - 1), \\ 1 \le i_1, \quad j_1, \quad i_2, \quad j_2 \le n. \end{gathered}$$

Correctly identify $\otimes$ with $\underset{\cdot}{\times}$, $\times\cdot$, $\dot{\times}$, or $\cdot\times$ and write similar definitions for the remaining three direct products.

By definition the upper direct product $C = A \dot{\times} B$ of two matrices $A = (a_{ij})$ and $B = (b_{kl})$ that are of size m by n and r by s, respectively, is given by

$$(c_{kj}) = \begin{pmatrix} a_{1j} \\ a_{2j} \\ \cdot \\ \cdot \\ \cdot \\ a_{nj} \end{pmatrix} (b_{k1}, b_{k2}, \cdots b_{ks})$$

or

$$C = A \dot{\times} B = \begin{pmatrix} a_{11} & a_{12} \cdots & a_{1n} \\ a_{21} & a_{22} \cdots & a_{2n} \\ \cdot & \cdot \quad \cdot & \cdot \\ \cdot & \cdot \quad \cdot & \cdot \\ \cdot & \cdot \quad \cdot & \cdot \\ a_{m1} & a_{m2} \cdots & a_{mn} \end{pmatrix} \dot{\times} \begin{pmatrix} b_{11} & b_{12} \cdots & b_{1s} \\ b_{21} & b_{22} \cdots & b_{2s} \\ \cdot & \cdot \quad \cdot & \cdot \\ \cdot & \cdot \quad \cdot & \cdot \\ \cdot & \cdot \quad \cdot & \cdot \\ b_{r1} & b_{r2} & b_{rs} \end{pmatrix}$$

$$= \begin{pmatrix} \alpha^1\beta_1 & \alpha^2\beta_1 \cdots & \alpha^n\beta_1 \\ \alpha^1\beta_2 & \alpha^2\beta_2 \cdots & \alpha^n\beta_2 \\ \cdot & \cdot & \cdot \\ \cdot & \cdot & \cdot \\ \cdot & \cdot & \cdot \\ \alpha^1\beta_r & \alpha^2\beta_r \cdots & \alpha^n\beta_r \end{pmatrix} = (\alpha^1, \alpha^2, \cdots, \alpha^n) \begin{pmatrix} \beta_1 \\ \beta_2 \\ \cdot \\ \cdot \\ \cdot \\ \beta_r \end{pmatrix} \tag{20}$$

(20 *continued*)

$$
\left(\begin{array}{c:c:c:c}
\begin{pmatrix} a_{11} \\ a_{21} \\ \vdots \\ a_{m1} \end{pmatrix}(b_{11}, b_{12}, \cdots, b_{1s}) & \begin{pmatrix} a_{12} \\ a_{22} \\ \vdots \\ a_{m2} \end{pmatrix}(b_{11}, b_{12}, \cdots, b_{1s}) & \cdots & \begin{pmatrix} a_{1m} \\ a_{2n} \\ \vdots \\ a_{mn} \end{pmatrix}(b_{11}, b_{12}, \cdots, b_{1s}) \\
\hdashline
\begin{pmatrix} a_{11} \\ a_{21} \\ \vdots \\ a_{m1} \end{pmatrix}(b_{21}, b_{22}, \cdots, b_{2s}) & \begin{pmatrix} a_{12} \\ a_{22} \\ \vdots \\ a_{m2} \end{pmatrix}(b_{21}, b_{22}, \cdots, b_{2s}) & \cdots & \begin{pmatrix} a_{1n} \\ a_{2n} \\ \vdots \\ a_{mn} \end{pmatrix}(b_{21}, b_{22}, \cdots, b_{2s}) \\
\hdashline
\vdots & \vdots & \ddots & \vdots \\
\hdashline
\begin{pmatrix} a_{11} \\ a_{21} \\ \vdots \\ a_{m1} \end{pmatrix}(b_{r1}, b_{r2}, \cdots, b_{rs}) & \begin{pmatrix} a_{12} \\ a_{22} \\ \vdots \\ a_{m2} \end{pmatrix}(b_{r1}, b_{r2}, \cdots, b_{rs}) & \cdots & \begin{pmatrix} a_{1n} \\ a_{2n} \\ \vdots \\ a_{mn} \end{pmatrix}(b_{r1}, b_{r2}, \cdots, b_{rs})
\end{array}\right)
$$

$$
C = \begin{array}{c}
\\ 1 \\ \\ \\ \\ 2 \\ \\ \\ k \\ \\ \\ r \\ \\
\end{array}
\left(\begin{array}{ccc:ccc:c:ccc}
 & 1 & & & 2 & & j & & n & \\
a_{11}b_{11} & a_{11}b_{12}\cdots & a_{11}b_{1s} & a_{12}b_{11} & a_{12}b_{12}\cdots & a_{12}b_{1s} & & a_{1n}b_{11} & a_{1n}b_{12}\cdots & a_{1n}b_{1s} \\
a_{21}b_{11} & a_{21}b_{12}\cdots & a_{21}b_{1s} & a_{22}b_{11} & a_{22}b_{12}\cdots & a_{22}b_{1s} & \cdots & a_{2n}b_{11} & a_{2n}b_{12}\cdots & a_{2n}b_{1s} \\
\vdots & \vdots & \vdots & \vdots & \vdots & \vdots & & \vdots & \vdots & \vdots \\
a_{m1}b_{11} & a_{m1}b_{12}\cdots & a_{m1}b_{1s} & a_{m2}b_{11} & a_{m2}b_{12}\cdots & a_{m2}b_{1s} & & a_{mn}b_{11} & a_{mn}b_{12}\cdots & a_{mn}b_{1s} \\
\hdashline
a_{11}b_{21} & a_{11}b_{22}\cdots & a_{11}b_{2s} & a_{12}b_{21} & a_{12}b_{22}\cdots & a_{12}b_{2s} & & a_{1n}b_{21} & a_{1n}b_{22}\cdots & a_{1n}b_{2s} \\
a_{21}b_{21} & a_{21}b_{22}\cdots & a_{21}b_{2s} & a_{22}b_{21} & a_{12}b_{22}\cdots & a_{22}b_{22} & \cdots & a_{2n}b_{21} & a_{2n}b_{22}\cdots & a_{2n}b_{2s} \\
\vdots & \vdots & \vdots & \vdots & \vdots & \vdots & & \vdots & \vdots & \vdots \\
a_{m1}b_{21} & a_{m1}b_{22}\cdots & a_{m1}b_{2s} & a_{m2}b_{21} & a_{m2}b_{22}\cdots & a_{m2}b_{2s} & & a_{mn}b_{21} & a_{mn}b_{22}\cdots & a_{mn}b_{2s} \\
\hdashline
 & \vdots & & & \vdots & & \ddots & & \vdots & \\
\hdashline
a_{11}b_{r1} & a_{11}b_{r2} & a_{11}b_{rs} & a_{12}b_{r1} & a_{12}b_{r2}\cdots & a_{12}b_{rs} & & a_{1n}b_{r1} & a_{1n}b_{r2}\cdots & a_{1n}b_{rs} \\
a_{21}b_{r1} & a_{21}b_{r2}\cdots & a_{21}b_{rs} & a_{22}b_{r1} & a_{22}b_{r2}\cdots & a_{22}b_{rs} & \cdots & a_{2n}b_{r1} & a_{2n}b_{r2}\cdots & a_{2n}b_{rs} \\
\vdots & \vdots & \vdots & \vdots & \vdots & \vdots & & \vdots & \vdots & \vdots \\
a_{m1}b_{r1} & a_{m1}b_{r2}\cdots & a_{m1}b_{rs} & a_{m2}b_{r1} & a_{m2}b_{r2}\cdots & a_{m2}b_{rs} & & a_{mn}b_{r1} & a_{mn}b_{r2}\cdots & a_{mn}b_{rs}
\end{array}\right)
\tag{21}
$$

Thus the rule for upper direct multiplication forms a block matrix c_{kj} at the intersection of block row k and block column j by multiplying the jth column of A by the ith row of B by ordinary matrix multiplication (the outer product of two vectors). Hence upper direct multiplication is a column-by-row rule for multiplication analogous in reverse to the row-by-column rule for ordinary matrix multiplication.

For clarity, it is advantageous to define several additional row and column vectors closelt related to a matrix $C = (c_{gh})$, whose dimensionality for the time being will be written (γ, ζ). Let

$$\mathbf{C} = \Lambda\left(\bigvee_{g=1}^{\gamma} \bigwedge_{h=1}^{\zeta}\right)^{-1} C = \Lambda \begin{pmatrix} C_{11} \rightarrow C_{12} \rightarrow \cdots C_{1\zeta} \\ \downarrow \\ C_{21} \rightarrow C_{22} \rightarrow \cdots C_{2\zeta} \\ \downarrow \\ \cdot \quad \cdot \quad \cdot \\ \cdot \quad \cdot \quad \cdot \\ \cdot \quad \cdot \quad \cdot \\ \downarrow \\ C_{\gamma 1} \rightarrow C_{\gamma 2} \rightarrow \cdots C_{\gamma\zeta} \end{pmatrix} \tag{22}$$

$$= (C_{11}, C_{12}, \cdots, C_{1\zeta}, C_{21}, C_{22}, \cdots, C_{2\zeta}, \cdots, C_{\zeta 1}, C_{\zeta 2}, \cdots, C_{\gamma\zeta}) = \Lambda C,$$

and

$$\mathbf{Ç} = \Lambda\left(\bigwedge_{h=1}^{\zeta} \bigvee_{g=1}^{\gamma}\right) C = \Lambda \begin{pmatrix} C_{11} & \rightarrow C_{12} & \rightarrow \cdots & \rightarrow C_{1\zeta} \\ \downarrow & \downarrow & & \downarrow \\ C_{21} & C_{22} & \cdots & C_{2\zeta} \\ \downarrow & \downarrow & \cdot & \downarrow \\ \cdot & \cdot & \cdot & \cdot \\ \cdot & \cdot & \cdot & \cdot \\ \downarrow & \downarrow & \cdot & \downarrow \\ C_{\gamma 1} & C_{\gamma 2} & \cdots & C_{\gamma\zeta} \end{pmatrix} \tag{23}$$

$$= C_{11}, C_{21}, \cdots, C_{\gamma 1}, C_{12}\ C_{22}, \cdots, C_{\gamma 2}, \cdots, C_{1\zeta}, C_{2\zeta}, \cdots, C_{\gamma\zeta}) = \Lambda C^T.$$

Vector (22) is called the "row elongation of C by rows" and vector (23) is called the "row elongation of C by columns." The corresponding column elongations, by rows and by columns, respectively, are

$$\mathbf{C}^T = \mathrm{V}C \tag{24}$$

and

$$\mathbf{Ç}^T = \mathrm{V}C^T. \tag{25}$$

The simple case afforded by a 3 square matrix will clarify the situation considerably. Let

$$A = \begin{pmatrix} a & \rightarrow b & \rightarrow c \\ d & e & f \\ g & h & j \end{pmatrix}.$$

Then

$$\left(\bigvee_{i=1}^{3}\bigwedge_{j=1}^{3}\right)^{-1} A = \begin{pmatrix} a \rightarrow b \rightarrow c \\ \downarrow \\ d \rightarrow e \rightarrow f \\ \downarrow \\ g \rightarrow h \rightarrow j \end{pmatrix} \quad \text{and} \quad \left(\bigwedge_{j=1}^{3}\bigvee_{i=1}^{3}\right)^{-1} A = \begin{pmatrix} a & \rightarrow b & \rightarrow c \\ \downarrow & \downarrow & \downarrow \\ d & e & f \\ \downarrow & \downarrow & \downarrow \\ g & h & j \end{pmatrix}$$

Collecting the results in a small table:

	By Rows	By Columns
Row Elongation	$\underline{\Lambda A} = \Lambda\left(\bigvee_{i=1}^{3}\bigwedge_{j=1}^{3}\right)^{-1} A$ $= (abcdefghj)$	$\underline{\Lambda A^T} = \Lambda\left(\bigwedge_{j=1}^{3}\bigvee_{i=1}^{3}\right)^{-1} A$ $= (adgbehcfj)$
Column Elongation	$\underline{\mathrm{V}A} = \mathrm{V}\left(\bigvee_{i=1}^{3}\bigwedge_{j=1}^{3}\right)^{-1} A$ $= \begin{pmatrix} a \\ b \\ c \\ d \\ e \\ f \\ g \\ h \\ j \end{pmatrix}$	$\underline{\mathrm{V}A^T} = \mathrm{V}\left(\bigwedge_{j=1}^{3}\bigvee_{i=1}^{3}\right)^{-1} A$ $= \begin{pmatrix} a \\ d \\ g \\ b \\ e \\ h \\ c \\ f \\ j \end{pmatrix}$

$$M \times N = \begin{pmatrix} M_{11}N & M_{12}N \\ M_{21}N & M_{22}N \end{pmatrix}$$

$$\begin{pmatrix} M_{11}N_{11} & M_{11}N_{12} & M_{12}N_{11} & M_{12}N_{12} \\ M_{11}N_{21} & M_{11}N_{22} & M_{12}N_{21} & M_{12}N_{22} \\ M_{21}N_{11} & M_{21}N_{12} & M_{22}N_{11} & M_{22}N_{12} \\ M_{21}N_{21} & M_{21}N_{22} & M_{22}N_{21} & M_{22}N_{22} \end{pmatrix}$$

$$M \times N = \begin{pmatrix} MN_{11} & MN_{12} \\ MN_{21} & MN_{22} \end{pmatrix}$$

$$\begin{pmatrix} M_{11}N_{11} & M_{12}N_{11} & M_{11}N_{12} & M_{12}N_{12} \\ M_{21}N_{11} & M_{22}N_{11} & M_{21}N_{12} & M_{22}N_{12} \\ M_{11}N_{21} & M_{12}N_{21} & M_{11}N_{22} & M_{12}N_{22} \\ M_{21}N_{21} & M_{22}N_{21} & M_{21}N_{22} & M_{22}N_{22} \end{pmatrix}$$

$$\begin{pmatrix} a & b \\ c & d \end{pmatrix} \dot{\times} \begin{pmatrix} e & f \\ g & h \end{pmatrix} =$$

$$\left(\begin{array}{c:c} \begin{pmatrix} a \\ c \end{pmatrix} (e \quad f) & \begin{pmatrix} b \\ d \end{pmatrix} (e \quad f) \\ \hdashline \begin{pmatrix} a \\ c \end{pmatrix} (g \quad h) & \begin{pmatrix} b \\ d \end{pmatrix} (g \quad h) \end{array}\right)$$

$$\begin{pmatrix} a & b \\ c & d \end{pmatrix} \underset{\cdot}{\times} \begin{pmatrix} e & f \\ g & h \end{pmatrix} = \begin{pmatrix} a \\ b \\ c \\ d \end{pmatrix} (e \quad f \quad g \quad h)$$

Figure 10. The four basic direct products.

H. Matriplication

Assume that $A = (a_{ij})$ and $B = (b_{kl})$ are of size m by n and r by s, respectively. Furthermore, suppose that the elements a_{ij} and b_{kl} commute with the elements c_{gh} of a matrix C of size γ by ζ. If, for convenience, the four direct products $\cdot\times$, $\dot{\times}$, $\times\cdot$, and $\underset{\cdot}{\times}$ (shown in Fig. 12) collectively are symbolized temporarily by $\cdot\dot{\underset{\cdot}{\times}}\cdot$, then interesting results are obtained by forming the 16 double products indicated by:

$$P_1 = \mathbf{C}(A \cdot\dot{\underset{\cdot}{\times}}\cdot B), \tag{1}$$

$$P_2 = \mathbf{Ç}(A \cdot\dot{\underset{\cdot}{\times}}\cdot B), \tag{2}$$

$$P_3 = (A \cdot\dot{\underset{\cdot}{\times}}\cdot B)\mathbf{C}^T, \tag{3}$$

and

$$P_4 = (A \cdot\dot{\underset{\cdot}{\times}}\cdot B)\mathbf{Ç}^T. \tag{4}$$

To reinforce the terminology, these double products, which are vectors, will be termed "biproducts" because they entail two multiplications. It is often convenient to reshape these vectors as matrices, and for reasons that will become evident later, the resulting matrices, denoted by

$$C{:}A \cdot\dot{\underset{\cdot}{\times}}\cdot B = \mathrm{V}\Lambda P_1, \tag{5}$$

$$C^T{:}A \cdot\dot{\underset{\cdot}{\times}}\cdot B = \mathrm{V}\Lambda P_2, \tag{6}$$

$$A \cdot\dot{\underset{\cdot}{\times}}\cdot B{:}C = \mathrm{V}\Lambda P_3, \tag{7}$$

and

$$A \cdot\dot{\underset{\cdot}{\times}}\cdot B{:}C^T = \mathrm{V}\Lambda P_4, \tag{8}$$

will be termed "triproducts" (see Fig. 11).

EXERCISES

127. Verify:

(*a*) $I \cdot\times I \cdot A = I \times\cdot I{:}A = A.$

(*b*) $I \dot{\times} I{:}A = A^T.$

(*c*) $I \underset{\cdot}{\times} I{:}A = (\mathrm{tr}\, A)I.$

128. Prove:

(*a*) $C{:}A \cdot\times B = B^TCA.$

(*b*) $C{:}A \dot{\times} B = A^TC^TB.$

(*c*) $C{:}A \times\cdot B = A^TCB.$

(*d*) $A \cdot\times B{:}C = BCA^T.$

(*e*) $A \dot{\times} B{:}C = BC^TA^T.$

(*f*) $A \times\cdot B.C = ACB^T.$

$$A \cdot\!\times B{:}C = A \dot{\times} B{:}C^T = B \times\!\cdot A{:}C = \underset{\substack{\uparrow\\ \text{Ordinary product}}}{BCA^T} = C{:}A^T \cdot\!\times B^T$$

$$= C^T{:}B^T \dot{\times} A^T = C{:}B^T \times\!\cdot A^T$$

Some algebraic rules:

$$C{:} = ({:}C)^T \qquad \times\!\cdot = (\cdot\!\times)^T \qquad \dot{\times} = (\dot{\times})^T$$

$${:}C = \begin{pmatrix} 1 \rightharpoondown 2 \\ 3 \hookrightarrow 4 \end{pmatrix} \to \begin{pmatrix} 1 \\ 2 \\ 3 \\ 4 \end{pmatrix} \qquad {:}C^T = \begin{pmatrix} 1 \rightharpoondown 3 \\ 2 \hookrightarrow 4 \end{pmatrix} \to \begin{pmatrix} 1 \\ 3 \\ 2 \\ 4 \end{pmatrix}$$

$$C{:} = \begin{pmatrix} 1 \rightharpoondown 2 \\ 3 \hookrightarrow 4 \end{pmatrix} = (1 \quad 2 \quad 3 \quad 4)$$

	$\cdot\times$	$\dot{\times}$	$\times\cdot$
$:R$	$A \cdot\!\times B{:}C$	$A \dot{\times} B{:}C^T$	$B \times\!\cdot A{:}C$
$L:$	$C{:}A^T \cdot\!\times B^T$	$C^T{:}B^T \dot{\times} A^T$	$C{:}B^T \times\!\cdot A^T$

Multiplication table

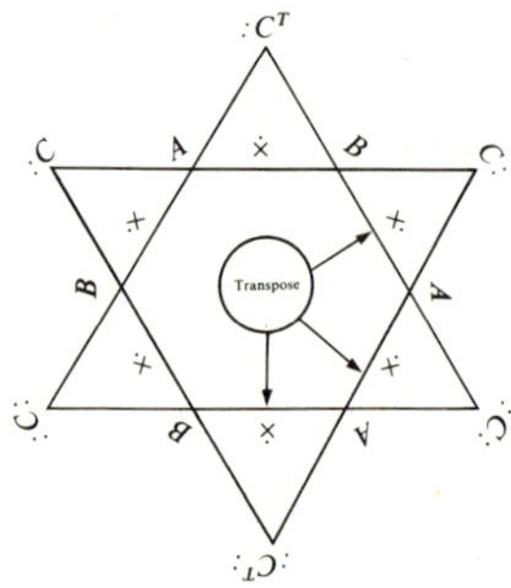

Figure 13. Matrix triproducts.

129. (*a*) Express A^2 as a triproduct in many ways.
(*b*) Represent $A(\text{tr}\, B)$ and $(\text{tr}\, A)B$ as triproducts.

130. Prove:
(*a*) $C{:}A \underset{\cdot}{\times} B = [L(C \,\textcircled{H}\, A)L^T]B = A{:}C \times B.$
(*b*) $A \underset{\cdot}{\times} B{:}C = A[L(B \,\textcircled{H}\, C)L^T] = A \times C{:}B.$

131. Let A and B be of size m by n and r by s, respectively. Verify the following relationships:
(*a*) $I_m \cdot\!\times I_n = I_m \times\!\cdot I_n = I_{mn}.$
(*b*) $K_m \cdot\!\times K_n = K_m \times\!\cdot K_n = K_{mn}.$
(*c*) $A \underset{\cdot}{\times} B = VA\,\Lambda B.$
(*d*) $A \cdot\!\times B = B \times\!\cdot A.$
(*e*) $(A \dot{\times} A^T)^T = A \dot{\times} A^T.$
(*f*) $(A \dot{\times} B)^T = B^T \dot{\times} A^T.$

(*g*) $(A \underset{\cdot}{\times} B)^T = B \underset{\cdot}{\times} A.$

(*h*) $\operatorname{tr}(A \cdot\!\times B) = \operatorname{tr}(A \times\!\cdot B) = \operatorname{tr} A \operatorname{tr} B;\ {}_{m=n,\ r=s}.$

(*i*) $\operatorname{tr}(A \underset{\cdot}{\times} B) = \underset{m=r,\ n=s}{L(A \circledH B)L^T} = \underset{mn=rs}{\Lambda A V B}.$

(*j*) $(A \cdot\!\times B{:}C)M = M^TA \cdot\!\times B{:}C.$

(*k*) $M(A \cdot\!\times B{:}C) = A \cdot\!\times MB{:}C.$

(*l*) $A \cdot\!\times B{:}(C \cdot\!\times D{:}E) = AC \cdot\!\times BD{:}E.$

(*m*) $A \dot{\times} B{:}(C \dot{\times} D{:}E) = AD \dot{\times} BC{:}E.$

132. Establish the following formulas:

(*a*) $(C^T \cdot\!\times A{:}B)(F^T \cdot\!\times D{:}E) = (EF)^T \cdot\!\times (AB){:}(CD).$

(*b*) $(A \dot{\times} B)(C \dot{\times} D) = AD \cdot\!\times BC,$

(*c*) $(A \cdot\!\times B)(C \dot{\times} D) = (B \dot{\times} A)(C \cdot\!\times D) = AC \dot{\times} BD.$

(*d*) $(A \cdot\!\times B)^{-1} = A^{-1} \cdot\!\times B^{-1}.$

(*e*) $(A \dot{\times} B)^{-1} = B^{-1} \dot{\times} A^{-1}.$

(*f*) $(A \times\!\cdot B)^{-1} = A^{-1} \times\!\cdot B^{-1}.$

(*g*) $A \cdot\!\times B{:}C = B \times\!\cdot A{:}C.$

(*h*) $A \cdot\!\times B{:}C = A \dot{\times} B{:}C^T.$

(*i*) $A \cdot\!\times B{:}C = (A \times\!\cdot B{:}C^T)^T.$

(*j*) $A \cdot\!\times B{:}C = C^T{:}B^T \times\!\cdot A^T.$

(*k*) $A \cdot\!\times B{:}C + D \cdot\!\times E{:}F = (A \cdot\!\times B + D \cdot\!\times E){:}F.$

(*l*) $A \dot{\times} B{:}C + D \dot{\times} E{:}F = (A \dot{\times} B + D \dot{\times} E){:}F.$

133. Let the variables x and y appearing in the "bilinear form"

$$a_{ij}x_iy_j = Axy,$$

with matrix A, be subjected to the following linear transformations:

$$x_i = b_{im}x'_m, \qquad \text{and} \quad y_j = C_{jn}y'_n$$

with matrices B and C, respectively. Show that the resulting linear form in the primed variables has the matrix

$$A{:}B \times\!\cdot C.$$

134. According to C. C. MacDuffee, the general linear equation

$$A_1 \times B_1 + A_2 \times B_2 + \cdots + A_h \times B_h = C$$

was first studied by Sylvester, who called the matrix

$$B_1{}^T \times A_1 + B_2{}^T \times A_2 + \cdots + B_h{}^T \times A_h$$

the *nivellateur* or "level seeker," although he did not identify it as a sum of direct products. Let A denote a 4-square matrix. Express the fact that A is symmetric by a triproduct equation and evaluate the resulting nivellateur.

135. (*a*) Do the following candidates have sufficiently nice properties to qualify for the name "Kronecker sums"?

(1) $A \cdot\!\oplus B = I \cdot\!\times A + B^T \times I.$

(2) $A \dot{\oplus} B = I \dot{\times} A + B^T \dot{\times} I.$

(3) $A \oplus\!\cdot B = I \times A + B^T \times\!\cdot I.$

(4) $A \underset{\cdot}{\oplus} B = I \underset{\cdot}{\times} A + B^T \underset{\cdot}{\times} I.$

(*b*) Evaluate $(A \cdot\!\oplus B)^T$, $(A \dot{\oplus} B)^T$, $(A \oplus\!\cdot B)^T$, and $(A \underset{\cdot}{\oplus} B)^T$.

136. Using triproducts, the solution of the linear matrix equation $A \times B + \sum C_i\{X_i\}D_i = E$, where $\{X\}_i$ denotes a subset of the seven remaining elements of the octic set of X can be expressed as the following system of simultaneous linear equations in the n^2 unknowns x_{ij}:

$$\left(B^T \cdot\!\times A + \sum\{C_i\} \dot{\times} \{D_i\}\right):X = :E,$$

where the expression in parentheses represents the coefficient matrix, the colon indicates that the matrices X and E temporarily are elongated into column vectors, and $\dot{\times}$ denotes either the left or upper direct product, whichever is appropriate. Express the equation

$$A_1 \text{F} B_1 + A_2 \text{╥} B_2 + A_3 \text{Ǝ} B_3 + A_4 \text{╨} B_4 + A_5 \text{╦} B_5 + A_6 \text{ꟻ} B_6 + A_7 \text{╩} B_7 + A_8 \text{E} B_8 = C$$

as such a system.

137. (*a*) Express the Lyapunov transformation,

$$L(X) = AX + (AX)^T = B;\ A, B, X \text{ real},$$

which arises in stability theory, as a triproduct equation in the unknown symmetric matrix X.

(*b*) How does the nivellateur change when X is no longer assumed to be symmetric?

(*c*) Express the complex Lyapunov transformation

$$L(Z) = AZ + (AZ)^\dagger = B;\ A, B, Z \text{ complex},$$

as a real triproduct equation in the unknown hermitian matrix $Z = X + iV$.

(*d*) How does the nivellateur change when X is no longer taken to be hermitian?

(*e*) Show that $L(X) =$ alternator of $AX = (A \dot{\oplus} A^T):X^T$.

Next consider two general 2 by 2 matrices:

$$A = \begin{pmatrix} A_{11} & A_{12} \\ A_{21} & A_{22} \end{pmatrix}, \qquad B = \begin{pmatrix} B_{11} & B_{12} \\ B_{21} & B_{22} \end{pmatrix}. \tag{9}$$

Their direct or outer product, $A \times B$, is not the same as their more familiar inner or contracted product, AB, but rather is defined to be

$$A \times\cdot B = A_{ij}B_{kl} = C_{ik,jl} = \begin{pmatrix} A_{11} & A_{12} \\ A_{21} & A_{22} \end{pmatrix} \times\cdot \begin{pmatrix} B_{11} & B_{12} \\ B_{21} & B_{22} \end{pmatrix}$$

$$= \begin{pmatrix} A_{11}\begin{pmatrix} B_{11} & B_{12} \\ B_{21} & B_{22} \end{pmatrix} & A_{12}\begin{pmatrix} B_{11} & B_{12} \\ B_{21} & B_{22} \end{pmatrix} \\ A_{21}\begin{pmatrix} B_{11} & B_{12} \\ B_{21} & B_{22} \end{pmatrix} & A_{22}\begin{pmatrix} B_{11} & B_{12} \\ B_{21} & B_{22} \end{pmatrix} \end{pmatrix}$$

$$= \begin{pmatrix} A_{11}B_{11} & A_{11}B_{12} & A_{12}B_{11} & A_{12}B_{12} \\ A_{11}B_{21} & A_{11}B_{22} & A_{12}B_{21} & A_{12}B_{22} \\ A_{21}B_{11} & A_{21}B_{12} & A_{22}B_{11} & A_{22}B_{12} \\ A_{21}B_{21} & A_{21}B_{22} & A_{22}B_{21} & A_{22}B_{22} \end{pmatrix} \tag{10}$$

$$= \begin{pmatrix} C_{11,11} & C_{11,12} & C_{12,11} & C_{12,12} \\ C_{11,21} & C_{11,22} & C_{12,21} & C_{12,22} \\ C_{21,11} & C_{21,12} & C_{22,11} & C_{22,12} \\ C_{21,21} & C_{21,22} & C_{22,21} & C_{22,22} \end{pmatrix} = \begin{pmatrix} d_{11} & d_{12} & d_{13} & d_{14} \\ d_{21} & d_{22} & d_{23} & d_{24} \\ d_{31} & d_{32} & d_{33} & d_{34} \\ d_{41} & d_{42} & d_{43} & d_{44} \end{pmatrix}.$$

Although the product $C_{ij,kl}$ can be displayed in the form of a 4 by 4 matrix D, it really is a tensor operator of higher rank than either of the two original matrices A or B.

EXERCISE 138

Show that the vector-matrix relationship

$$\begin{pmatrix} A \\ B \end{pmatrix} = \begin{pmatrix} M_{11} & M_{12} \\ M_{21} & M_{22} \end{pmatrix} \begin{pmatrix} C \\ D \end{pmatrix}$$

is equivalent to

$$\begin{pmatrix} B \\ A \end{pmatrix} = \begin{pmatrix} M_{22} & M_{21} \\ M_{12} & M_{11} \end{pmatrix} \begin{pmatrix} D \\ C \end{pmatrix}$$

by using the counter identity transformation

$$M' = KMK = KM\tilde{K} = KMK^{-1} = (K \times K)M, \qquad K^2 = I$$

instead of the correct but less elegant straight algebraic approach.

Lexographical order is the arrangement that words have in a dictionary, where *aa* would come before *bb* and before *ab*, while *ba* would come after *ab* and before *bb*. Therefore, the dictionary entries *aa*, *ab*, *ba*, *bb* are listed in lexographical or dictionary order.

The rules of dictionary order for index couplets are: (ik) comes before $(i'k')$ *if* $i' > i$ and $k' > k$ *or if* $i = i$ and $k' > k$. When applied to two 2 by 2 matrices A and B these rules yield:

$$A \times B = \begin{array}{c|cc:cc|} & (11) & (12) & (21) & (22) \\ \hline (11) & A_{11}B_{11} & A_{11}B_{12} & A_{12}B_{11} & A_{12}B_{12} \\ (12) & A_{11}B_{21} & A_{11}B_{22} & A_{12}B_{21} & A_{12}B_{22} \\ \hdashline (21) & A_{21}B_{11} & A_{21}B_{12} & A_{22}B_{11} & A_{22}B_{12} \\ (22) & A_{21}B_{21} & A_{21}B_{22} & A_{22}B_{21} & A_{22}B_{22} \\ \hline \end{array}$$

EXERCISES

139. Try to find out about the lexographical rules that govern the arrangement of ideograms in a Japanese dictionary.

140. Let the indices i, j, k, l range as follows: $1 \le i \le m$, $1 \le j \le n$, $1 \le k \le r$, $1 \le l \le s$. Complete the following table:

Dictionary Order

(i) 1, 2, $\cdots$, n.

(ij) 11, 12, $\cdots$, $1n$, ——.

(ijk) 111, 112, $\cdots$, $11n$, ——.

$(ijkl)$ 1111, 1112, $\cdots$, $111n$, ——.

Certain direct products, especially those involving the identity or counter-identity matrix, reoccur so frequently that they warrant special emphasis. Among the most powerful of these direct products are those bearing the suggestive names given in Fig. 12. The application of these operators is explained in Fig. 13. A subset of the basic operators, when combined as in Fig. 14, generate a dihedral group of order 8 that recreates the octic set of A.

Identifier:

$$I_n \cdot\times I_n = I_n \times\cdot I_n = I_{n^2}$$

$$\begin{pmatrix} 1 & 0 \\ 0 & 1 \end{pmatrix} \times \begin{pmatrix} 1 & 0 \\ 0 & 1 \end{pmatrix} = \begin{pmatrix} 1 & 0 & 0 & 0 \\ 0 & 1 & 0 & 0 \\ 0 & 0 & 1 & 0 \\ 0 & 0 & 0 & 1 \end{pmatrix}$$

Transpose-cisposer:

$$K_n \cdot\times K_n = K_n \times\cdot K_n = K_{n^2}$$

$$\begin{pmatrix} 0 & 1 \\ 1 & 0 \end{pmatrix} \times \begin{pmatrix} 0 & 1 \\ 1 & 0 \end{pmatrix} = \begin{pmatrix} 0 & 0 & 0 & 1 \\ 0 & 0 & 1 & 0 \\ 0 & 1 & 0 & 0 \\ 1 & 0 & 0 & 0 \end{pmatrix}$$

Transposer:

$$I_n \dot{\times} I_n = T_{n^2}$$

$$\begin{pmatrix} 1 & 0 \\ 0 & 1 \end{pmatrix} \underset{\cdot}{\times} \begin{pmatrix} 1 & 0 \\ 0 & 1 \end{pmatrix} = \begin{pmatrix} 1 & 0 & 0 & 0 \\ 0 & 0 & 1 & 0 \\ 0 & 1 & 0 & 0 \\ 0 & 0 & 0 & 1 \end{pmatrix}$$

Cisposer:

$$K_n \dot{\times} K_n - C_{n^2}$$

$$\begin{pmatrix} 0 & 1 \\ 1 & 0 \end{pmatrix} \dot{\times} \begin{pmatrix} 0 & 1 \\ 1 & 0 \end{pmatrix} = \begin{pmatrix} 0 & 0 & 0 & 1 \\ 0 & 1 & 0 & 0 \\ 0 & 0 & 1 & 0 \\ 1 & 0 & 0 & 0 \end{pmatrix}$$

Tracer:

$$I \underset{\cdot}{\times} I$$

$$\begin{pmatrix} 1 & 0 \\ 0 & 1 \end{pmatrix} \underset{\cdot}{\times} \begin{pmatrix} 1 & 0 \\ 0 & 1 \end{pmatrix} = \begin{pmatrix} 1 & 0 & 0 & 1 \\ 0 & 0 & 0 & 0 \\ 0 & 0 & 0 & 0 \\ 1 & 0 & 0 & 1 \end{pmatrix}$$

Countertracer:

$$K \underset{\cdot}{\times} K$$

$$\begin{pmatrix} 0 & 1 \\ 1 & 0 \end{pmatrix} \underset{\cdot}{\times} \begin{pmatrix} 0 & 1 \\ 1 & 0 \end{pmatrix} = \begin{pmatrix} 0 & 0 & 0 & 0 \\ 0 & 1 & 1 & 0 \\ 0 & 1 & 1 & 0 \\ 0 & 0 & 0 & 0 \end{pmatrix}$$

Figure 14. Some basic direct-product operators.

Transpose:

$$
\begin{aligned}
I \dot{\times} I : A &= \begin{pmatrix} 1 & 0 \\ 0 & 1 \end{pmatrix} \dot{\times} \begin{pmatrix} 1 & 0 \\ 0 & 1 \end{pmatrix} : \begin{pmatrix} a_{11} & a_{12} \\ a_{21} & a_{22} \end{pmatrix} \\
&= \begin{pmatrix} 1 & 0 & 0 & 0 \\ 0 & 0 & 1 & 0 \\ 0 & 1 & 0 & 0 \\ 0 & 0 & 0 & 1 \end{pmatrix} \begin{pmatrix} a_{11} \\ a_{12} \\ a_{21} \\ a_{22} \end{pmatrix} \\
&= \begin{pmatrix} a_{11} \\ a_{21} \\ a_{12} \\ a_{22} \end{pmatrix} = \begin{pmatrix} a_{11} & a_{21} \\ a_{12} & a_{22} \end{pmatrix}
\end{aligned}
$$

Cispose:

$$
\begin{aligned}
K \dot{\times} K : A &= \begin{pmatrix} 0 & 1 \\ 1 & 0 \end{pmatrix} \dot{\times} \begin{pmatrix} 0 & 1 \\ 1 & 0 \end{pmatrix} \begin{pmatrix} a_{11} & a_{12} \\ a_{21} & a_{22} \end{pmatrix} \\
&= \begin{pmatrix} a_{22} & a_{12} \\ a_{21} & a_{11} \end{pmatrix} \\
&= A^C = KA^TK
\end{aligned}
$$

Trace:

$$
\begin{aligned}
I \underset{\cdot}{\times} I : A &= \begin{pmatrix} 1 & 0 \\ 0 & 1 \end{pmatrix} \underset{\cdot}{\times} \begin{pmatrix} 1 & 0 \\ 0 & 1 \end{pmatrix} : \begin{pmatrix} a_{11} & a_{12} \\ a_{21} & a_{22} \end{pmatrix} \\
&= \begin{pmatrix} 1 & 0 & 0 & 1 \\ 0 & 0 & 0 & 0 \\ 0 & 0 & 0 & 0 \\ 1 & 0 & 0 & 1 \end{pmatrix} \begin{pmatrix} a_{11} \\ a_{12} \\ a_{21} \\ a_{22} \end{pmatrix} \\
&= \begin{pmatrix} a_{11} + a_{22} \\ 0 \\ 0 \\ a_{11} + a_{22} \end{pmatrix} \\
&= (a_{11} + a_{22}) \begin{pmatrix} 1 & 0 \\ 0 & 1 \end{pmatrix}
\end{aligned}
$$

Cispose Transpose:
Consider

$$
\begin{aligned}
U &= MV \\
KU &= KMV = KMKKV \\
U' &= M'V'
\end{aligned}
$$

thus

$$
\begin{matrix}
\begin{pmatrix} a \\ b \end{pmatrix} & = & \begin{pmatrix} M_{11} & M_{12} \\ M_{21} & M_{22} \end{pmatrix} & \begin{pmatrix} c \\ d \end{pmatrix} \\
\downarrow & & \downarrow & \downarrow \\
\begin{pmatrix} b \\ a \end{pmatrix} & & \begin{pmatrix} M_{22} & M_{21} \\ M_{12} & M_{11} \end{pmatrix} & \begin{pmatrix} d \\ c \end{pmatrix}
\end{matrix}
$$

Figure 15. Application of birpoducts to obtain useful triproducts.

For	Use on A:
A	$(I \cdot\!\times I) = (I \times\!\cdot I) = (I \dot{\times} I)(I \dot{\times} I)$ $= (K \dot{\times} K)(K \dot{\times} K)$
A^T	$(I \dot{\times} I)$
KA^TK	$(K \dot{\times} K)$
KAK	$(K \cdot\!\times K) = (K \times\!\cdot K) = (I \dot{\times} I)(K \dot{\times} K)$ $= (K \dot{\times} K)(I \dot{\times} I)$
KA^T	$(I \dot{\times} K)$
A^TK	$(K \dot{\times} I)$
KA	$(K \times I) = (I \dot{\times} I)(K \dot{\times} I) = (K \dot{\times} I)(I \dot{\times} I)$
AK	$(I \times K) = (I \dot{\times} I)(I \dot{\times} K) = (I \dot{\times} K)(I \dot{\times} I)$
	(Also represents dihedral group of order 8.)

Figure 16. The set of octic operators $\{O_i\}:A = \{A\}$.

Before we take up the subject of vector and tensor analysis, a few closing remarks are appropriate concerning those matrix transformations that are RST relations on certain classes of matrices over various number fields. The table below gives an aerial view of the ground to be covered (the P and Q are nonsingular).

Transformation	Definition	Related Triproduct
Equivalence	$B = PAQ$	$B = Q^T \cdot\!\times P:A$
Similarity	$B = P^{-1}AP$	$B = P^T \cdot\!\times P^{-1}:A$
Congruence	$B = P^TAP$	$B = P^T \cdot\!\times P^T:A$
Hermitian congruence	$B = P^*AP$	$B = P^T \cdot\!\times P^*:A$

These transformations are called RST relations because they are reflexive, symmetric, and transitive. To illustrate, the equivalence relation $B = PAQ$ satisfies the following conditions:

Reflexive: $A = PAQ$. Let $P = I_m$ and $Q = I_n$.
Symmetric: $B = PAQ \Rightarrow A = P'BQ'$. Let $P' = P^{-1}$ and $Q' = Q^{-1}$
Transitive: $B = P_1AQ_1$ and $C = P_2BQ_2 \Rightarrow C = P^1AQ^1$. Let $P' = P_2P_1$ and $Q' = Q_1Q_2$.

The other transformations are shown to be RST relations in the same way.

The following chart shows what happens when two such transformations occur at the same time:

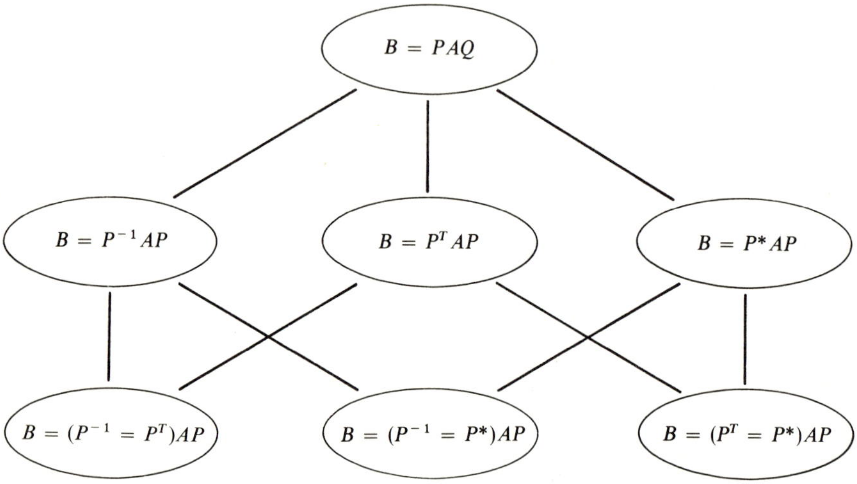

This chart permits an extension of the preceding table (O = orthogonal, U = unitary, and R = real):

Transformation	Definition	Related Triproduct
Orthogonal similarity	$B = (O^{-1} = O^T)AQ$	$B = O^T \times (O^{-1} = O^T):A$
Unitary similarity	$B = (U^{-1} = U^*)AU$	$B = U^T \times (U^{-1} = U^*):A$
Real	$B = (R^T = R^*)AR$	$B = R^T \times (R^T = R^*):A$

Certain classes of matrices are reduced by these transformations to standard forms. For example, real symmetric matrices under similarity over a field F are reduced to a diagonal matrix $N(\lambda_1, \lambda_2, \cdots, \lambda_n)$ exhibiting the eigenvalues along the main diagonal.

The best that can be said about arbitrary matrices under similarity is that they are reduced to a direct sum of Jordan matrices.

Such considerations have led to the chart shown in Fig. 17. This chart is intentionally incomplete, since it is intended to provide insight rather than to add to the confusion. By working the exercises and through collateral reading* the student may obtain a more complete conception.

* S. Perlis, *Theory of Matrices*, Addison-Wesley Publishing Co., 1958.

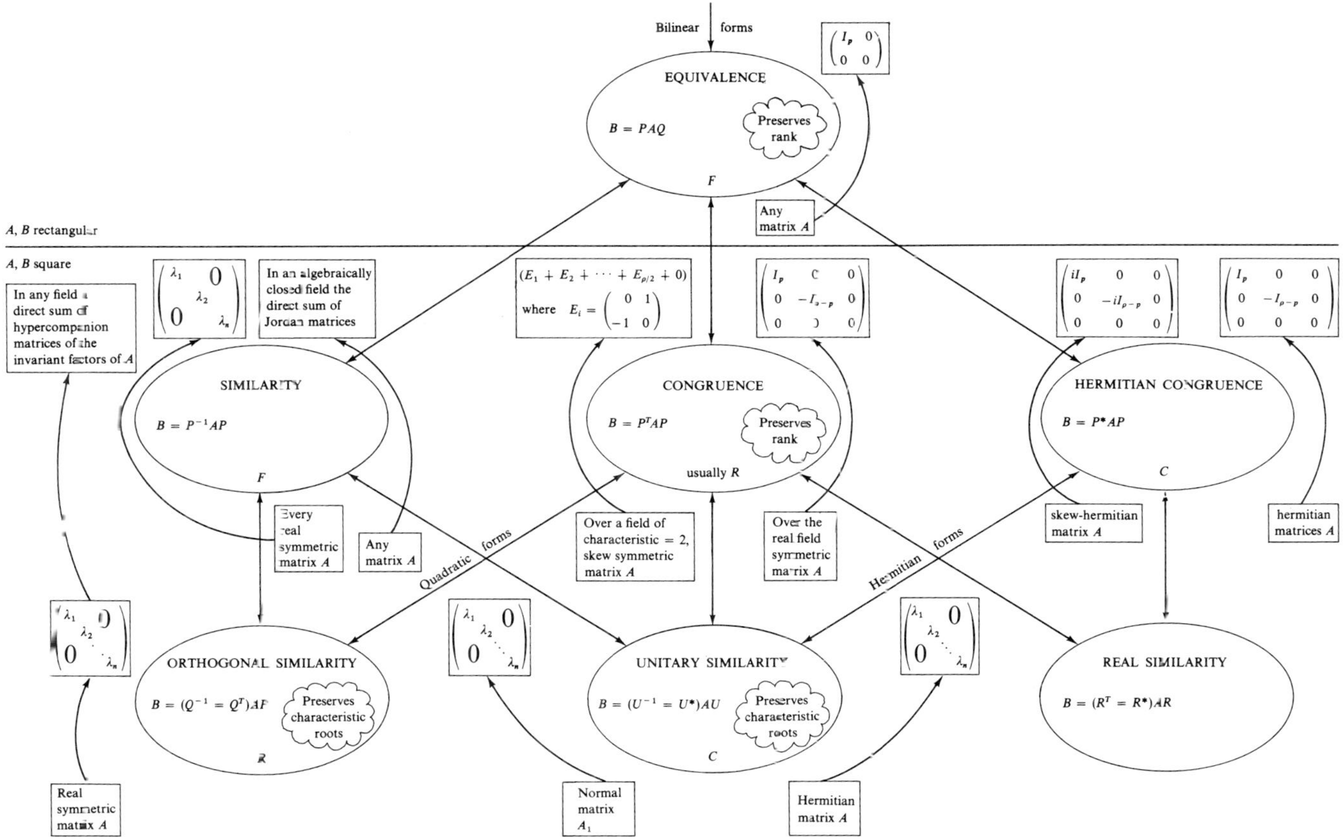

P, Q and O, U, R nonsingular $\quad$ R real field $\quad$ C complex field $\quad$ F arbitrary field

Figure 17. The grand conception.

EXERCISES

141. Show that two matrices over the scalar field F are equivalent over F if and only if they have the same rank.

142. (*a*) Prove that if matrices A and B are similar, then they have the same eigenvalues. (*b*) Is the converse true? If not, give an example of two matrices with the same eigenvalues that are dissimilar.

143. Prove that the trace and determinant are similarity invariants.

144. Prove that $\lambda(AB) = \lambda(BA)$.

145. (*a*) Prove that if A has linearly independent eigenvectors, then A is similar to a diagonal matrix. (*b*) Prove that if two matrices over F are similar over F, then they both have the same minimum equation.

146. Let

$$U = \begin{pmatrix} 1 & 0 & \cdots & 0 \\ r_1 & 1 & \cdots & 0 \\ \cdot & \cdot & \cdot & \cdot \\ \cdot & \cdot & \cdot & \cdot \\ \cdot & \cdot & \cdot & \cdot \\ r_{n-1} & 0 & \cdots & 1 \end{pmatrix}.$$

Take a general n-square matrix $T = \mathbf{c} \cdot \mathbf{x}^T = (c_i x_j)$ of rank 1. Determine a particular U such that

$$U^{-1}TU = \begin{pmatrix} a_{11} & a_{12} & \cdots & a_{1n} \\ 0 & 0 & \cdots & 0 \\ \cdot & \cdot & \cdot & \cdot \\ \cdot & \cdot & \cdot & \cdot \\ \cdot & \cdot & \cdot & \cdot \\ 0 & 0 & \cdots & 0 \end{pmatrix}.$$

The following lemma will be used immediately. Let λ be an eigenvalue of

$$M = \begin{pmatrix} A & \vdots & B \\ \cdots & \vdots & \cdots \\ -B & \vdots & -A \end{pmatrix} \tag{1}$$

and let $\mathbf{x}$ be the corresponding eigenvector. Then the equation

$$M\mathbf{x} = \lambda\mathbf{x} \tag{2}$$

implies that

$$-M\mathbf{x} = -\lambda\mathbf{x}. \tag{3}$$

So, $-\lambda$ is seen to be an eigenvalue of $-M$. But

$$-M = \begin{pmatrix} -A & \vdots & -B \\ \cdots & \cdots & \cdots \\ B & \vdots & A \end{pmatrix} = \begin{pmatrix} 0 & \vdots & I \\ \cdots & \cdots & \cdots \\ I & \vdots & 0 \end{pmatrix} \begin{pmatrix} A & \vdots & B \\ \cdots & \cdots & \cdots \\ -B & \vdots & -A \end{pmatrix} \begin{pmatrix} 0 & \vdots & I \\ \cdots & \cdots & \cdots \\ I & \vdots & 0 \end{pmatrix} \cdot K^{-1}MK \quad (4)$$

is a similarity transformation. Consequently,

LEMMA. If λ is an eigenvalue of M, then so is $-\lambda$.

Now set $C = (A + B)(A - B)$ and assume that C has the eigenvalues $\lambda_1, \lambda_2, \cdots, \lambda_n$. On the basis of this skimpy information it is possible to find the eigenvalues $\lambda(M)$ of M. Let the unitary transformation $U^{-1}CU$ reduce C to the easily recognizable Jordan canonical form:

$$N = U^{-1}CU = \begin{pmatrix} \lambda_1 & \epsilon_1 & 0 & \cdots & 0 \\ 0 & \lambda_2 & \epsilon_2 & \cdots & 0 \\ 0 & 0 & \lambda_3 & \cdots & 0 \\ \cdot & \cdot & \cdot & \cdot & \cdot \\ \cdot & \cdot & \cdot & \cdot & \cdot \\ \cdot & \cdot & \cdot & \cdot & \cdot \\ 0 & 0 & 0 & \cdots & \epsilon_{n-1} \\ 0 & 0 & 0 & \cdots & \lambda_n \end{pmatrix}. \quad (5)$$

where ϵ_i are either 1 or 0. Consequently,

$$\begin{pmatrix} U^{-1} & 0 \\ 0 & U^{-1} \end{pmatrix} \begin{pmatrix} C & 0 \\ 0 & C \end{pmatrix} \begin{pmatrix} U & 0 \\ 0 & U \end{pmatrix}$$

$$= (U^{-1} \dotplus U^{-1})(C \dotplus C)(U \dotplus U) = N \dotplus N, \quad (6)$$

where the eigenvalues of C, taken with their proper multiplicity, appear twice.

In conclusion, if a matrix V can be found such that

$$V^{-1}M^2V = C \dotplus C, \quad (7)$$

then M^2 will have the eigenvalues $\lambda_1, \lambda_2, \cdots, \lambda_n, \lambda_1, \lambda_2, \cdots, \lambda_n$; so, applying the preceding lemma,

$$\lambda(M) = \pm\sqrt{\lambda_1}, \qquad \pm\sqrt{2}, \cdots, \pm\sqrt{\lambda_n}. \quad (8)$$

EXERCISES

147. Find the required V.

148. Assume that A and B are square matrices. Prove that the proper values of $\begin{matrix} A & -B \\ B & A \end{matrix}$ coincide with those of $A \pm iB$.

149. Give a short account of the matrix representations* of a finite group explaining what is meant by the terms "equivalent," "reducible," and "irreducible."

150. Investigate the concepts of character and class in group theory and create character tables for various simple groups.

151. (*a*) Generate the two-dimensional matrix representation of the group of operations that transform an equilateral triangle into itself.

(*b*) Generate the three-dimensional matrix representation for this group.

(*c*) Show that this representation is reducible.

152. Find a suitable unitary S-matrix that reduces the three-dimensional representation given above. That is, if Γ represents a member of the group, then

$$S^{-1}\Gamma S = \Gamma_D$$

or

$$\begin{pmatrix} \text{3 by 3} \\ \text{unknown} \\ S\text{-matrix} \end{pmatrix}^{-1} \begin{pmatrix} \text{3 by 3} \\ \text{known} \\ \text{matrix} \end{pmatrix} \begin{pmatrix} \text{3 by 3} \\ \text{unknown} \\ S\text{-matrix} \end{pmatrix} = \begin{pmatrix} 1 & 0 \quad 0 \\ 0 & \text{unknown 2 by 2} \\ 0 & \text{matrix} \end{pmatrix}.$$

153. (*a*) Prove that the matrices in the two- and three-dimensional representations obey the group properties under multiplication.

(*b*) Generate the group multiplication table.

(*c*) Show that the unitary S-matrix can be formed from a suitable set of orthogonal vectors.

(*d*) What other representations can be found for this group?

154. (*a*) Generate the three-dimensional group representation for the permutation of four objects. Show that the generator of this representation may be taken as the basis vector:

$$\begin{pmatrix} x \\ y \\ z \end{pmatrix}.$$

*Morris Newman, *Matrix Representations of Groups* (NBS Applied Math Series 60, for sale by the Superintendent of Documents, U.S. Government Printing Office, Washington D.C. 20402).

(*b*) How is this group modified if the generator is taken as the basis vector?

$$\begin{pmatrix} x^2 \\ y^2 \\ z^2 \end{pmatrix}.$$

(*c*) Find a suitable unitary S-matrix to reduce this representation.

155. Find the four-dimensional representation for the permutation group of four objects.

156. Reduce the four-dimensional representation with a suitable S-matrix.

157. Show that a four-dimensional unitary transformation matrix S_4 may be formed as the right direct product or "direct square" of a two-dimensional unitary transformation matrix S_2:

$$S_4 = S_2 \times_r S_2.$$

158. What representations for a group of order 24 are generated by the following basis vectors:

$$(xyz) \sim \text{volume},$$
$$(xyz)^2,$$

and

$$(x^2 + y^2 + z^2) = r^2.$$

159. Relate the concept of a basis or generator of a group representation with quantum-mechanical wave functions for appropriate atomic or molecular systems in chemistry and solid state physics.

160. Show that a similarity transformation $S^{-1}MS = D$ reducing a matrix M to diagonal form may be thought of differently, as follows. First, S is formed from the eigenvectors of M. Next, the matrix M operates on S, bringing forth the eigenvalues. Then S^{-1} operates transforming these vectors to a new form, or alternatively, producing an n-square array of all possible products of the n-orthonormal eigenvectors. Thus the whole process represents a manipulation of eigenvectors that incidentally happen to be placed together in S-matrix form.

Part II

Vector and Tensor Analysis

3 Vector Analysis

A. Reciprocal Lattices

To find the reciprocal lattice for two-dimensional polar coordinates r, θ, it is best to work with the full three-dimensional cylindrical coordinate system. Accordingly, introduce the three unitary vectors $\mathbf{a}_r$, $\mathbf{a}_\theta$, $\mathbf{a}_z$, defined in the next section and form their triple scalar product:

$$V = \mathbf{a}_r \cdot (\mathbf{a}_\theta \times \mathbf{a}_z) = r, \tag{1}$$

which also can be interpreted as the volume of the parallelopiped defined by the three unitary vectors. Now define two new unitary vectors $\mathbf{a}^r$, $\mathbf{a}^\theta$, by the following two vector cross products:

$$\begin{aligned} \mathbf{a}^r &= \frac{\mathbf{a}_\theta \times \mathbf{a}_z}{V} = \frac{r(-\hat{\mathbf{i}}\sin\theta + \hat{\mathbf{j}}\cos\theta) \times \hat{\mathbf{k}}}{r} \\ &= (\hat{\mathbf{i}}\cos\theta + \hat{\mathbf{j}}\sin\theta) = \hat{\mathbf{e}}_r \equiv \mathbf{a}_r, \end{aligned} \tag{2}$$

and

$$\begin{aligned} \mathbf{a}^\theta &= \frac{\mathbf{a}_z \times \mathbf{a}_r}{V} = \frac{\hat{\mathbf{k}} \times (\hat{\mathbf{i}}\cos\theta + \hat{\mathbf{j}}\sin\theta)}{r} \\ &= \frac{1}{r}(\hat{\mathbf{j}}\cos\theta - \hat{\mathbf{i}}\sin\theta) = \frac{1}{r^2}\mathbf{a}_\theta \neq \hat{\mathbf{e}}_\theta. \end{aligned} \tag{3}$$

Notice that an orthogonal set of unitary base vectors generates a second set of orthogonal base vectors (Fig. 1). One of the base vectors was not a unit vector, but

$$\mathbf{a}_\theta \cdot \mathbf{a}^\theta = 1 \tag{4}$$

For any vector $\mathbf{V}$,

$$\begin{aligned} \mathbf{V} &= V_r\hat{\mathbf{e}}_r + V_\theta\hat{\mathbf{e}}_\theta = v_r\mathbf{a}^r + v_\theta\mathbf{a}^\theta \\ &= v^r\mathbf{a}_r + v^\theta\mathbf{a}_\theta = V_x\hat{\mathbf{i}} + V_y\hat{\mathbf{j}}. \end{aligned} \tag{5}$$

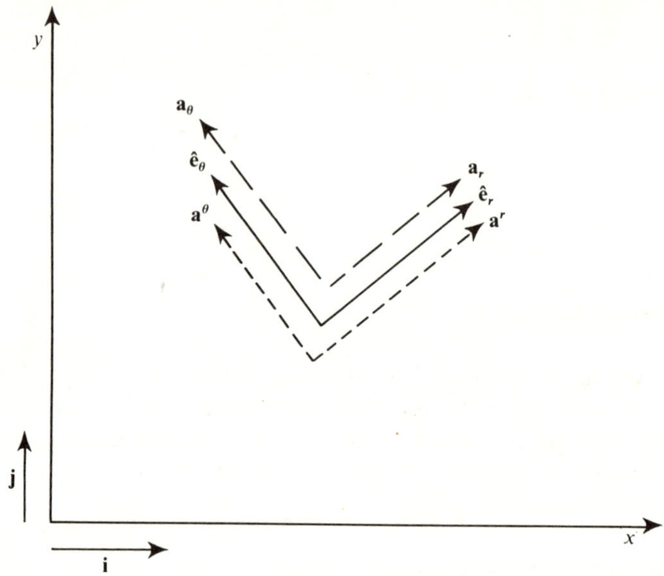

Figure 1. The parallel unit and unitary vectors for an orthogonal coordinate system.

Remember that

$$\begin{aligned} \mathbf{a}^r \cdot \mathbf{a}_r &= 1, \qquad & \mathbf{a}^\theta \cdot \mathbf{a}_\theta &= 1, \\ \mathbf{a}_r \cdot \mathbf{a}^\theta &= 0, \qquad & \mathbf{a}_\theta \cdot \mathbf{a}^r &= 0. \end{aligned} \tag{6}$$

Thus

$$\begin{aligned} \mathbf{a}^r \cdot \mathbf{V} &= v^r, \qquad & \mathbf{a}_r \cdot \mathbf{V} &= v_r, \\ \mathbf{a}^\theta \cdot \mathbf{V} &= v^\theta, \qquad & \mathbf{a}_\theta \cdot \mathbf{V} &= v_\theta. \end{aligned} \tag{7}$$

Further,

$$v^r = v_r = V_r \tag{8}$$

since

$$\hat{\mathbf{e}}_r = \mathbf{a}_r = \mathbf{a}^r. \tag{9}$$

But

$$v^\theta = \frac{1}{r} V_\theta, \qquad v_\theta = r V_\theta, \tag{10}$$

since

$$r\mathbf{a}^\theta = \hat{\mathbf{e}}_\theta = \frac{1}{r}\mathbf{a}_\theta. \tag{11}$$

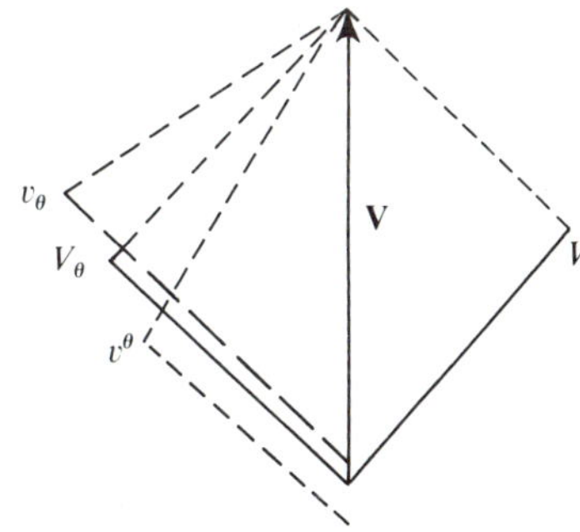

Figure 2. Covariant and contravariant components of a vector.

Notice that

$$\begin{aligned} \mathbf{a}_\theta v^\theta &= (\hat{\mathbf{e}}_\theta r)\left(\frac{1}{r} V_\theta\right) \leftrightarrow \hat{\mathbf{e}}_\theta \frac{r}{r} V_\theta \leftrightarrow \hat{\mathbf{e}}_\theta V_\theta \\ \hat{\mathbf{e}}_\theta V_\theta &\leftrightarrow \hat{\mathbf{e}}_\theta \frac{r}{r} V_\theta \leftrightarrow \left(\hat{\mathbf{e}}_\theta \frac{1}{r}\right)(r V_\theta) \leftrightarrow \mathbf{a}^\theta v_\theta, \end{aligned} \tag{12}$$

as suggested by Fig. 2.

EXERCISES

1. Show that the reciprocal lattice for a solid possessing a body-centered cubic crystal structure is a face-centered cubic crystal structure.

2. Describe Bragg reflection from a solid crystal in terms of Ewald's construction in the reciprocal lattice.

3. Prove that the position vector to a point (h, k, l) in a reciprocal lattice is perpendicular to the lattice planes in the original solid crystal characterized by the Miller indices (hkl).

4. Prove that the length of the position vector to the point (h, k, l) in the reciprocal lattice is inversely proportional to the distance between adjacent parallel planes characterized by the Miller indices (hkl) in the original solid.

B. Curvilinear Coordinate Systems and Transformations

In a given three-dimensional region, let

$$u^1 = f_1(x, y, z), \quad u^2 = f_2(x, y, z), \quad u^3 = f_3(x, y, z) \tag{1}$$

be three independent, continuous, single-valued functions of the rectangular coordinates x, y, z. That equations (1) can be solved to obtain x, y, and z as functions of u^1, u^2, and u^3 may be seen as follows:

$$x = \phi_1(u^1, u^2, u^3), \quad y = \phi_2(u^1, u^2, u^3), \quad z = \phi_3(u^1, u^2, u^3). \tag{2}$$

In general, the functions in both sets (1) and (2) are continuously differentiable, except possibly at certain singular points.

To each point $P(x, y, z)$ there corresponds sets of values x, y, z and u^1, u^2, u^3, and, inversely, for each such set of values there is a point within the region. The functions u^1, u^2, u^3 are called "general" or "curvilinear coordinates." Through each point, there pass three surfaces:

$$u^1 = \text{constant}, \quad u^2 = \text{constant}, \qquad \text{and} \quad u^3 = \text{constant}, \tag{3}$$

called the "coordinate surfaces." On each of these surfaces one coordinate is constant and two are variable—the constant is used to define the surface. When n is the dimensionality of space, there are n surfaces. Each pair of surfaces meets in a curve and there are $(n/2)(n - 1)$ of these curves; three for $n = 3$. On these curves two coordinates are constant, so that it is convenient to denote the curves by the variable coordinate (see Fig. 3).

Let $\mathbf{r}$ denote a position vector to a variable point $P(x, y, z)$. In curvilinear coordinates, u^1, u^2, u^3 either

$$P = P(x, y, z), \quad \mathbf{r} = \mathbf{r}(x, y, z) \tag{4a}$$

or

$$P = P(u^1, u^2, u^3), \quad \mathbf{r} = \mathbf{r}(u^1, u^2, u^3). \tag{4b}$$

A differential change in $\mathbf{r}$ due to small displacements along the coordinates may be written

$$d\mathbf{r} = \frac{\partial \mathbf{r}}{\partial u^1} du^1 + \frac{\partial \mathbf{r}}{\partial u^2} du^2 + \frac{\partial \mathbf{r}}{\partial u^3} du^3. \tag{5}$$

But

$$\mathbf{r} = x\hat{\mathbf{i}} + y\hat{\mathbf{j}} + z\hat{\mathbf{k}} \tag{6}$$

and

$$d\mathbf{r} = dx\hat{\mathbf{i}} + dy\hat{\mathbf{j}} + dz\hat{\mathbf{k}}, \tag{7}$$

so that

$$d\mathbf{r} = du^1\mathbf{a}_1 + du^2\mathbf{a}_2 + du^3\mathbf{a}_3, \tag{8}$$

where the vectors

$$\mathbf{a}_1 = \frac{\partial \mathbf{r}}{\partial u^1}, \quad \mathbf{a}_2 = \frac{\partial \mathbf{r}}{\partial u^2}, \quad \mathbf{a}_3 = \frac{\partial \mathbf{r}}{\partial u^3} \tag{9}$$

are the unitary vectors associated with the point P, which form a local base system of reference for all other vectors associated with that point.

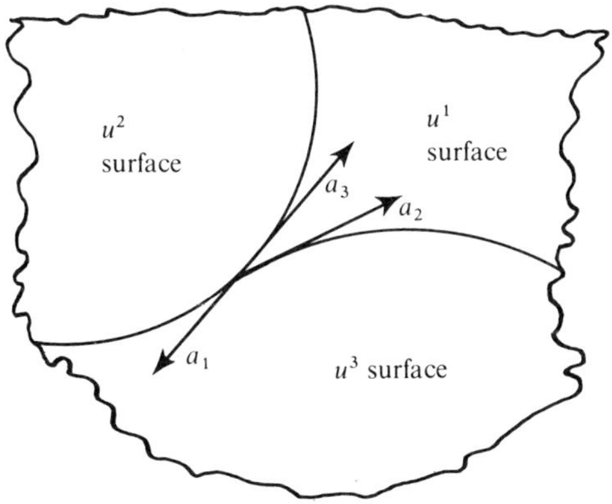

Figure 3. The coordinate surfaces for curvilinear coordinates.

For cartesian coordinates **r** and $d\mathbf{r}$, use the same unit vectors $\hat{\mathbf{i}}$, $\hat{\mathbf{j}}$, $\mathbf{k}$; for curvilinear coordinates **r** and $d\mathbf{r}$, however, use different unitary vectors depending on position.

It must be remembered that unitary vectors are not necessarily of unit length. Consequently, the three vectors $\mathbf{a}_1, \mathbf{a}_2, \mathbf{a}_3$ define a small parallelopiped with volume

$$V = \mathbf{a}_1 \cdot (\mathbf{a}_2 \times \mathbf{a}_3) = \mathbf{a}_2 \cdot (\mathbf{a}_3 \times \mathbf{a}_1) = \mathbf{a}_3 \cdot (\mathbf{a}_1 \times \mathbf{a}_2). \tag{10}$$

Now form three new vectors:

$$\mathbf{a}^1 = \frac{\mathbf{a}_2 \times \mathbf{a}_3}{V}, \quad \mathbf{a}^2 = \frac{\mathbf{a}_3 \times \mathbf{a}_1}{V}, \quad \text{and} \quad \mathbf{a}^3 = \frac{\mathbf{a}_1 \times \mathbf{a}_2}{V}, \tag{11}$$

or:

$$\mathbf{a}^1 = \frac{\mathbf{a}_2 \times \mathbf{a}_3}{\mathbf{a}_1 \cdot (\mathbf{a}_2 \times \mathbf{a}_3)}, \quad \mathbf{a}^2 = \frac{\mathbf{a}_3 \times \mathbf{a}_1}{\mathbf{a}_1 \cdot (\mathbf{a}_2 \times \mathbf{a}_3)}, \quad \text{and} \quad \mathbf{a}^3 = \frac{\mathbf{a}_1 \times \mathbf{a}_2}{\mathbf{a}_1 \cdot (\mathbf{a}_2 \times \mathbf{a}_3)} \tag{12}$$

normal to each u^i plane. These new vectors can be written succinctly in combination as

$$\mathbf{a}^i = \frac{\mathbf{a}_j \times \mathbf{a}_k}{V} = \frac{\mathbf{a}_j \times \mathbf{a}_k}{\mathbf{a}_i \cdot (\mathbf{a}_j \times \mathbf{a}_k)}. \tag{13}$$

It is easy to determine all possible scalar products of the form $\mathbf{a}^i \cdot \mathbf{a}_j$; in fact,

$$\mathbf{a}^i \cdot \mathbf{a}_j = \delta_{ij}. \tag{14}$$

Also, it is easy to obtain the inverse relationships:

$$\mathbf{a}_1 = \frac{\mathbf{a}^2 \times \mathbf{a}^3}{V}, \quad \mathbf{a}_2 = \frac{\mathbf{a}^3 \times \mathbf{a}^1}{V}, \quad \text{and} \quad \mathbf{a}_3 = \frac{\mathbf{a}^1 \times \mathbf{a}^2}{V}. \tag{15}$$

Any two sets of vectors related by equations like (13), (14), and (15) are said to constitute a "reciprocal system." Vectors $\mathbf{a}^1$, $\mathbf{a}^2$, $\mathbf{a}^3$, which are called "reciprocal unitary vectors," may be used as a new basis in terms of which $d\mathbf{r}$ is written

$$d\mathbf{r} = \mathbf{a}^1 du_1 + \mathbf{a}^2 du_2 + \mathbf{a}^3 du_3. \tag{16}$$

Thus

$$d\mathbf{r} = \sum_{i=1}^{3} \mathbf{a}_i du^i = \sum_{j=1}^{3} \mathbf{a}^j du_j. \tag{17}$$

Multiply equation (17) by $\mathbf{a}^i$ and by $\mathbf{a}_j$ in turn:

$$\begin{aligned} \mathbf{a}^i \cdot d\mathbf{r} &= \mathbf{a}^i \cdot \sum \mathbf{a}_i du^i = \mathbf{a}^i \cdot \sum \mathbf{a}^j du_j \\ &= du^i = \sum \mathbf{a}^i \cdot \mathbf{a}_i du^i = \sum \mathbf{a}^i \cdot \mathbf{a}^j du_j \end{aligned} \tag{18}$$

and

$$\begin{aligned} \mathbf{a}_j \cdot d\mathbf{r} &= \mathbf{a}_j \cdot \sum \mathbf{a}_i du^i = \mathbf{a}_j \cdot \sum \mathbf{a}_i du^i \\ &= du_j = \sum \mathbf{a}_j \cdot \mathbf{a}_i du^i = \sum \mathbf{a}_j a^j du_j. \end{aligned} \tag{19}$$

The scalar products, $\mathbf{a}_i \cdot \mathbf{a}_j$, customarily are designated by

$$g_{ij} = \mathbf{a}_i \cdot \mathbf{a}_j = g_{ji} \tag{20}$$

and

$$g^{ij} = \mathbf{a}^i \cdot \mathbf{a}^j = g^{ji}.$$

Then

$$du_j = g_{ji} du^i \qquad \text{and} \quad du^i = g^{ij} du_j. \tag{22}$$

A vector $\mathbf{F}$ at the point P may be resolved into components in either the original or reciprocal system:

$$\mathbf{F} = f^i \mathbf{a}_i = f_j \mathbf{a}^j. \tag{23}$$

Multiply $\mathbf{F}$ by $\mathbf{a}^i$ to obtain

$$f^i = \mathbf{a}^i \cdot \mathbf{a}^j f_j = g^{ij} f_j, \tag{24}$$

or multiply $\mathbf{F}$ by $\mathbf{a}_j$ to obtain

$$f_j = \mathbf{a}_j \cdot \mathbf{a}_i f^i = g_{ji} f^i. \tag{25}$$

Since $\mathbf{a}^i$ is perpendicular to both $\mathbf{a}_j$ and $\mathbf{a}_k$,

$$f^i = \mathbf{F} \cdot \mathbf{a}^i, \quad f_j = \mathbf{F} \cdot \mathbf{a}_j, \tag{26}$$

or

$$\mathbf{F} = (\mathbf{F} \cdot \mathbf{a}^i)\mathbf{a}_i = (\mathbf{F} \cdot \mathbf{a}_j)\mathbf{a}^j. \tag{27}$$

In this expression the f^i are called the "contravariant components of the vector **F**" and the f_j are called the "covariant components of the vector **F**." The letters F_1, F_2, F_3 refer to a system in which the base vectors are of unit length. The magnitude and dimensions of the base vectors depend on the particular system of coordinates in use. To construct a set of unit vectors tangent to the u^i curves, proceed as follows:

$$\hat{\mathbf{i}}_1 = \frac{\mathbf{a}_1}{\sqrt{\mathbf{a}_1 \cdot \mathbf{a}_1}}, \quad \hat{\mathbf{i}}_2 = \frac{\mathbf{a}_2}{\sqrt{\mathbf{a}_2 \cdot \mathbf{a}_2}}, \quad \hat{\mathbf{i}}_3 = \frac{\mathbf{a}_3}{\sqrt{\mathbf{a}_3 \cdot \mathbf{a}_3}}. \tag{28}$$

Then

$$\mathbf{F} = F_1\hat{\mathbf{i}}_1 + F_2\hat{\mathbf{i}}_2 + F_3\hat{\mathbf{i}}_3, \tag{29}$$

but

$$\begin{aligned} \mathbf{F} &= f^1\mathbf{a}_1 + f^2\mathbf{a}_2 + f^3\mathbf{a}_3 \\ &= \frac{F_1\mathbf{a}_1}{\sqrt{\mathbf{a}_1 \cdot \mathbf{a}_1}} + \frac{F_2\mathbf{a}_2}{\sqrt{\mathbf{a}_2 \cdot \mathbf{a}_2}} + \frac{F_3\mathbf{a}_3}{\sqrt{\mathbf{a}_3 \cdot \mathbf{a}_3}}. \end{aligned} \tag{30}$$

So

$$\frac{F_i}{\sqrt{\mathbf{a}_i \cdot \mathbf{a}_i}} = f^i \tag{31}$$

or

$$F_i = \sqrt{g_{ii}}f^i \qquad \text{(No sum intended.)} \tag{32}$$

Compare this to

$$f_j = g_{ji}f^i \qquad \text{(Sum intended.)} \tag{33}$$

The F_i have the same physical dimensions as does F.

Now let $d\mathbf{r}$ represent an infinitesimal displacement from a point $P(u^1, u^2, u^3)$ to another point $P'(u^1 + du^1, u^2 + du^2, u^3 + du^3)$ having a line element ds whose square is given by

$$\begin{aligned} ds^2 &= d\mathbf{r} \cdot d\mathbf{r} = \mathbf{a}_i \cdot \mathbf{a}_j du^i du^j = \mathbf{a}^i \cdot \mathbf{a}^j du_i du_j \\ &= g_{ij} du^i du^j = g^{ij} du_i du_j. \end{aligned} \tag{34}$$

The g_{ij} and g^{ij} are called "metrical coefficients"; line elements are expressed in terms of them.

It is possible to express arcs, surfaces, and volume elements in a system of curvilinear coordinates. Let $d\mathbf{s}_1$ correspond to an infinitesimal displacement at P; then

$$d\mathbf{s}_1 = \mathbf{a}_1 du^1 = \sqrt{g_{11}}\, du^1\hat{\mathbf{i}}, \quad ds_1 = \sqrt{g_{11}}\, du^1. \tag{35}$$

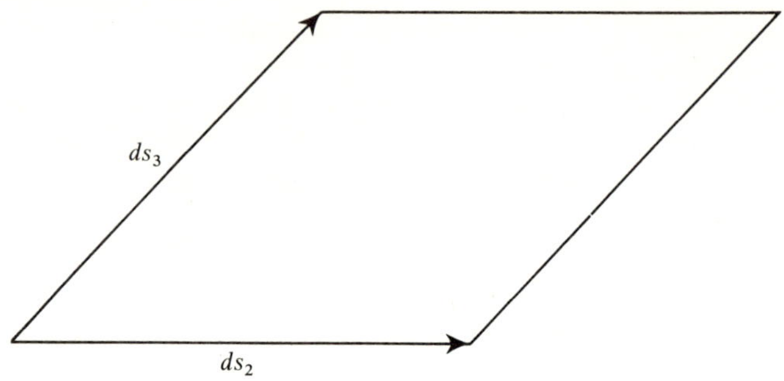

Figure 4. The U^1 surface.

And similarly,

$$ds_2 = \sqrt{g_{22}}\,du^2, \quad ds_3 = \sqrt{g_{33}}\,du^3. \tag{36}$$

Form an area element as follows. Consider the u^1 "plane" (really a surface) and draw vectors $\mathbf{a}_2$ and $\mathbf{a}_3$ as shown in Fig. 4. Calculate:

$$d\mathbf{A}_1 = d\mathbf{s}_2 \times d\mathbf{s}_3 = \mathbf{a}_2 \times \mathbf{a}_3\, du^2\, du^3. \tag{37}$$

Or, in terms of absolute values,

$$dA_1 = |\mathbf{a}_2 \times \mathbf{a}_3| du^2 du^3 = [(\mathbf{a}_2 \times \mathbf{a}_3) \cdot (\mathbf{a}_2 \times \mathbf{a}_3)]^{1/2} du^2 du^3. \tag{38}$$

Since

$$(\mathbf{a} \times \mathbf{b}) \cdot (\mathbf{c} \times \mathbf{d}) = (\bar{a} \cdot \bar{c})(\bar{b} \cdot \bar{d}) - (\bar{a} \cdot \bar{d})(\bar{b} \cdot \bar{c}), \tag{39}$$

it follows that

$$\begin{aligned}(\mathbf{a}_2 \times \mathbf{a}_3) \cdot (\mathbf{a}_2 \times \mathbf{a}_3) &= (a_2 \cdot a_2)(a_3 \cdot a_3) - (a_2 \cdot a_3)(a_3 \cdot a_2) \\ &= g_{22} g_{33} - (g_{23})^2.\end{aligned} \tag{40}$$

Thus

$$dA_1 = \sqrt{g_{22} g_{33} - (g_{23})^2}\, du^2\, du^3. \tag{41}$$

Similarly,

$$dA_2 = \sqrt{g_{11} g_{33} - (g_{13})^2}\, du^1\, du^3, \tag{42}$$

and

$$dA_3 = \sqrt{g_{11} g_{22} - (g_{12})^2}\, du^1\, du^2. \tag{43}$$

Finally, for a volume element,

$$dV = d\mathbf{s}_1 \cdot (d\mathbf{s}_2 \times d\mathbf{s}_3) = \mathbf{a}_1 \cdot (\mathbf{a}_2 \times \mathbf{a}_3) du^1\, du^2\, du^3. \tag{44}$$

Remember that

$$\mathbf{F} = (\mathbf{F} \cdot \mathbf{a}^i)\mathbf{a}_i, \tag{45}$$

and let

$$\mathbf{F} = \mathbf{a}_2 \times \mathbf{a}_3. \tag{46}$$

Then

$$\begin{aligned}
\mathbf{a}_1 \cdot (\mathbf{a}_2 \times \mathbf{a}_3) &= \mathbf{a}_1 \cdot \mathbf{F} = \mathbf{a}_1 \cdot (\mathbf{F} \cdot \mathbf{a}^i)\mathbf{a}_i \\
&= \mathbf{a}_1 \cdot [(\mathbf{a}_2 \times \mathbf{a}_3) \cdot \mathbf{a}^i]\mathbf{a}_i = [(\mathbf{a}_2 \times \mathbf{a}_3) \cdot \mathbf{a}^i]g_{1i} \\
&= \frac{(\mathbf{a}_2 \times \mathbf{a}_3) \cdot (\mathbf{a}_2 \times \mathbf{a}_3)}{\mathbf{a}_1 \cdot (\mathbf{a}_2 \times \mathbf{a}_3)} g_{11} + \frac{(\mathbf{a}_2 \times \mathbf{a}_3) \cdot (\mathbf{a}_3 \times \mathbf{a}_1)}{\mathbf{a}_1 \cdot (\mathbf{a}_2 \times \mathbf{a}_3)} g_{12} \\
&\quad + \frac{(\mathbf{a}_2 \times \mathbf{a}_3) \cdot (\mathbf{a}_1 \times \mathbf{a}_2)}{\mathbf{a}_1 \cdot (\mathbf{a}_2 \times \mathbf{a}_3)} g_{13}
\end{aligned} \tag{47}$$

or

$$\begin{aligned}
[\mathbf{a}_1 \cdot (\mathbf{a}_2 \times \mathbf{a}_3)]^2 &= (\mathbf{a}_2 \times \mathbf{a}_3) \cdot (\mathbf{a}_2 \times \mathbf{a}_3)g_{11} + (\mathbf{a}_2 \times \mathbf{a}_3) \cdot (\mathbf{a}_3 \times \mathbf{a}_1)g_{12} \\
&\quad + (\mathbf{a}_2 \times \mathbf{a}_3) \cdot (\mathbf{a}_1 \times \mathbf{a}_2)g_{13} \\
&= (\mathbf{a}_2 \cdot \mathbf{a}_2)(\mathbf{a}_3 \cdot \mathbf{a}_3)g_{11} - (\mathbf{a}_2 \cdot \mathbf{a}_3)(\mathbf{a}_3 \cdot \mathbf{a}_2)g_{11} \\
&\quad + (\mathbf{a}_2 \cdot \mathbf{a}_3)(\mathbf{a}_3 \cdot \mathbf{a}_1)g_{12} - (\mathbf{a}_2 \cdot \mathbf{a}_1)(\mathbf{a}_3 \cdot \mathbf{a}_3)g_{12} \\
&\quad + (\mathbf{a}_2 \cdot \mathbf{a}_1)(\mathbf{a}_3 \cdot \mathbf{a}_2)g_{13} - (\mathbf{a}_2 \cdot \mathbf{a}_2)(\mathbf{a}_3 \cdot \mathbf{a}_1)g_{13} \\
&= g_{11}g_{22}g_{33} + g_{23}g_{31}g_{12} + g_{21}g_{32}g_{13} \\
&\quad - g_{23}g_{32}g_{11} - g_{21}g_{33}g_{12} - g_{22}g_{31}g_{13} \\
&= g = \begin{vmatrix} g_{11} & g_{12} & g_{13} \\ g_{21} & g_{22} & g_{23} \\ g_{31} & g_{32} & g_{33} \end{vmatrix}.
\end{aligned} \tag{48}$$

Therefore

$$dV = \sqrt{g}\, du^1\, du^2\, du^3. \tag{49}$$

To employ the reciprocal base system, replace the g_{ij} with g^{ij}.

Clearly the coefficients g_{ij} are sufficient to characterize completely the geometrical properties of space with respect to any coordinate system. To determine these coefficients, represent the rectangular coordinates of a point P by the letters x^1, x^2, x^3; then

$$ds^2 = (dx^1)^2 + (dx^2)^2 + (dx^3)^2. \tag{50}$$

In this coordinate system the g_{ij} become δ_{ij} and the coalescent reciprocal and original systems are indistinguishable.

Now

$$x^1 = x^1(u^1, u^2, u^3), \quad x^2 = x^2(u^1, u^2, u^3), \quad x^3 = x^3(u^1, u^2, u^3),$$

$$\begin{aligned}
dx^1 &= \frac{\partial x^1}{\partial u^1} du^1 + \frac{\partial x^1}{\partial u^2} du^2 + \frac{\partial x^1}{\partial u^3} du^3 \\
dx^2 &= \frac{\partial x^2}{\partial u^1} du^1 + \frac{\partial x^2}{\partial u^2} du^2 + \frac{\partial x^2}{\partial u^3} du^3 \\
dx^3 &= \frac{\partial x^3}{\partial u^1} du^1 + \frac{\partial x^3}{\partial u^2} du^2 + \frac{\partial x^3}{\partial u^3} du^3.
\end{aligned} \tag{51}$$

Thus

$$\begin{aligned}
ds^2 = {} & \left(\frac{\partial x^1}{\partial u^1}\right)^2 du^{1^2} + \left(\frac{\partial x^1}{\partial u^2}\right)^2 du^{2^2} + \left(\frac{\partial x^1}{\partial u^3}\right)^2 du^{3^2} \\
& + 2\left[\frac{\partial x^1}{\partial u^1}\frac{\partial x^1}{\partial u^2} du^1 du^2 + \frac{\partial x^1}{\partial u^1}\frac{\partial x^1}{\partial u^3} du^1 du^3 + \frac{\partial x^1}{\partial u^2}\frac{\partial x^1}{\partial u^3} du^2 du^3\right] \\
& + \left(\frac{\partial x^2}{\partial u^1}\right)^2 du^{1^2} + \left(\frac{\partial x^2}{\partial u^2}\right)^2 du^{2^2} + \left(\frac{\partial x^2}{\partial u^3}\right)^2 du^{3^2} \\
& + 2\left[\frac{\partial x^2}{\partial u^1}\frac{\partial x^2}{\partial u^2} du^1 du^2 + \frac{\partial x^2}{\partial u^1}\frac{\partial x^2}{\partial u^3} du^1 du^3 + \frac{\partial x^2}{\partial u^2}\frac{\partial x^2}{\partial u^3} du^2 du^3\right] \\
& + \left(\frac{\partial x^3}{\partial u^1}\right)^2 du^{1^2} + \left(\frac{\partial x^3}{\partial u^2}\right)^2 du^{2^2} + \left(\frac{\partial x^3}{\partial u^3}\right)^2 du^{3^2} \\
& + 2\left[\frac{\partial x^3}{u^1}\frac{\partial x^3}{\partial u^2} du^1 du^2 + \frac{\partial x^3}{\partial u^1}\frac{\partial x^3}{\partial u^3} du^1 du^3 + \frac{\partial x^3}{\partial u^2}\frac{\partial x^3}{\partial u^3} du^2 du^3\right].
\end{aligned} \tag{52}$$

Therefore

$$\begin{aligned}
g_{11} &= \left(\frac{\partial x^1}{\partial u^1}\right)^2 + \left(\frac{\partial x^2}{\partial u^1}\right)^2 + \left(\frac{\partial x^3}{\partial u^1}\right)^2, \\
g_{22} &= \left(\frac{\partial x^1}{\partial u^2}\right)^2 + \left(\frac{\partial x^2}{\partial u^2}\right)^2 + \left(\frac{\partial x^3}{\partial u^2}\right)^2, \\
g_{12} &= \frac{\partial x^1}{\partial u^1}\frac{\partial x^1}{\partial u^2} + \frac{\partial x^2}{\partial u^1}\frac{\partial x^2}{\partial u^2} + \frac{\partial x^3}{\partial u^1}\frac{\partial x^3}{\partial u^2},
\end{aligned} \tag{53}$$

or, in general,

$$\begin{aligned} g_{ij} &= \frac{\partial x^1}{\partial u^i}\frac{\partial x^1}{\partial u^j} + \frac{\partial x^2}{\partial u^i}\frac{\partial x^2}{\partial u^j} + \frac{\partial x^3}{\partial u^i}\frac{\partial x^3}{\partial u^j} \\ &= \frac{\partial x^k}{\partial u^i}\frac{\partial x^k}{\partial u^j}. \end{aligned} \tag{54}$$

A cursory examination of the differential relationship between coordinate systems is informative. Consider two sets of coordinates:

$$x^1, x^2, x^3 \equiv x^\mu; \qquad \mu = 1, 2, 3; \tag{55}$$

and

$$x'^1, x'^2, x'^3 \equiv x'^\nu; \qquad \nu = 1, 2, 3. \tag{56}$$

where x is not to be associated with the first cartesian coordinate in an (x, y, z) system. The integer superscripts μ and ν range from 1 to n inclusive, where n is the dimensionality of the space. These coordinates are not necessarily the components of a readily identifiable vector, such as the position vector $\mathbf{r} \equiv \mathbf{x}$.

The two sets of coordinates are related as follows:

$$\begin{aligned} x^\mu &= f^\mu(x'^\nu), \\ x'^\nu &= f'^\nu(x^\mu). \end{aligned} \tag{57}$$

Assume that the coordinates are generalized coordinates like q, curvilinear, and not necessarily orthogonal. Take differential lengths of each coordinate and relate them to the other system, as follows:

$$\begin{aligned} dx'^1 &= \frac{\partial x'^1}{\partial x^1}dx^1 + \frac{\partial x'^1}{\partial x^2}dx^2 + \frac{\partial x'^1}{\partial x^3}dx^3, \\ dx'^2 &= \frac{\partial x'^2}{\partial x^1}dx^1 + \frac{\partial x'^2}{\partial x^2}dx^2 + \frac{\partial x'^2}{\partial x^3}dx^3, \end{aligned} \tag{58}$$

and

$$dx'^3 = \frac{\partial x'^3}{\partial x^1}dx^1 + \frac{\partial x'^3}{\partial x^2}dx^2 + \frac{\partial x'^3}{\partial x^3}dx^3;$$

or:

$$dx'^\nu - \frac{\partial x'^\nu}{\partial x^\mu}dx^\mu = M_\mu{}^\nu dx^\mu. \tag{59}$$

Looking at the equations the other way around,

$$
\begin{aligned}
dx^1 &= \frac{\partial x^1}{\partial x'^1}dx'^1 + \frac{\partial x^1}{\partial x'^2}dx'^2 + \frac{\partial x^1}{\partial x'^3}dx'^3 \\
dx^2 &= \frac{\partial x^2}{\partial x'^1}dx'^1 + \frac{\partial x^2}{\partial x'^2}dx'^2 + \frac{\partial x^2}{\partial x'^3}dx'^3 \\
dx^3 &= \frac{\partial x^3}{\partial x'^1}dx'^1 + \frac{\partial x^3}{\partial x'^2}dx'^2 + \frac{\partial x^3}{\partial x'^3}dx'^3,
\end{aligned}
\tag{60}
$$

or:

$$
dx^\nu = \frac{\partial x^\mu}{\partial x'^\nu}dx'^\nu = N_\nu{}^\mu\, dx'^\nu, \tag{61}
$$

where the matrices M and N are different.

Now consider a scalar point function

$$
\phi(x^\mu) = \phi(x'^\nu) = \phi; \tag{62}
$$

form the gradient and express it in a different system:

$$
\begin{aligned}
\frac{\partial\phi}{\partial x^1} &= \frac{\partial\phi}{\partial x'^1}\frac{\partial x'^1}{\partial x^1} + \frac{\partial\phi}{\partial x'^2}\frac{\partial x'^2}{\partial x^1} + \frac{\partial\phi}{\partial x'^3}\frac{\partial x'^3}{\partial x^1}, \\
\frac{\partial\phi}{\partial x^2} &= \frac{\partial\phi}{\partial x'^1}\frac{\partial x'^1}{\partial x^2} + \frac{\partial\phi}{\partial x'^2}\frac{\partial x'^2}{\partial x^2} + \frac{\partial\phi}{\partial x'^3}\frac{\partial x'^3}{\partial x^2} \\
\frac{\partial\phi}{\partial x^3} &= \frac{\partial\phi}{\partial x'^1}\frac{\partial x'^1}{\partial x^3} + \frac{\partial\phi}{\partial x'^2}\frac{\partial x'^2}{\partial x^3} + \frac{\partial\phi}{\partial x'^3}\frac{\partial x'^3}{\partial x^3},
\end{aligned}
\tag{63}
$$

or:

$$
\frac{\partial\phi}{\partial x^\mu} = \frac{\partial x'^\nu}{\partial x^\mu}\frac{\partial\phi}{\partial x'^\nu} = M_\mu{}^\nu \frac{\partial\phi}{\partial x'^\nu}, \tag{64}
$$

or:

$$
\nabla_\mu\phi = \frac{\partial x'^\nu}{\partial x^\mu}\nabla'_\nu\phi = M_\mu{}^\nu\nabla'_\nu\phi. \tag{65}
$$

The next three relationships are inverse:

$$
\begin{aligned}
\frac{\partial\phi}{\partial x'^1} &= \frac{\partial\phi}{\partial x^1}\frac{\partial x^1}{\partial x'^1} + \frac{\partial\phi}{\partial x^2}\frac{\partial x^2}{\partial x'^1} + \frac{\partial\phi}{\partial x^3}\frac{\partial x^3}{\partial x'^1} \\
\frac{\partial\phi}{\partial x'^2} &= \frac{\partial\phi}{\partial x^1}\frac{\partial x^1}{\partial x'^2} + \frac{\partial\phi}{\partial x^2}\frac{\partial x^2}{\partial x'^2} + \frac{\partial\phi}{\partial x^3}\frac{\partial x^3}{\partial x'^2},
\end{aligned}
\tag{66}
$$

and

$$\frac{\partial \phi}{\partial x'^3} = \frac{\partial \phi}{\partial x^1}\frac{\partial x^1}{\partial x'^3} + \frac{\partial \phi}{\partial x^2}\frac{\partial x^2}{\partial x'^3} + \frac{\partial \phi}{\partial x^3}\frac{\partial x^3}{\partial x'^3},$$

or:

$$\frac{\partial \phi}{\partial x'^\nu} = \frac{\partial x^\mu}{\partial x'^\nu}\frac{\partial \phi}{\partial x^\mu} = N_\nu{}^\mu \frac{\partial \phi}{\partial x^\mu}, \tag{67}$$

or:

$$\nabla'_\nu \phi = \frac{\partial x^\nu}{\partial x'^\nu} \nabla_\mu \phi = N_\nu{}^\mu \nabla_\mu \phi. \tag{68}$$

Consider all four relationships in symbolic form:

$$\begin{aligned}
&(1) \quad dx^\mu = N_\nu{}^\mu dx'^\nu, \quad \text{or} \quad x^\mu = N_\nu{}^\mu x'^2, \quad \text{or} \quad X = NX'.\\
&(2) \quad dx'^\nu = M_\mu{}^\nu dx^\mu, \quad \text{or} \quad x'^\nu = M_\mu{}^\nu x^\nu, \quad \text{or} \quad X' = MX.\\
&(3) \quad \nabla_\mu \phi = M_\mu{}^\nu \nabla'_\nu \phi, \quad \text{or} \quad \nabla \phi = M \nabla' \phi.\\
&(4) \quad \nabla'_\nu \phi = N_\nu{}^\mu \nabla_\mu \phi, \quad \text{or} \quad \nabla' \phi = N \nabla \phi.
\end{aligned} \tag{69}$$

Now (1) and (4) are cogredient to each other and (2) and (3) are cogredient to each other. Furthermore, if $M = N^{-1}$, then (1) is contragredient to (3) and (2) is contragredient to (4).

C. Covariance and Contravariance

Mechanics, electromagnetic theory, relativity, and quantum mechanics would be either sterile or severely limited without such things as coordinates, scalars, vectors, and tensors. Created by the more mathematically minded physicists, these quantities describe the laws of physics in a simple and powerful manner.

To be correct, a physical theory must be expressible in laws that are covariant with respect to Lorentz transformations. The meaning of this statement will be explored in greater detail later. Let it suffice for now to say that a physical law must be invariant in form for any observer when expressed in his coordinate system. Scalar numbers, the measurable quantities in physics, are examples of other invariants.

Covariance and contravariance refer to the transformation properties of vectors and tensors. First introduce the two- and three-dimensional cartesian coordinate systems shown in Fig. 5. Next, introduce vectors into these coordinate systems, as in Fig. 6. The vector components may be shown explicitly (see Fig. 7). The concept of a vector in cartesian coordinates may not

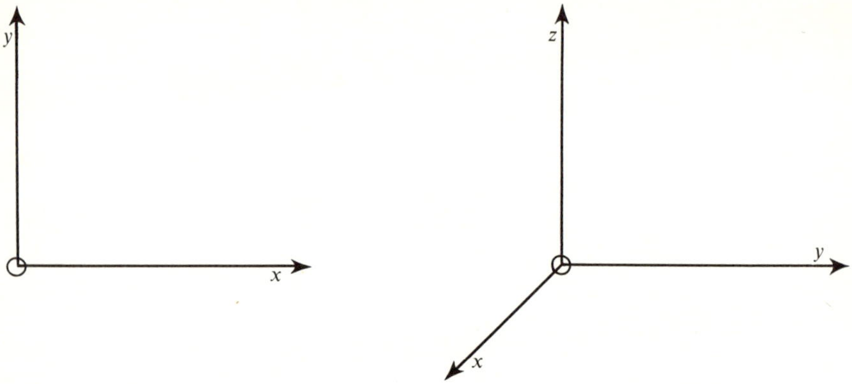

(*a*) Two-dimensional x, y coordinate system.

(*b*) Three-dimensional x, y, z coordinate system.

Figure 5

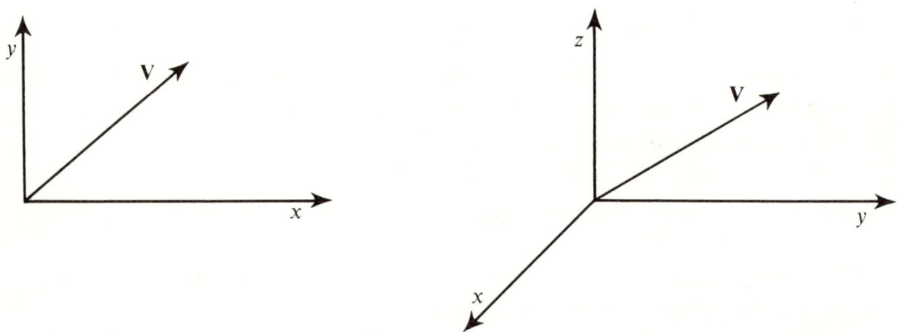

(*a*) Vector **V** in two-dimensional space.

(*b*) Vector **V** in three-dimensional space.

Figure 6

be hard to understand. Vectors are not always this simple but may have a complicated physical interpretation. Moreover, the coordinate system may not always be cartesian. Some noncartesian coordinate systems are linear (but not orthogonal), curvilinear (orthogonal), and curvilinear (but not orthogonal). General coordinate systems have any number of dimensions N (see Fig. 8).

In order to reduce the concepts of covariance and contravariance to their simplest terms, consider two-dimensional linear coordinates. The same vector **V** may be expressed in both rectangular and oblique coordinate systems, as shown in Fig. 9. If unit vectors parallel to the coordinates are introduced in

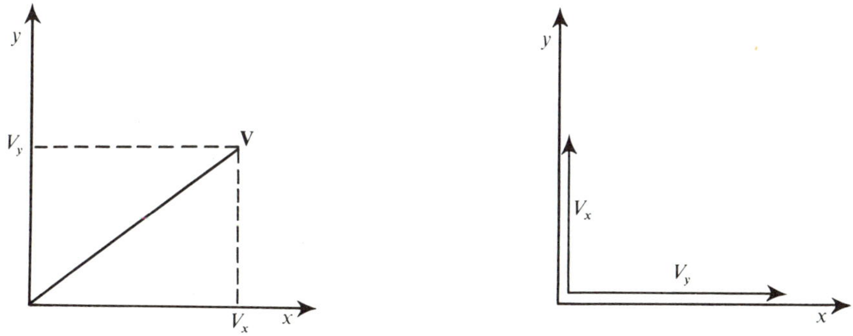

(a) Resolution of **V** into two components.

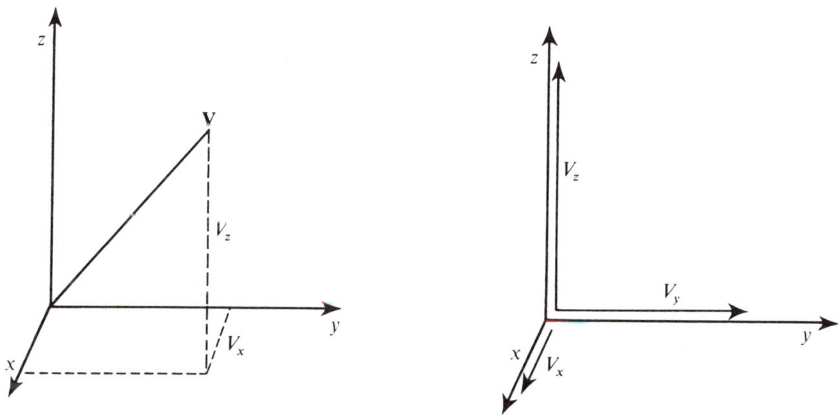

(b) Resolution of **V** into three components.

Figure 7

each of these systems, then:

$$
\left.\begin{aligned}
&\hat{\mathbf{i}}\cdot\hat{\mathbf{i}}=\hat{\mathbf{j}}\cdot\hat{\mathbf{j}}=\hat{\mathbf{k}}\cdot\hat{\mathbf{k}}=1\\
&\hat{\mathbf{i}}\cdot\hat{\mathbf{j}}=\hat{\mathbf{i}}\cdot\hat{\mathbf{k}}=\hat{\mathbf{j}}\cdot\hat{\mathbf{k}}=0\\
&\hat{\mathbf{i}}\times\hat{\mathbf{i}}=\hat{\mathbf{j}}\times\hat{\mathbf{j}}=\hat{\mathbf{k}}\times\hat{\mathbf{k}}=0\\
&\hat{\mathbf{i}}\times\hat{\mathbf{j}}=-(\hat{\mathbf{j}}\times\hat{\mathbf{i}})=\mathbf{k}\\
&\hat{\mathbf{j}}\times\mathbf{k}=-(\hat{\mathbf{k}}\times\hat{\mathbf{j}})=\hat{\mathbf{i}}\\
&\hat{\mathbf{k}}\times\hat{\mathbf{i}}=-(\hat{\mathbf{i}}\times\hat{\mathbf{k}})=\hat{\mathbf{j}}
\end{aligned}\right\}\ \text{in the } (x, y, z) \text{ system}
\qquad
\text{rules to be defined later for } \hat{\mathbf{e}}_1, \hat{\mathbf{e}}_2, \hat{\mathbf{e}}_3 \ \text{in the } (u_1 u_2 u_3) \text{ system.}
\tag{2}
$$

As a consequence, the same vector **V** will have different components in the

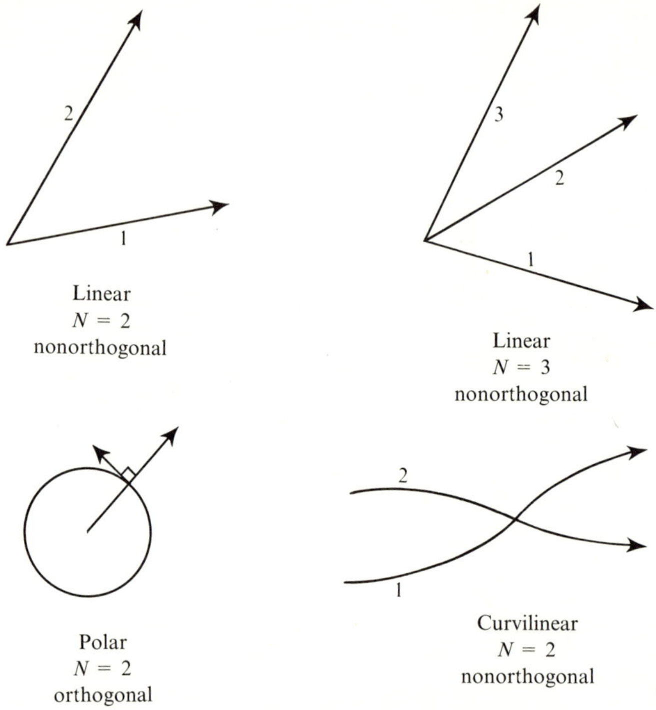

Figure 8. Types of coordinate systems.

two coordinate systems, shown in Fig. 10. Several important factors in the present treatment are:

1. The base vectors $\hat{\mathbf{e}}_i$ need not be unit vectors.

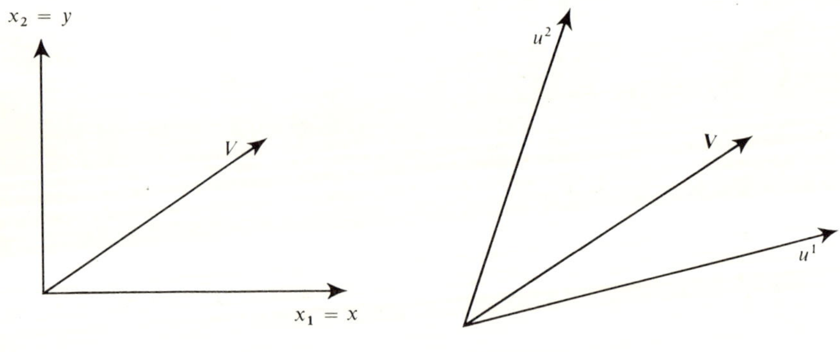

Figure 9

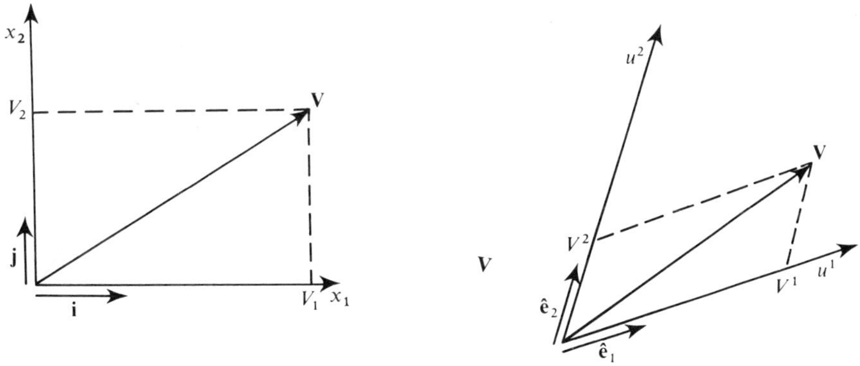

Figure 10

2. The third, or $x_3 = u^3 = z$, coordinate is taken perpendicular to the page for convenience in defining some vector products. Notice that this produces a partially orthogonal coordinate system u^1, u^2, u^3. This is necessary to keep all of the vector products in the plane of the paper.

3. The vector components in the u coordinate system have superscripts instead of subscripts.

When cartesian coordinates are used, it is obvious that the scalar product of **V** with itself is

$$V^2 = \mathbf{V} \cdot \mathbf{V} = V_x{}^2 + V_y{}^2 = V_1{}^2 + V_2{}^2 = \sum_{i=1}^{2} V_i^2 = V_i^2. \tag{3}$$

But when nonorthogonal coordinates are used, it is equally obvious that the scalar product is *not* simply the sum of squares:

$$V^2 = \mathbf{V} \cdot \mathbf{V} \neq V^{1^2} + V^{2^2} = \sum_{j=1}^{2} V^{j^2} = V^{j^2}. \tag{4}$$

The components V^j are called the contravariant components of the vector **V**. Elsewhere **V** will be defined using nonunit base vectors, introduced in a formal way.

In order to define a scalar product in a nonorthogonal coordinate system, it is convenient to define what is called a "reciprocal lattice system." For this purpose consider the volume of a small elementary parallelopiped in the original system:

$$v = \hat{\mathbf{e}}_1 \cdot (\hat{\mathbf{e}}_2 \times \hat{\mathbf{e}}_3) = \hat{\mathbf{e}}_2 \cdot (\hat{\mathbf{e}}_3 \times \hat{\mathbf{e}}_1) = \hat{\mathbf{e}}_3 \cdot (\hat{\mathbf{e}}_1 \times \hat{\mathbf{e}}_2). \tag{5}$$

Also consider the following vector products:

$$\mathbf{a}^1 = \frac{\hat{\mathbf{e}}_2 \times \hat{\mathbf{e}}_3}{v} = \frac{\hat{\mathbf{e}}_2 \times \hat{\mathbf{e}}_3}{\hat{\mathbf{e}}_1 \cdot (\hat{\mathbf{e}}_2 \times \hat{\mathbf{e}}_3)},$$

$$\mathbf{a}^2 = \frac{\hat{\mathbf{e}}_3 \times \hat{\mathbf{e}}_1}{v} = \frac{\hat{\mathbf{e}}_3 \times \hat{\mathbf{e}}_1}{\hat{\mathbf{e}}_1 \cdot (\hat{\mathbf{e}}_2 \times \hat{\mathbf{e}}_3)}, \tag{6}$$

and

$$\mathbf{a}^3 = \frac{\hat{\mathbf{e}}_1 \times \hat{\mathbf{e}}_2}{v} = \frac{\hat{\mathbf{e}}_1 \times \hat{\mathbf{e}}_2}{\hat{\mathbf{e}}_1 \cdot (\hat{\mathbf{e}}_2 \times \hat{\mathbf{e}}_3)}.$$

Notice that the new vectors $\mathbf{a}^1$, $\mathbf{a}^2$, $\mathbf{a}^3$ have superscripts. Although called a "set of contravariant base vectors" in the dual space or reciprocal lattice, they are not, in general, unit vectors. Their inner product with the original unit vectors is

$$\hat{\mathbf{e}}_1 \cdot \mathbf{a}^1 = \frac{\hat{\mathbf{e}}_1 \cdot (\hat{\mathbf{e}}_2 \times \hat{\mathbf{e}}_3)}{v} = \frac{\hat{\mathbf{e}}_1 \cdot (\hat{\mathbf{e}}_2 \times \hat{\mathbf{e}}_3)}{\hat{\mathbf{e}}_1 \cdot (\hat{\mathbf{e}}_2 \times \hat{\mathbf{e}}_3)} = 1. \tag{7}$$

Similarly,

$$\hat{\mathbf{e}}_2 \cdot \mathbf{a}^2 = 1, \qquad \hat{\mathbf{e}}_3 \cdot \mathbf{a}^3 = 1. \tag{8}$$

The original unit vectors $\hat{\mathbf{e}}_i$ are called "covariant unit vectors." Rewrite equation (8) as

$$\hat{\mathbf{e}}_i \cdot \mathbf{a}^i = 1, \tag{9}$$

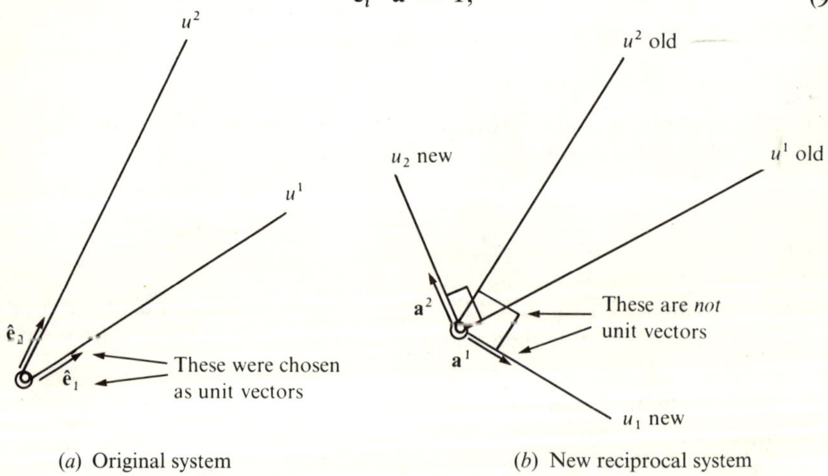

(*a*) Original system (*b*) New reciprocal system

Figure 11. Relationship between a coordinate system and its dual or reciprocal coordinate system.

no Einstein sum convention intended! Notice that $\mathbf{a}^1$ is perpendicular to $\hat{\mathbf{e}}_2$ and that $\mathbf{a}^2$ is perpendicular to $\hat{\mathbf{e}}_1$. Both $\mathbf{a}^1$ and $\mathbf{a}^2$ are generally in the direction of $\hat{\mathbf{e}}_1$ and $\hat{\mathbf{e}}_2$, respectively. Figure 11 shows the situation existing between the base vectors and the *two* nonorthogonal coordinate systems.

A corresponding relationship exists for the base vectors of the cartesian coordinate system. Remember that for the given initial unit base vectors $\hat{\mathbf{i}}, \hat{\mathbf{j}}, \hat{\mathbf{k}}$,

$$v = \hat{\mathbf{i}} \cdot (\hat{\mathbf{j}} \times \hat{\mathbf{k}}) = \hat{\mathbf{j}} \cdot (\hat{\mathbf{k}} \times \hat{\mathbf{i}}) = \hat{\mathbf{k}} \cdot (\hat{\mathbf{i}} \times \hat{\mathbf{j}}) = 1, \tag{10}$$

and that the reciprocal base vectors (6) are also unit vectors:

$$\hat{\mathbf{i}} = \frac{(\hat{\mathbf{j}} \times \hat{\mathbf{k}})}{v} = \hat{\mathbf{j}} \times \hat{\mathbf{k}}, \quad \hat{\mathbf{j}} = \frac{(\hat{\mathbf{k}} \times \hat{\mathbf{i}})}{v} = \hat{\mathbf{k}} \times \hat{\mathbf{i}}, \quad \hat{\mathbf{k}} = \frac{(\hat{\mathbf{i}} \times \hat{\mathbf{j}})}{v} = \hat{\mathbf{i}} \times \hat{\mathbf{j}}. \tag{11}$$

Thus the coordinate system is identical with its dual or reciprocal system. Since it maps into itself, a cartesian system is called "self-dual" and is a special case of Fig. 11:

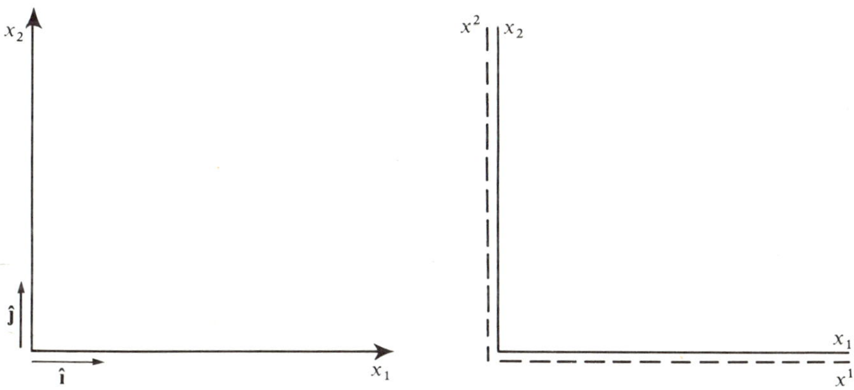

The distinction between covariant and contravariant vector components vanishes for cartesian coordinates. It does not really matter whether a superscript or a subscript notation is used.

The components of a vector in nonorthogonal coordinates are shown in Fig. 12. The proper way to form a scalar product in this situation is as follows: In a linear space u^i and its reciprocal space u_i define contravariant and covariant vector components v^i and v_i. Now multiply:

$$\begin{aligned} V^2 &= \mathbf{V} \cdot \mathbf{V} = v^1 v_1 + v^2 v_2 = v_1 v^1 + v_2 v^2 \\ &= \sum_{i=1}^{2} v_i v^i = \sum_{i=1}^{2} v^i v_i = v^i v_i = v_i v^i, \end{aligned} \tag{12}$$

Einstein summation convention intended.

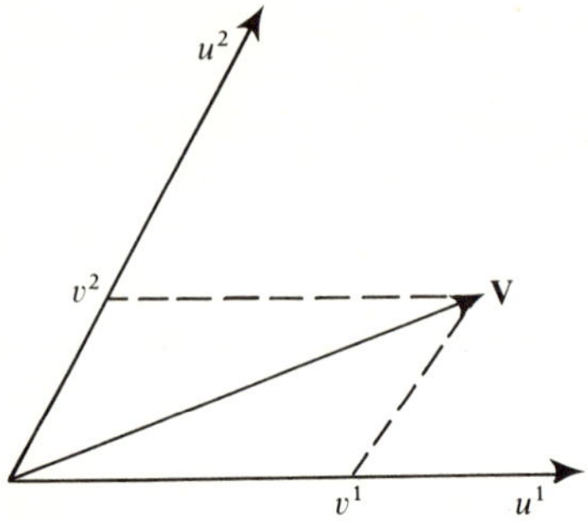

(*a*) Contravariant vector components in original space.

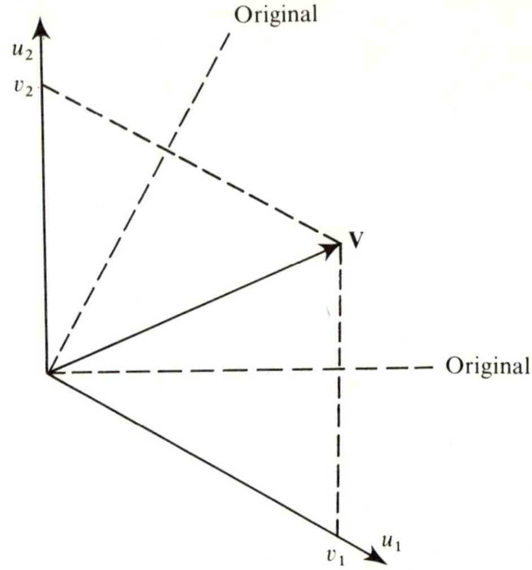

(*b*) Covariant vector components in dual or reciprocal coordinate space.

Figure 12. Covariant and contravariant decompositions of a vector **V**. (The dotted lines in the contravariant components are parallel to the u^i directions because the $\hat{\mathbf{e}}_i$ are unit vectors; thus $v^i = V_i$.)

To clarify Fig. 12, Fig. 13 shows the components of the vector another way. The contravariant components are defined by the dotted sides of a parallelogram and the covariant components are defined by the sides of a

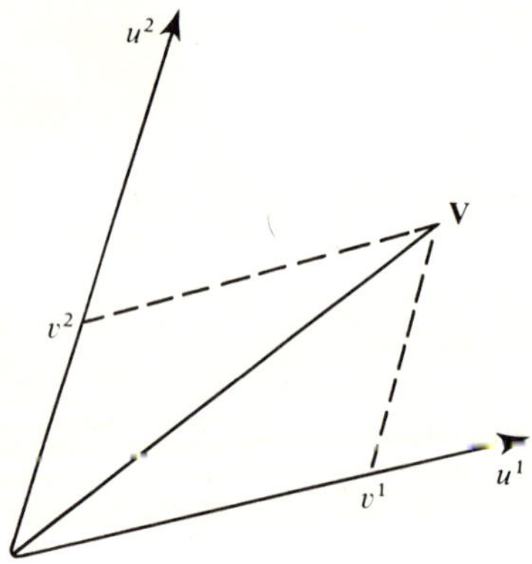

(*a*) Contravariant components

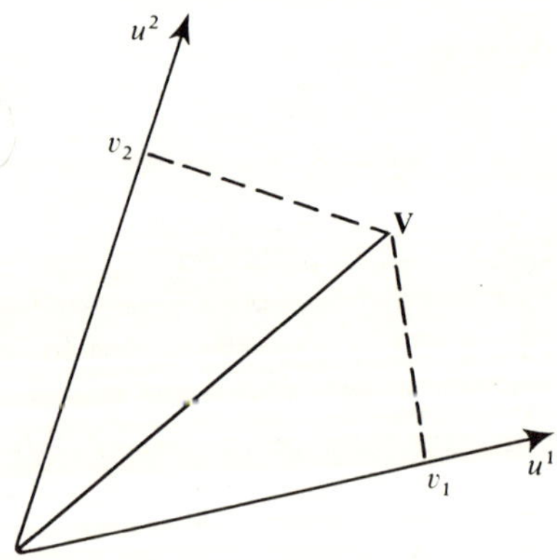

(*b*) Covariant components

Figure 13. Covariant and contravariant components of a vector superimposed on the original coordinate system.

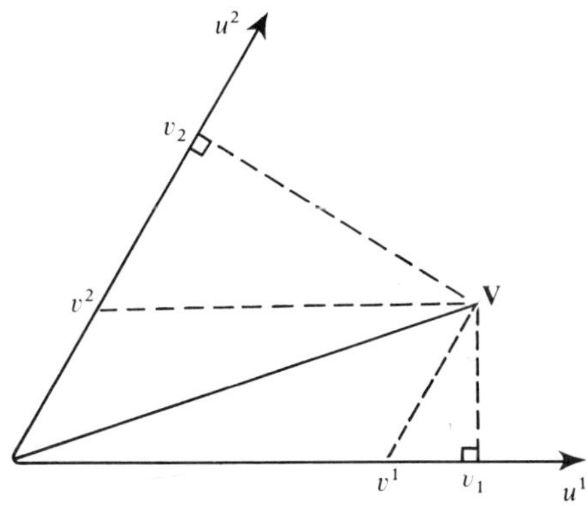

Figure 14. Covariant and contravariant components superimposed.

trapezoid with two right angles. It is instructive to superimpose the components (see Fig. 14).

The vector **V** in the cartesian system x_i and in the nonorthogonal system u^i are compared in Fig. 15. The cartesian component $V_2 = V^2$ lies somewhere intermediate to the covariant and contravariant component. In the symmetric example shown in Fig. 15, **V** appears to be at a 45° angle to x and y; it also is symmetric with respect to u^1 and u^2. If the vector **V** is taken to be of unit length, then

$$V_1 = V^1 = V_2 = V^2 = 0{\cdot}707 \tag{13}$$

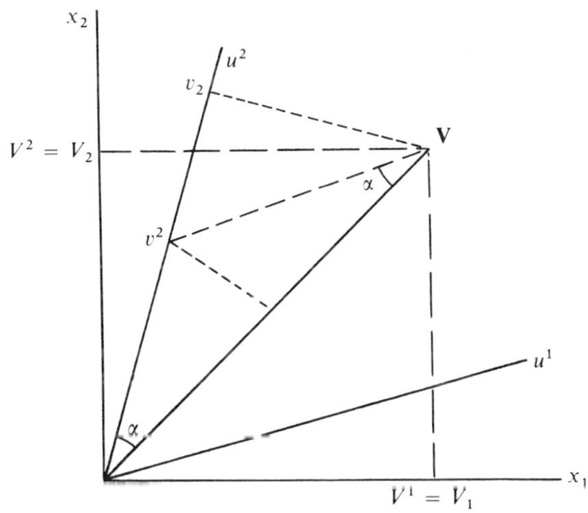

Figure 15. Comparison of three different components for **V**.

and

$$V_1{}^2 = V_2{}^2 = 0{\cdot}5. \tag{14}$$

From the isosceles triangle rule, it follows that

$$\frac{0{\cdot}5}{v^2} = \cos\alpha \qquad \text{or} \quad v^2 = \frac{0{\cdot}5}{\cos\alpha} \tag{15}$$

and

$$\frac{v_2}{1} = \cos\alpha \qquad \text{or} \quad v_2 = \cos\alpha, \tag{16}$$

which lead to the product

$$v_2 v^2 = 0{\cdot}5\,\frac{\cos\alpha}{\cos\alpha} = 0{\cdot}5. \tag{17}$$

In a like manner,

$$v_1 v^1 = 0{\cdot}5, \tag{18}$$

so that

$$\underbrace{V_1{}^2 + V_2{}^2}_{\text{cartesian}} = \underbrace{v_1 v^1 + v_2 v^2}_{\substack{\text{nonorthogonal}\\ \text{but linear}}} = 1, \tag{19}$$

a formula transcending this special case.

Finally, to improve on the definition of the symmetry between the dual coordinate systems, the inverse of equation (6) is

$$\begin{aligned}
\hat{\mathbf{e}}_1 &= \frac{\mathbf{a}^2 \times \mathbf{a}^3}{v^1} = \frac{\mathbf{a}^2 \times \mathbf{a}^3}{\mathbf{a}^1 \cdot (\mathbf{a}^2 \times \mathbf{a}^3)},\\
\hat{\mathbf{e}}_2 &= \frac{\mathbf{a}^3 \times \mathbf{a}^1}{v^1} = \frac{\mathbf{a}^3 \times \mathbf{a}^1}{\mathbf{a}^1 \cdot (\mathbf{a}^2 \times \mathbf{a}^3)},\\
&\text{and}\\
\hat{\mathbf{e}}_3 &= \frac{\mathbf{a}^1 \times \mathbf{a}^2}{v^1} = \frac{\mathbf{a}^1 \times \mathbf{a}^2}{\mathbf{a}^1 \cdot (\mathbf{a}^2 \times \mathbf{a}^3)}.
\end{aligned} \tag{20}$$

The two coordinate systems are equivalent *and* reciprocal. Therefore, either could have been defined in terms of the other.

The meaning of covariant and contravariant can be interchanged. But once a choice is made, there is a distinct physical difference between the two decompositions. This difference will be discussed later.

The objects of interest in physics are moving or changing. Being able to express a vector or tensor in a proper coordinate system requires accurate expressions for its covariant and contravariant components. Motion or change is often expressed by a coordinate transformation, and one such expression is the Lorentz transformation of relativity. This transformation may be thought of either as a rotation or as a distortion of a coordinate system, depending on the choice of real or imaginary time as the fourth dimension.

1. Manifestly covariant form of a 4-vector with *ict*:

$$\mathbf{r} = (x, y, z, t) \Rightarrow (x, y, z, ict) = (x_1, x_2, x_3, x_4), \tag{21}$$

and

$$\begin{aligned} r^2 &= \mathbf{r} \cdot \mathbf{r} = x_1{}^2 + x_2{}^2 + x_3{}^2 + x_4{}^2 = x^2 + y^2 + z^2 + (ict)^2 \\ &= x^2 + y^2 + z^2 - c^2t^2 = r^2 - c^2t^2. \end{aligned} \tag{22}$$

2. Bivariant form with *ct*:

$$\begin{aligned} \mathbf{r}^i &= (x, y, z, t) \Rightarrow (x, y, z, ct) = (x_1, x_2, x_3, x_0), && \text{contravariant;} \\ \mathbf{r}_i &= (x, y, z, -ct) = (x_1, x_2, x_3, -x_0), && \text{covariant.} \end{aligned} \tag{23}$$

Thus, in both cases (1) and (2), the scalar product is the same:

$$\mathbf{r}_i\mathbf{r}^i = x_1{}^2 + x_2{}^2 + x_3{}^2 - x_0{}^2 = x^2 + y^2 + z^2 - c^2t^2 = r^2 - c^2t^2. \tag{24}$$

Next consider the covariance and contravariance for nonorthogonal coordinates. For the unprimed and primed coordinate systems I and II shown in Fig. 16, the position vector $\mathbf{r}$ may be written

$$\mathbf{r} = \hat{\mathbf{i}}x + \hat{\mathbf{j}}y = \hat{\mathbf{i}}\left(x' - \frac{1}{\sqrt{2}}y'\right) + \hat{\mathbf{j}}y' = \hat{\mathbf{i}}x' + \left(\hat{\mathbf{j}} - \frac{1}{\sqrt{2}}\hat{\mathbf{i}}\right)y'. \tag{25}$$

The derivative,

$$\frac{\partial \mathbf{r}}{\partial x'} = \hat{\mathbf{i}} = \mathbf{a}_{x'}, \tag{26}$$

a unitary vector, is also a unit vector. Since

$$\frac{\partial \mathbf{r}}{\partial y'} = \left(\hat{\mathbf{j}} - \frac{1}{\sqrt{2}}\hat{\mathbf{i}}\right) = \mathbf{a}_{y'}, \tag{27}$$

it follows that

$$d\mathbf{r} = \mathbf{a}_{x'}\,dx' + \mathbf{a}_{y'}\,dy'. \tag{28}$$

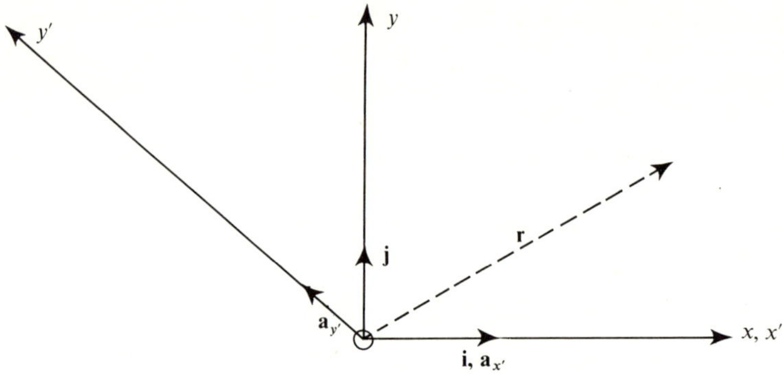

Figure 16. System I: x, y. System II: $x' = x$, y'.

Introduce the unit vector **k** in the z direction. Then

$$\begin{aligned} \hat{\imath} \times \hat{\jmath} &= \hat{k}, & \hat{\jmath} \times \hat{k} &= \hat{\imath}, & \hat{k} \times \hat{\imath} &= \hat{\jmath}, \\ \hat{\jmath} \times \hat{\imath} &= -\hat{k}, & \hat{k} \times \hat{\jmath} &= -\hat{\imath}, & \hat{\imath} \times \hat{k} &= -\hat{\jmath}. \end{aligned} \tag{29}$$

Form

$$\hat{\imath} \cdot \left[\left(\hat{\jmath} - \frac{\hat{\imath}}{\sqrt{2}}\right) \times \hat{k}\right] = V = \hat{\imath} \cdot \left(\hat{\imath} + \frac{\hat{\jmath}}{\sqrt{2}}\right) = 1. \tag{30}$$

Call

$$\mathbf{a}_1 = \hat{\imath}, \qquad \mathbf{a}_2 = \left(\hat{\jmath} - \frac{\hat{\imath}}{\sqrt{2}}\right), \tag{31}$$

and

$$\mathbf{a}_3 = \hat{k}. \tag{32}$$

Then

$$\mathbf{a}^1 = \frac{\mathbf{a}_2 \times \mathbf{a}_3}{V} = \frac{\left(\hat{\jmath} - \frac{\hat{\imath}}{\sqrt{2}}\right) \times \hat{k}}{1} = \left(\hat{\imath} + \frac{\hat{\jmath}}{\sqrt{2}}\right). \tag{33}$$

The length of a is 1·5.

$$\mathbf{a}^2 = \frac{\mathbf{a}_3 \times \mathbf{a}_1}{V} = \frac{\hat{k} \times \hat{\imath}}{1} = \hat{\jmath},$$

$$\mathbf{a}^3 = \frac{\mathbf{a}_1 \times \mathbf{a}_2}{V} = \frac{\hat{\imath} \times \left(\hat{\jmath} - \frac{\hat{\imath}}{\sqrt{2}}\right)}{1} = \hat{k}, \tag{34}$$

$$\mathbf{a}_i \cdot \mathbf{a}^j = \delta_i^{\,j}.$$

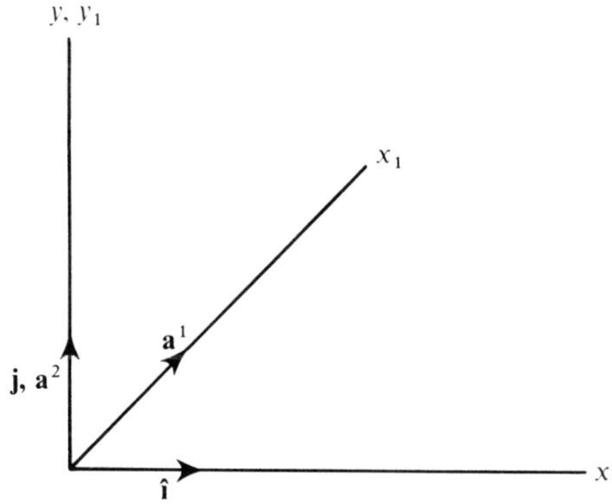

Figure 17. System I: x, y. System III: x_1, $y_1 = y$.

For two sets of unitary vectors

$\mathbf{a}_1$	$\mathbf{a}_2$	$\mathbf{a}_3$
$\hat{\mathbf{i}}$		$\mathbf{k}$

(35)

and

$\mathbf{a}^1$	$\mathbf{a}^2$	$\mathbf{a}^3$
	$\hat{\mathbf{j}}$	$\hat{\mathbf{k}}$

, (36)

write the vector

$$\begin{aligned} \mathbf{F} &= F_x\hat{\mathbf{i}} + F_y\hat{\mathbf{j}} + F_z\mathbf{k} = F_1\hat{\mathbf{e}}_1 + F_2\hat{\mathbf{e}}_2 + F_3\hat{\mathbf{e}}_3 \\ &= F^1\hat{\mathbf{e}}^1 + F^2\hat{\mathbf{e}}^2 + F^3\hat{\mathbf{e}}^3 \\ &= f^1\mathbf{a}_1 + f^2\mathbf{a}_2 + f^3\mathbf{a}_3 = f_1\mathbf{a}^1 + f_2\mathbf{a}^2 + f_3\mathbf{a}^3 \end{aligned} \tag{37}$$

where f^i are the contravariant components of $\mathbf{F}$ for the unitary basis $\mathbf{a}_i$, and f_i are the covariant components of $\mathbf{F}$ for the reciprocal unitary basis $\mathbf{a}^i$. The components F^i or F_i are "semi-," "co-," or "contra-" components for unit sets $\hat{\mathbf{e}}_i$ or $\hat{\mathbf{e}}^i$. Now the components:

$$f^i - \mathbf{F}\cdot\mathbf{a}^i \tag{38}$$

and

$$f_i = \mathbf{F}\cdot\mathbf{a}_i \tag{39}$$

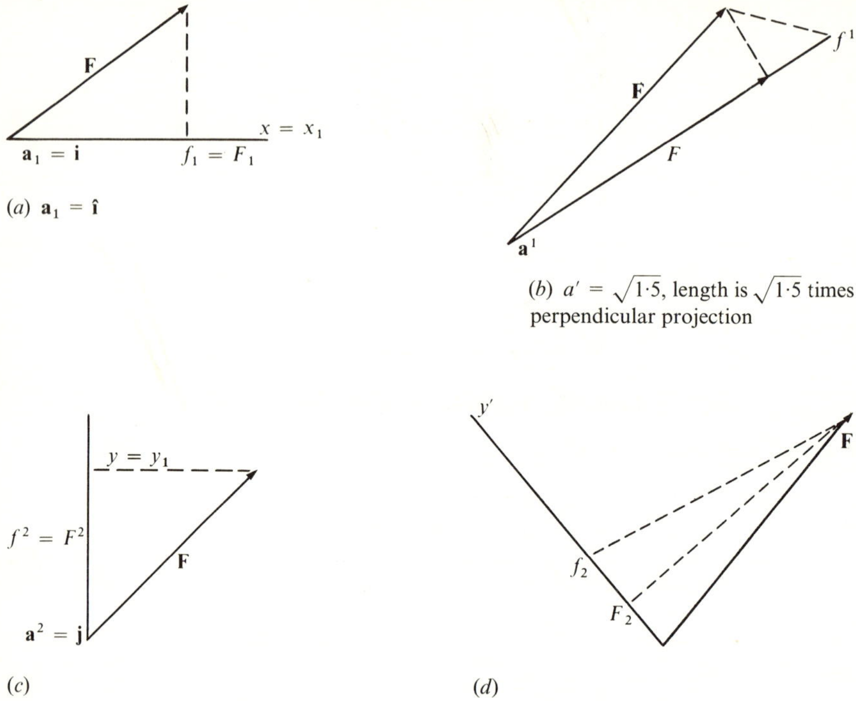

Figure 18. Four views of the vector F.

are obtained from the same class basis vector, but they are then

$$\mathbf{F} = f^i\mathbf{a}_i = f_i\mathbf{a}^i \tag{40}$$

used with the opposite class basis, as shown in Fig. 18.

In cartesian coordinates, the length of the two-dimensional vector **F** is given by

$$|\mathbf{F}| = \sqrt{\mathbf{F}\cdot\mathbf{F}}, \tag{41}$$

the square root of the vector product of **F** and itself. Thus

$$\mathbf{F}\cdot\mathbf{F} = F_x{}^2 + F_y{}^2 = (F_1)^2 + (F_2{}^2)^2. \tag{42}$$

In nonorthogonal coordinates, however, the length is given by

$$\mathbf{F}\cdot\mathbf{F} = f_if^i = f_1f^1 + f_2f^2. \tag{43}$$

To see this, first consider the dot products of the unitary vectors:

$$\mathbf{a}_1 \cdot \mathbf{a}^1 = (\hat{\imath}) \cdot \left(\hat{\imath} + \frac{\hat{\jmath}}{\sqrt{2}}\right) = 1 = \mathbf{a}_2 \cdot \mathbf{a}^2 = \left(\hat{\jmath} - \frac{\hat{\imath}}{\sqrt{2}}\right) \cdot (\hat{\jmath}),$$

$$\mathbf{a}_1 \cdot \mathbf{a}^2 = \hat{\imath} \cdot \hat{\jmath} = 0 = \mathbf{a}_2 \cdot \mathbf{a}^1 = \left(\hat{\jmath} - \frac{\hat{\imath}}{\sqrt{2}}\right) \cdot \left(\hat{\imath} + \frac{\hat{\jmath}}{\sqrt{2}}\right),$$

$$\mathbf{a}_1 \cdot \mathbf{a}_1 = \hat{\imath} \cdot \hat{\imath} = 1 = \mathbf{a}^2 \cdot \mathbf{a}^2 = \hat{\jmath} \cdot \hat{\jmath},$$

$$\mathbf{a}_2 \cdot \mathbf{a}_2 = \left(\hat{\jmath} - \frac{\hat{\imath}}{\sqrt{2}}\right) \cdot \left(\hat{\jmath} - \frac{\hat{\imath}}{\sqrt{2}}\right) = 1.5 = \mathbf{a}^1 \cdot \mathbf{a}^1 \tag{44}$$

$$= \left(\hat{\imath} + \frac{\hat{\jmath}}{\sqrt{2}}\right) \cdot \left(\hat{\imath} + \frac{\hat{\jmath}}{\sqrt{2}}\right),$$

$$\mathbf{a}_1 \cdot \mathbf{a}_2 = (\hat{\imath}) \cdot \left(\hat{\jmath} - \frac{\hat{\imath}}{\sqrt{2}}\right) = -\frac{1}{\sqrt{2}} = \mathbf{a}_2 \cdot \mathbf{a}_1,$$

$$\mathbf{a}^1 \cdot \mathbf{a}^2 = \mathbf{a}^2 \cdot \mathbf{a}^1 = \left(\hat{\imath} + \frac{\hat{\jmath}}{\sqrt{2}}\right) \cdot (\hat{\jmath}) = \frac{1}{\sqrt{2}},$$

$$F^2 = \tilde{f}gf = \tilde{F}GF.$$

The metric tensors have the following values:

$$g_{ij} = \begin{pmatrix} a_1 a_1 & a_1 a_2 \\ a_2 a_1 & a_2 a_2 \end{pmatrix} = \begin{pmatrix} 1 & -\frac{1}{\sqrt{2}} \\ -\frac{1}{\sqrt{2}} & \frac{3}{2} \end{pmatrix} = \begin{pmatrix} 1 & -0.707 \\ -0.707 & 1.5 \end{pmatrix} \tag{45}$$

and

$$g^{ij} = \begin{pmatrix} a^1 a^1 & a^1 a^2 \\ a^2 a^1 & a^2 a^2 \end{pmatrix} = \begin{pmatrix} \frac{3}{2} & \frac{1}{\sqrt{2}} \\ \frac{1}{\sqrt{2}} & 1 \end{pmatrix} = \begin{pmatrix} 1.5 & 0.707 \\ 0.707 & 1 \end{pmatrix}. \tag{46}$$

Consequently the determinants of g_{ij} and g^{ij} are both unity:

$$\det g_{ij} = \det g^{ij} = 1. \tag{47}$$

Moreover, the ordinary matrix product is the identity matrix:

$$(g_{ij})(g^{ij}) = \begin{pmatrix} 1 & -\frac{1}{\sqrt{2}} \\ -\frac{1}{\sqrt{2}} & \frac{3}{2} \end{pmatrix}\begin{pmatrix} \frac{3}{2} & \frac{1}{\sqrt{2}} \\ \frac{1}{\sqrt{2}} & 1 \end{pmatrix} = \begin{pmatrix} 1 & 0 \\ 0 & 1 \end{pmatrix} = \delta_j^{\ i}. \tag{48}$$

Calculate the scalar product:

$$\begin{aligned} \mathbf{F}\cdot\mathbf{F} &= (\mathbf{a}_1 f^1 + \mathbf{a}_2 f^2)\cdot(\mathbf{a}^1 f_1 + \mathbf{a}^2 f_2) \\ &= (\mathbf{a}_1\cdot\mathbf{a}^1) f_1 f^1 + (\mathbf{a}_2\cdot\mathbf{a}^2) f_2 f^2 + (\mathbf{a}_1\cdot\mathbf{a}^2) f^1 f_2 + (\mathbf{a}^1\cdot\mathbf{a}_2) f_1 f^2 \\ &= f_1 f^1 + f_2 f^2 = f_i f^i. \end{aligned} \tag{49}$$

Consider:

$$\mathbf{F} = F_x\hat{\mathbf{i}} + F_y\hat{\mathbf{j}}, \tag{50}$$

and calculate the two scalar products:

$$\begin{aligned} f_i &= \mathbf{a}_i\cdot\mathbf{F}, \\ f_1 &= \mathbf{a}_1\cdot\mathbf{F} = \hat{\mathbf{i}}\cdot\mathbf{F} = F_x, \\ f_2 &= \mathbf{a}_2\cdot\mathbf{F} = \left(\hat{\mathbf{j}} - \frac{\hat{\mathbf{i}}}{\sqrt{2}}\right)\cdot\mathbf{F} = -\frac{F_x}{\sqrt{2}} + F_y, \end{aligned} \tag{51}$$

and

$$\begin{aligned} f^i &= \mathbf{a}^i\cdot\mathbf{F}, \\ f^1 &= \left(\hat{\mathbf{i}} + \frac{\hat{\mathbf{j}}}{\sqrt{2}}\right)\cdot\mathbf{F} = \mathbf{a}^1\cdot\mathbf{F} = \left(F_x + \frac{F_y}{\sqrt{2}}\right), \\ f^2 &= \mathbf{a}^2\cdot\mathbf{F} = \hat{\mathbf{j}}\cdot\mathbf{F} = F_y. \end{aligned} \tag{52}$$

Thus

$$f_1 f^1 + f_2 f^2 = (F_x)\left(F_x + \frac{F_y}{\sqrt{2}}\right) + \left(-\frac{F_x}{\sqrt{2}} + F_y\right)(F_y) = F_x^{\ 2} + F_y^{\ 2}. \tag{53}$$

Raise and lower indices with the metric tensor:

$$\begin{aligned} f_i &= g_{ij} f^j \qquad f_i f^j = g_{ij} f^j f^i \\ f_1 &= f_{11} f^1 + g_{12} f^2 = (1)\left(F_x + \frac{F_y}{\sqrt{2}}\right) + \left(-\frac{1}{\sqrt{2}}\right)(F_y) = F_x \\ f_2 &= g_{21} f^1 + g_{22} f^2 = \left(-\frac{1}{\sqrt{2}}\right)\left(F_x + \frac{F_y}{\sqrt{2}}\right) + 1{\cdot}5(F_y) \\ &= \left(F_y - \frac{F_x}{\sqrt{2}}\right) \end{aligned} \tag{54}$$

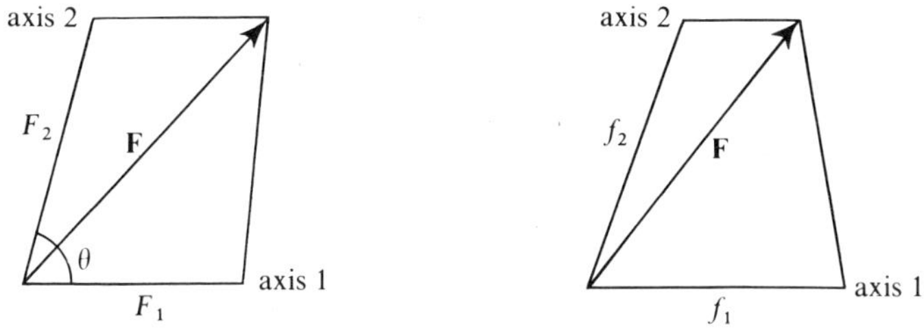

Figure 19. The decomposition of vector F.

$$\begin{aligned}\mathbf{F}\cdot\mathbf{F} &= f_i f^i = g_{ij} f^j f^i = g_{1j} f^j f^1 + g_{2j} f^j f^2 \\ &= g_{11} f^1 f^1 + g_{12} f^2 f^1 + g_{21} f^1 f^2 + g_{22} f^2 f^2 \\ &= (\sqrt{g_{11}} f^1)^2 + (\sqrt{g_{22}} f^2)^2 + 2 g_{12} f^1 f^2,\end{aligned}$$

which essentially is a law of cosines, as is inevitable if the components are taken from one coordinate system (see Fig. 19).

Because of the use of unit vectors $\hat{\mathbf{e}}^i$, the law of cosines may be stated:

$$F^2 = F_1{}^2 + F_2{}^2 + 2\cos\theta\, F_1 F_2. \tag{55}$$

Several examples of covariance and contravariance will now be discussed. Consider the cartesian system I with coordinates x, y that coincides with the cartesian system II with coordinates x', y' after rotation about the origin through an angle θ. For system II in terms of I:

$$\begin{aligned} x' &= x\cos\theta - y\sin\theta, \\ y' &= x\sin\theta + y\cos\theta. \end{aligned} \tag{56a}$$

For system I in terms of II:

$$\begin{aligned} x &= x'\cos\theta + y'\sin\theta, \\ y &= -\,x'\sin\theta + y'\cos\theta. \end{aligned} \tag{56}$$

The derivatives of II in terms of I are

$$\begin{aligned} \frac{\partial x'}{\partial x} &= \cos\theta, & \frac{\partial x'}{\partial y} &= -\sin\theta, \\ \frac{\partial y'}{\partial x} &= \sin\theta, \quad \text{and} & \frac{\partial y'}{\partial y} &= \cos\theta. \end{aligned} \tag{57}$$

The derivatives of system I in terms of II are

$$\frac{\partial x}{\partial x'} = \cos\theta, \qquad \frac{\partial x}{\partial y'} = \sin\theta,$$
$$\frac{\partial y}{\partial x'} = -\sin\theta, \quad \text{and} \quad \frac{\partial y}{\partial y'} = \cos\theta. \tag{57}$$

For the total derivatives of system II with respect to system I:

$$dx' = \frac{\partial x'}{\partial x}dx + \frac{\partial x'}{\partial y}dy = \cos\theta\,dx - \sin\theta\,dy,$$
$$dy' = \frac{\partial y'}{\partial x}dx + \frac{\partial y'}{\partial y}dy = \sin\theta\,dx + \cos\theta\,dy. \tag{58}$$

For the total derivatives of system I with respect to system II:

$$dx = \frac{\partial x}{\partial x'}dx' + \frac{\partial x}{\partial y'}dy' = \cos\theta\,dx' + \sin\theta\,dy',$$
$$dy = \frac{\partial y}{\partial x'}dx' + \frac{\partial y}{\partial y'}dy' = -\sin\theta\,dx' + \cos\theta\,dy'. \tag{58}$$

In matrix form equations (58) become

$$\begin{pmatrix} dx' \\ dy' \end{pmatrix} = \begin{pmatrix} \cos\theta & -\sin\theta \\ \sin\theta & \cos\theta \end{pmatrix}\begin{pmatrix} dx \\ dy \end{pmatrix}, \qquad \begin{pmatrix} dx \\ dy \end{pmatrix} = \begin{pmatrix} \cos\theta & \sin\theta \\ -\sin\theta & \cos\theta \end{pmatrix}\begin{pmatrix} dx' \\ dy' \end{pmatrix}, \tag{59}$$

or, more simply,

$$\mathbf{dr}' = M\,\mathbf{dr}, \qquad \mathbf{dr} = M^{-1}\,\mathbf{dr}'. \tag{59}$$

Similarly, starting with $\nabla'\phi$, $\nabla\phi$:

$$\frac{\partial\phi}{\partial x'} = \frac{\partial\phi}{\partial x}\frac{\partial x}{\partial x'} + \frac{\partial\phi}{\partial y}\frac{\partial y}{\partial x'} = \cos\theta\,\frac{\partial\phi}{\partial x} - \sin\theta\,\frac{\partial\phi}{\partial y},$$
$$\frac{\partial\phi}{\partial y'} = \frac{\partial\phi}{\partial x}\frac{\partial x}{\partial y'} + \frac{\partial\phi}{\partial y}\frac{\partial y}{\partial y'} = \cos\theta\,\frac{\partial\phi}{\partial x} + \cos\theta\,\frac{\partial\phi}{\partial y}; \tag{60}$$

and

$$\frac{\partial\phi}{\partial x} = \frac{\partial\phi}{\partial x'}\frac{\partial x'}{\partial x} + \frac{\partial\phi}{\partial y'}\frac{\partial y'}{\partial x} = \cos\theta\,\frac{\partial\phi}{\partial x'} + \sin\theta\,\frac{\partial\phi}{\partial y'},$$
$$\frac{\partial\phi}{\partial y} = \frac{\partial\phi}{\partial x'}\frac{\partial x'}{\partial y} + \frac{\partial\phi}{\partial y'}\frac{\partial y'}{\partial y} = -\sin\theta\,\frac{\partial\phi}{\partial x'} + \cos\theta\,\frac{\partial\phi}{\partial y'}. \tag{60}$$

In matrix form equations (60) may be written:

$$\begin{pmatrix} \dfrac{\partial\phi}{\partial x'} \\ \\ \dfrac{\partial\phi}{\partial y'} \end{pmatrix} = \begin{pmatrix} \cos\theta & -\sin\theta \\ \\ \sin\theta & \cos\theta \end{pmatrix} \begin{pmatrix} \dfrac{\partial\phi}{\partial x'} \\ \\ \dfrac{\partial\phi}{\partial y'} \end{pmatrix}, \qquad \begin{pmatrix} \dfrac{\partial\phi}{\partial x} \\ \\ \dfrac{\partial\phi}{\partial y} \end{pmatrix} = \begin{pmatrix} \cos\theta & \sin\theta \\ \\ -\sin\theta & \cos\theta \end{pmatrix} \begin{pmatrix} \dfrac{\partial\phi}{\partial x'} \\ \\ \dfrac{\partial\phi}{\partial y'} \end{pmatrix} \tag{61}$$

or, more simply,

$$\nabla'\phi = M\nabla\phi, \qquad \nabla\phi = M^{-1}\nabla'\phi. \tag{61}$$

Thus, for cartesian coordinates, there is no distinction between covariant and contravariant transformation laws. Moreover, the scalar product of $\nabla'\phi$ and dr' is invariant. This scalar product is calculated as follows:

$$\begin{aligned} [\nabla'\phi]\cdot[d\mathbf{r}'] &= (\widetilde{\nabla'\phi})(dr') = d\phi \\ &= [M\bar{\nabla}\phi]\cdot[M\,dr] = (\widetilde{M\nabla\phi})(M\,dr) = \tilde{\nabla}\phi\tilde{M}M\,dr \\ &= \tilde{\nabla}\phi M^{-1}M\,dr = \widetilde{\nabla\phi}\,dr - [\nabla\phi]\cdot[d\mathbf{r}] = d\phi. \end{aligned} \tag{62}$$

The nonorthogonal transformation is

$$\begin{aligned} x' &= x + \frac{1}{\sqrt{2}}y, & x &= x' - \frac{1}{\sqrt{2}}y, \\ y' &= 0 + y, & y &= 0 + y', \end{aligned} \tag{63}$$

however, for which

$$\begin{aligned} dx' &= \frac{\partial x'}{\partial x}dx + \frac{\partial x'}{\partial y}dy = 1\,dx + \frac{1}{\sqrt{2}}dy, \\ dy' &= \frac{\partial y'}{\partial x}dx + \frac{\partial y'}{\partial y}dy = 0 + 1\,dy, \end{aligned} \tag{64}$$

has a different law of transformation.

In matrix form:

$$\begin{pmatrix} dx' \\ \\ dy' \end{pmatrix} - \begin{pmatrix} 1 & \dfrac{1}{\sqrt{2}} \\ \\ 0 & 1 \end{pmatrix} \begin{pmatrix} dx \\ \\ dy \end{pmatrix}, \tag{65}$$

$$d\mathbf{r}' = M\,d\mathbf{r}. \tag{66}$$

Similarly,

$$\frac{\partial\phi}{\partial x'} = \frac{\partial\phi}{\partial x}\frac{\partial x}{\partial x'} + \frac{\partial\phi}{\partial y}\frac{\partial y}{\partial x'} = 1\frac{\partial\phi}{\partial x} + 0\frac{\partial\phi}{\partial y} \tag{67}$$

and

$$\frac{\partial\phi}{\partial y'} = \frac{\partial\phi}{\partial x}\frac{\partial x}{\partial y'} + \frac{\partial\phi}{\partial y}\frac{\partial y}{\partial y'} = -\frac{1}{\phantom{\sqrt{2}}}\frac{\partial\phi}{\partial x} + 0\frac{\partial\phi}{\partial y} \tag{68}$$

In matrix form:

$$\begin{pmatrix} \dfrac{\partial\phi}{\partial x'} \\ \dfrac{\partial\phi}{\partial y'} \end{pmatrix} = \begin{pmatrix} 1 & 0 \\ -\dfrac{1}{\sqrt{2}} & 1 \end{pmatrix} \begin{pmatrix} \dfrac{\partial\phi}{\partial x} \\ \dfrac{\partial\phi}{\partial y} \end{pmatrix} \tag{69}$$

or:

$$\nabla'\phi = (\widetilde{M^{-1}})\nabla\phi. \tag{70}$$

The scalar product of $\nabla'\phi$ and dr',

$$\begin{aligned}
[\bar{\nabla}'\phi]\cdot[d\bar{r}'] &= (\widetilde{\nabla'\phi})(dr') = d\phi \\
&= [\tilde{M}^{-1}\nabla\phi]\cdot[Md\bar{r}] = (\widetilde{\tilde{M}^{-1}\nabla\phi})(Mdr) = \widetilde{\nabla\phi}(\widetilde{\tilde{M}^{-1}})Mdr \\
&= \widetilde{\nabla\phi}M^{-1}Mdr = \widetilde{\nabla\phi}\,dr = [\nabla\phi]\cdot[d\bar{r}] = d\phi \qquad (71) \\
= \widetilde{\nabla\phi} & \begin{pmatrix} 1 & -\dfrac{1}{\sqrt{2}} \\ 0 & 1 \end{pmatrix} \begin{pmatrix} 1 & \dfrac{1}{\sqrt{2}} \\ 0 & 1 \end{pmatrix} dr = \widetilde{\nabla\phi}\begin{pmatrix} 1 & 0 \\ 0 & 1 \end{pmatrix} dr,
\end{aligned}$$

is still invariant for the nonorthogonal linear transform.

To summarize:

Contravariance	Covariance
$dx'^{\nu} = \dfrac{\partial x'^{\nu}}{\partial x^{\mu}}dx^{\mu}$	$\nabla'_{\nu}\phi = \dfrac{\partial x^{\mu}}{\partial x'_{\nu}}\nabla_{\mu}\phi$
$d\mathbf{r}' = Md\mathbf{r}$	$\nabla'\phi = (\widetilde{M^{-1}})\nabla\phi$

In an orthogonal system:

$$\tilde{M} = M^{-1}, \quad M = \tilde{M}^{-1}, \quad M = \tilde{M}^{-1}, \tag{72}$$

and

$$dr' = Mdr \qquad \text{and} \quad \nabla'\phi = \widetilde{M\nabla\phi}, \tag{73}$$

$$\begin{aligned} \nabla'\phi \cdot dr' &= [\tilde{M}^{-1}\nabla\phi] \cdot Mdr = (\widetilde{M^{-1}\nabla\phi})(Mdr) \\ &= \widetilde{\nabla\phi}(\widetilde{\tilde{M}^{-1}}M)dr = \widetilde{\nabla\phi}\, dr = \nabla\phi \cdot d\mathbf{r}. \end{aligned} \tag{74}$$

To form the covariant derivative of a contravariant vector, start with a contravariant vector:

$$\bar{A}^i = \frac{\partial \bar{q}^i}{\partial q^\alpha} A^\alpha; \tag{75}$$

differentiate as follows:

$$\frac{\partial \bar{A}^i}{\partial \bar{q}^j} = \frac{\partial A^\alpha}{\partial q^\beta}\frac{\partial q^\beta}{\partial \bar{q}^j}\frac{\partial \bar{q}^i}{\partial q^\alpha} + A^\alpha \frac{\partial^2 \bar{q}^i}{\partial q^\alpha \partial q^\beta}\frac{\partial q^\beta}{\partial \bar{q}^j}, \tag{76}$$

and express the second derivative in terms of Christoffel symbols of the second kind:

$$\frac{\partial^2 \bar{q}^i}{\partial q^\alpha \partial q^\beta} = \begin{Bmatrix} s \\ k\beta \end{Bmatrix} \frac{\partial \bar{q}^i}{\partial q^s} - \begin{Bmatrix} \bar{i} \\ kl \end{Bmatrix} \frac{\partial \bar{q}^k}{\partial q^\alpha}\frac{\partial \bar{q}^l}{\partial q^\beta}. \tag{77}$$

Then:

$$\frac{\partial \bar{A}^i}{\partial \bar{q}^j} = \frac{\partial A^\alpha}{\partial q^\beta}\frac{\partial q^\beta}{\partial \bar{q}^j}\frac{\partial \bar{q}^i}{\partial q^\alpha} + A^\alpha \frac{\partial q^\beta}{\partial \bar{q}^i}\left[\begin{Bmatrix} s \\ \alpha\beta \end{Bmatrix} \frac{\partial \bar{q}^i}{\partial q^s} - \begin{Bmatrix} \bar{i} \\ kl \end{Bmatrix} \frac{\partial \bar{q}^k}{\partial q^\alpha}\frac{\partial \bar{q}^l}{\partial q^\beta}\right] \tag{78}$$

Rearrange terms to obtain

$$\frac{\partial \bar{A}^i}{\partial \bar{q}^j} + \begin{Bmatrix} \bar{i} \\ kl \end{Bmatrix} A^\alpha \frac{\partial \bar{q}^k}{\partial q^\alpha}\frac{\partial \bar{q}^l}{\partial \bar{q}^j} = \frac{\partial A^\alpha}{\partial q^\beta}\frac{\partial q^\beta}{\partial \bar{q}^j}\frac{\partial \bar{q}^i}{\partial q^\alpha} + \begin{Bmatrix} s \\ \alpha\beta \end{Bmatrix} A^\alpha \frac{\partial q^\beta}{\partial \bar{q}^j}\frac{\partial \bar{q}^i}{\partial q^s}. \tag{79}$$

Recognize a δ-function and A^k on the left to obtain

$$\frac{\partial \bar{A}^i}{\partial \bar{q}^j} + \begin{Bmatrix} \bar{i} \\ kj \end{Bmatrix} \bar{A}^k = \frac{\partial A^\alpha}{\partial q^\beta}\frac{\partial q^\beta}{\partial \bar{q}^j}\frac{\partial \bar{q}^i}{\partial q^\alpha} + \begin{Bmatrix} s \\ \alpha\beta \end{Bmatrix} A^\alpha \frac{\partial q^\beta}{\partial \bar{q}^j}\frac{\partial \bar{q}^i}{\partial q^s} \tag{80}$$

It is necessary also to switch some indices around on the right:

$$\frac{\partial \bar{A}^i}{\partial \bar{q}^j} + \begin{Bmatrix} \bar{i} \\ kj \end{Bmatrix} \bar{A}^k = \frac{\partial A^\alpha}{\partial q^\beta}\frac{\partial q^\beta}{\partial \bar{q}^j}\frac{\partial \bar{q}^i}{\partial q^\alpha} + \begin{Bmatrix} \alpha \\ s\beta \end{Bmatrix} A^s \frac{\partial q^\beta}{\partial \bar{q}^j}\frac{\partial \bar{q}^i}{\partial q^\alpha}. \tag{81}$$

Write the result:

$$\left[\frac{\partial \bar{A}^i}{\partial \bar{q}^j} + \begin{Bmatrix} i \\ kj \end{Bmatrix} \bar{A}^k\right] = \left[\frac{\partial A^\alpha}{\partial q^\beta} + \begin{Bmatrix} \alpha \\ s\beta \end{Bmatrix} A^s\right] \frac{\partial q^\beta}{\partial \bar{q}^j}\frac{\partial \bar{q}^i}{\partial q^\alpha}, \tag{82}$$

the "covariant derivative of a contravariant vector." It may be written also as:

$$A^{\alpha}{}_{,\beta} = \left[\frac{\partial A^{\alpha}}{\partial q^{\beta}} + \left\{ {\alpha \atop s\beta} \right\} A^{s}\right]. \tag{83}$$

This quantity transforms like a mixed tensor of rank two, but differs from the covariant derivative of a covariant vector, namely:

$$A_{\alpha,\beta} = \left[\frac{\partial A_{\alpha}}{\partial q^{\beta}} - \left\{ {k \atop \alpha\beta} \right\} A_{k}\right], \tag{84}$$

which will now be discussed.

To form the covariant derivative of a covariant vector, start with a covariant vector:

$$\bar{A}_i = \frac{\partial q^{\alpha}}{\partial \bar{q}^{i}} A_{\alpha}; \tag{85}$$

differentiate as follows:

$$\frac{\partial \bar{A}_i}{\partial \bar{q}^{j}} = \frac{\partial q^{\alpha}}{\partial \bar{q}^{i}}\frac{\partial q^{\beta}}{\partial \bar{q}^{j}}\frac{\partial A_{\alpha}}{\partial q^{\beta}} + \frac{\partial^2 q^{\alpha}}{\partial \bar{q}^{j}\partial \bar{q}^{i}} A_{\alpha}, \tag{86}$$

and express the second derivative in terms of Christoffel symbols of the second kind:

$$\frac{\partial^2 q^{\alpha}}{\partial \bar{q}^{j}\partial \bar{q}^{i}} = \left\{ {\bar{s} \atop ji} \right\}\frac{\partial q^{\alpha}}{\partial \bar{q}^{s}} - \left\{ {\alpha \atop kl} \right\}\frac{\partial q^{k}}{\partial \bar{q}^{j}}\frac{\partial q^{l}}{\partial \bar{q}^{i}}. \tag{87}$$

Then:

$$\frac{\partial \bar{A}_i}{\partial \bar{q}^{j}} = \frac{\partial q^{\alpha}}{\partial \bar{q}^{i}}\frac{\partial q^{\beta}}{\partial \bar{q}^{j}}\frac{\partial A_{\alpha}}{\partial q^{\beta}} + \left\{ {\bar{s} \atop ji} \right\} A_{\alpha}\frac{\partial q^{\alpha}}{\partial \bar{q}^{j}} - \left\{ {\alpha \atop kl} \right\} A_{\alpha}\frac{\partial q^{k}}{\partial \bar{q}^{j}}\frac{\partial q^{l}}{\partial \bar{q}^{i}}. \tag{88}$$

Rearrange the terms:

$$\left[\frac{\partial \bar{A}_i}{\partial \bar{q}^{j}} - \left\{ {\bar{s} \atop ji} \right\}\bar{A}_s\right] = \frac{\partial A_{\alpha}}{\partial q^{\beta}}\frac{\partial q^{\alpha}}{\partial \bar{q}^{i}}\frac{\partial q^{\beta}}{\partial \bar{q}^{j}} - \left\{ {\alpha \atop kl} \right\} A_{\alpha}\frac{\partial q^{k}}{\partial \bar{q}^{j}}\frac{\partial q^{l}}{\partial \bar{q}^{i}}. \tag{89}$$

Next interchange the indices in equation (89) and relable one of them to produce the result:

$$\left[\frac{\partial A_i}{\partial \bar{q}^{j}} - \left\{ {\bar{s} \atop ji} \right\}\bar{A}_s\right] = \left[\frac{\partial A_{\alpha}}{\partial q^{\beta}} - \left\{ {k \atop \alpha\beta} \right\} A_k\right]\frac{\partial q^{\alpha}}{\partial \bar{q}^{i}}\frac{\partial q^{\beta}}{\partial \bar{q}^{j}}, \tag{90}$$

called the "covariant derivative of a covariant vector," may be written as

$$A_{\alpha,\beta} = \left[\frac{\partial A_\alpha}{\partial q^\beta} - \begin{Bmatrix} k \\ \alpha\beta \end{Bmatrix} A_k\right]. \tag{91}$$

This quantity, which transforms as a covariant tensor of rank two, differs from the covariant derivative of a contravariant vector, discussed earlier, namely:

$$A^\alpha{}_{,\beta} = \left[\frac{\partial A^\alpha}{\partial q^\beta} + \begin{Bmatrix} \alpha \\ k\beta \end{Bmatrix} A^k\right]. \tag{92}$$

In summary, four words in the physicist's working vocabulary warrant the student's early attention: covariant, contravariant, cogredient, and contragredient. All refer to the transformation properties of vectors and tensors and all permit a certain amount of arbitrariness or relativeness in their definitions. What is covariant, or cogredient, to observer A perched on one coordinate frame might be relabled contravariant, or contragredient, by another equally keen observer B perched on another frame—with no change in the physics—then, of course, all other quantities would have to be relabled.

Consider a transformation matrix M. If a relationship exists, such as

$$p' - Mp \qquad \text{or} \qquad p = M^{-1}p', \tag{93}$$

and a second relationship exists, such as

$$q = Mq' \qquad \text{or} \qquad q' = M^{-1}q, \tag{94}$$

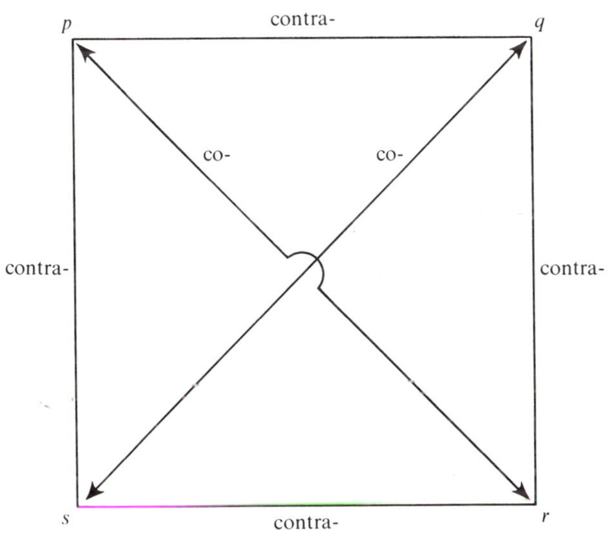

Figure 20. The interrelationship between coordinates and prefixes.

then p and q are said to "transform contragrediently with respect to each other" or simply to "transform contragrediently."

If a third relationship exists, such as

$$r' = Mr \qquad \text{or} \qquad r = M^{-1}r', \tag{95}$$

then r and p instead are said to "transform cogrediently," but r and q are said to "transform contragrediently."

If a fourth relationship exists, such as

$$s = Ms' \qquad \text{or} \qquad s' = M^{-1}s, \tag{96}$$

then r and s again are said to "transform contragrediently." This is clarified by Fig. 20, which is applicable to either suffix, -gredient or -variant.

EXERCISES

What transformation is contragredient to

$$\mathbf{x} = \begin{pmatrix} 1 & 0 & 1 \\ 0 & 1 & 1 \\ 1 & 1 & 1 \end{pmatrix} \mathbf{x}?$$

A study of the information contained in the two figures and the few related equations to follow will help make the meaning of covariance and contravariance clear and simple in a way no amount of words alone can convey.

Consider a vector $\mathbf{F}$ of length $\sqrt{2}$ as shown in Fig. 21; then we have:

$$\begin{aligned} \mathbf{F} &= F_x\hat{\mathbf{i}} + F_y\hat{\mathbf{j}} = F_1\hat{\mathbf{e}}_1 + F_2\hat{\mathbf{e}}_2 = F^1\hat{\mathbf{e}}^1 + F^2\hat{\mathbf{e}}^2 \\ &= f_1 a^1 + f_2 a^2 = f_a{}^1 + f^2 a_2. \end{aligned} \tag{97}$$

Also,

$$\begin{aligned} \mathbf{F}\cdot\mathbf{F} &= F_x{}^2 + F_y{}^2 + \underbrace{2\hat{\mathbf{i}}\cdot\hat{\mathbf{j}}}_{\cos 90^\circ = 0}F_xF_y \\ &= (F_1\hat{\mathbf{e}}_1 + F_2\hat{\mathbf{e}}_2)^2 = F_1{}^2 + F_1{}^2 + \underbrace{2\hat{\mathbf{e}}_1\cdot\hat{\mathbf{e}}_2}_{\cos\theta}\ F_1F_2 \\ &= F^1\hat{\mathbf{e}}^1 + F^2\hat{\mathbf{e}}^2)^2 = F^{12} + F^{22} + \underbrace{2\hat{\mathbf{e}}^1\cdot\hat{\mathbf{e}}^2}_{\cos\phi}F^1F^2 \\ &= F^1F_1 + F^2F_2 = f^1f_1 + f^2f_2 = \underbrace{(f_1)^2 + (f^2)^2}_{\text{special case}}. \end{aligned} \tag{98}$$

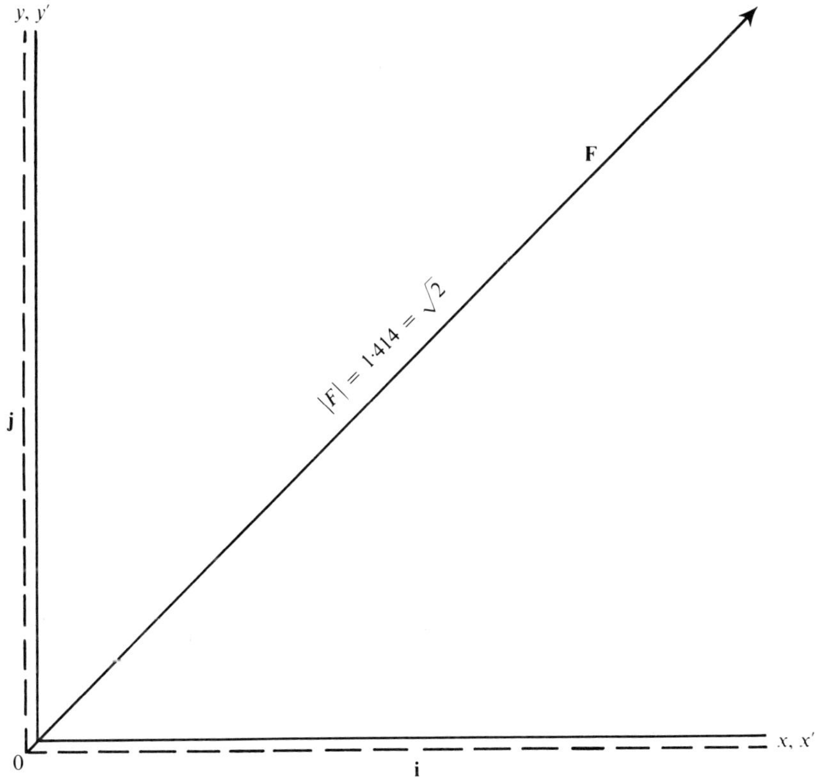

Figure 21. The vector F in cartesian coordinates.

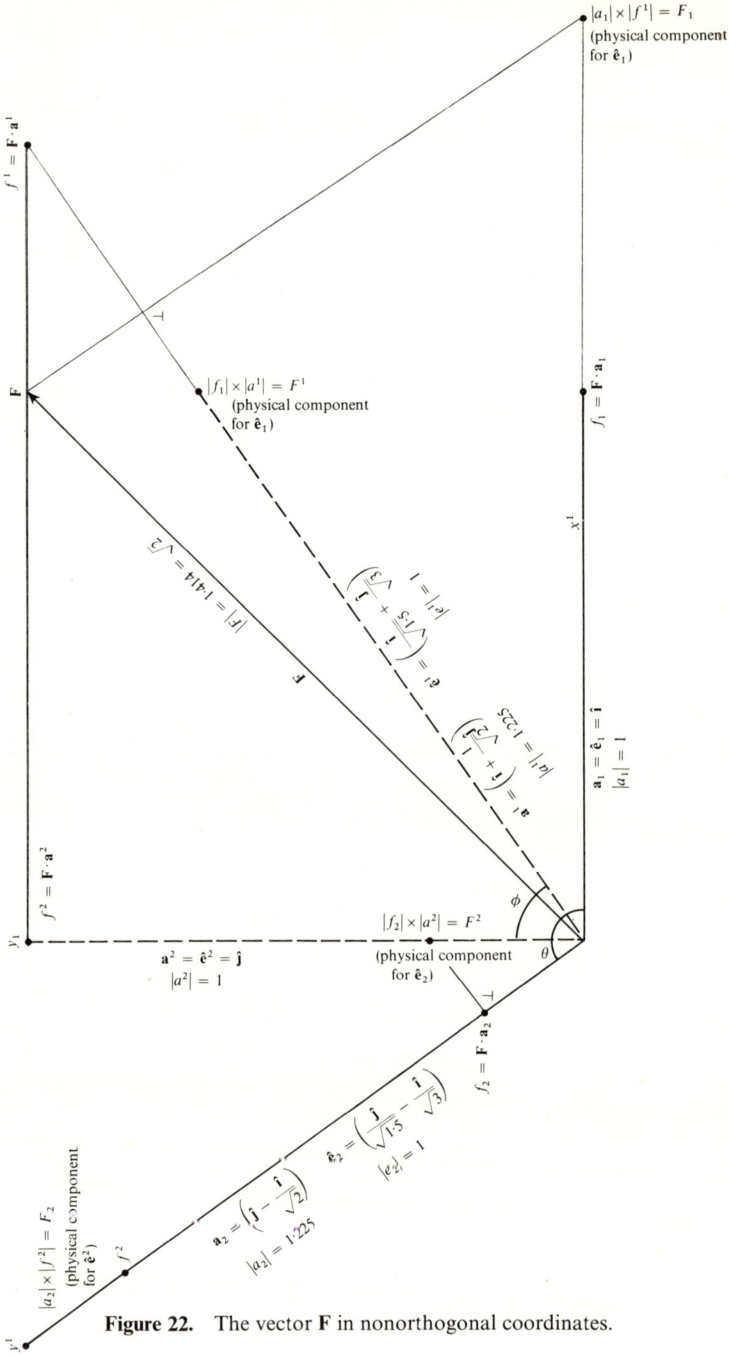

Figure 22. The vector **F** in nonorthogonal coordinates.

Table 1 lists some facts about covariance and contravariance for several elementary coordinate systems.

Table 1 A Summary of Properties of Elementary Coordinate Systems

Coordinate System	Basis Vector	Reciprocal System	Scale	Direction	Do Covariant and Contravariant Systems Differ?
X, Y, Z	$\hat{\mathbf{e}}$	same	same	same	no
x, y, z	$\mathbf{a}$	orthogonal superimposed	change	same	yes
r, θ	$\hat{\mathbf{e}}$	same	same	same	no
r, θ	$\mathbf{a}$	orthogonal superimposed	change	same	yes
r, θ, z	$\hat{\mathbf{e}}$	same	same	same	no
r, θ, z	$\mathbf{a}$	orthogonal superimposed	change	same	yes
r, θ, ϕ	$\hat{\mathbf{e}}$	same	same	same	no
r, θ, ϕ	$\mathbf{a}$	orthogonal superimposed	change	same	yes
Linear nonorthogonal:					
$Y, \left(X + \frac{Y}{\sqrt{2}}\right)$	$\hat{\mathbf{e}}$	NONorthogonal	change	change	yes
$Y, \left(X + \frac{Y}{\sqrt{2}}\right)$	$\mathbf{a}$	NONorthogonal	change	change	yes

4 Tensor Analysis

A. Tensors, Rank, and Dimensionality

The subject of tensor analysis possesses an undeserved aura of mystery, perhaps because various isolated fragments of knowledge pertaining to the simpler aspects of tensor algebra and calculus often contribute to the intellectual fallout in such undergraduate courses as matrix algebra and vector analysis without being given proper motivation. It is hoped that by collecting many of these fragments and additionally supplying a few important unifying concepts, this book will help the reader not only to use tensors as a theoretical research tool but also to develop a trust in his own ability to use them, since acquiring self-confidence is a key factor in the process of learning to solve problems successfully in mathematical physics.

Before wondering very much about the physical interpretation of a tensor —which depends entirely on the problem at hand—consider its mathematical description. As an entity, a tensor is somewhat like a pigeon-hole file, each compartment of which contains a number that represents some physical quantity. Such numbers are called "tensor components." Keeping track of these components individually is a big job, best handled by giving every component a unique designation. This assignment is reminiscent of the procedure through which every word in a computer memory is given a unique address, but differs from this procedure in that a computer memory is ordered serially, whereas a tensor is ordered spatially. That is, a word in computer memory has a single number as its address, but a tensor component may have a set of such numbers called "indices," which consist of either superscripts or subscripts or a combination thereof, the exact number of indices being termed the "tensor rank." In order to select a particular component of a tensor, these superscripts or subscripts must be given fixed numerical values, but to refer to tensor components generally, variable letter names are used.

The number of components in a given tensor depends on both its rank and on "tensor dimensionality." The lowest rank that a tensor may have is 0. The number of components possessed by a tensor of rank 0 is always 1, no matter what the dimensionality of the tensor may be. A common, possibly

Table 1 Maximum Number of Components for Arbitrary Tensor

Name	Rank	Dimension	Number of Components
Scalar	0	1	1
		2	1
		3	1
		.	.
		.	.
		.	.
		D	$D^0 = 1$
Vector	1	1	1
		2	2
		3	3
		.	.
		.	.
		.	.
		D	$D^1 = D$
Tensor of rank two	2	1	1
		2	4
		3	9
		.	.
		.	.
		.	.
		D	D^2
Tensor of rank three	3	1	1
		2	8
		3	27
		.	.
		.	.
		.	.
		D	D^3
Tensor of rank four	4	1	1
		2	16
		3	81
		.	.
		.	.
		.	.
		D	D^4
Tensor of rank R	R	1	1
		2	2^R
		3	3^R
		.	.
		.	.
		.	.
		D	D^R

unexpected name for such a tensor is "scalar." So an n-dimensional scalar is a one-component tensor of rank zero. An example of a scalar quantity is a thermometer reading representing temperature.

The next rank that a tensor may have is 1. The number of components possessed by a tensor of rank 1 is always n, the dimensionality of the tensor. Thus in two-dimensional space a tensor of rank 1 has 2 components; in three-dimensional space a tensor of rank 1 has 3 components; and so on. Again, a common name for such a tensor is "vector." So an n-dimensional vector is an n-component tensor of rank 1. The first example of a vector that comes to mind usually is velocity.

The next rank that a tensor may have is 2. The number of components possessed by a tensor of rank 2 is always n^2, the square of the dimensionality of the tensor. Thus in two-dimensional space a tensor of rank 2 has 4 components; in three-dimensional space a tensor of rank 2 has 9 components, and so on. A tensor of rank 2 is the first that normally would be called a true tensor.

Scalar tensors may have any rank. A rank-zero scalar tensor is a "scalar *scalar*," a rank-one scalar tensor is a "scalar *vector*," and a rank-two scalar tensor is a "scalar *matrix*." Higher rank scalar tensors have no special names.

Table 1 summarizes these results, making use of the general fact that the number of components in a tensor of rank R in a space of dimensionality D is given by the formula $Z = D^R$.

The last column of Table 1 lists the maximum number of terms or components in a complete tensor. One example of an incomplete tensor is called an "antisymmetric tensor." Such a tensor always has some zero components. The number of independent nonzero components is given in Table 2.

It is a small but convenient coincidence that in three dimensions an antisymmetric tensor of rank 2 has only three nonzero components. This

Table 2 Number of Nonzero Components for Antisymmetric Tensors

Special Name (If Any)	Rank	Dimension	Number of Components
Axial vector	2	1	0
pseudovector		2	1
		3	3
		.	.
		.	.
		.	.
		N	$\frac{1}{2}N(N-1)$

incidentally gives rise to this tensor's interpretation as a kind of vector; moreover, it is related to the fact that in three dimensions the number of planes (defined by pairs of coordinates x, y; x, z; z, y) is the same as the number of coordinates x, y, z. In any other dimension this does not happen, but the number of independent planes still is equal to the number of tensor components. Thus in two dimensions there is only one plane (1, 2) and one tensor component, while in three dimensions, there are six independent planes (1, 2; 1, 3; 1, 4; 2, 3; 2, 4; 3, 4) and six tensor components. Finally, Table 3 summarizes the number of components of simple tensors for spaces with dimension up to five. Notice that the number of components of the symmetric and antisymmetric tensors of a given rank sum to the maximum number of the rank.

Table 3 Number of Components of a Tensor

	Dimension				
	1	2	3	4	5
Scalar	1	1	1	1	1
Vector	1	2	3	4	5
Axial vector	1	2	3	4	5
Tensor—2nd rank	1	4	9	16	25
Symmetric tensor	1	3	6	10	15
Antisymmetric tensor	0	1	3	6	10
Tensor—3rd rank	1	8	27	64	125

All vectors are tensors but not all tensors are vectors.

A vector A_i has a transformation law:

$$A'_j = l_{ij} A_i, \tag{1}$$

whereas a tensor K_{ik} has a transformation law:

$$K'_{jl} = l_{ij} l_{kl} K_{ik}. \tag{2}$$

Here the symbol l_{ij} or l_{kl} represents the transformation matrix taking A_i or K_{ik} from the unprimed to the primed coordinate system. The n^2 elements of the matrix l_{ij} are derived from the direction cosines of the unit vectors in one system of coordinates with respect to the unit vectors in the other.

A tensor K_{ik} or K'_{jl} is a physical object and the double index ik or jl, while similar to ij or kl for the matrix l, refers to the axes in one system of coordinates. Thus they are different sorts of object even though they may have the same form.

Examples of quantities that can be and are represented by tensors include stress, strain, moment of inertia, electric and magnetic field strengths, and angular momentum.

A tensor of rank zero has a single scalar component and may be visualized in a space having any number of dimensions as shown in Fig. 1. A tensor of

(a) Tensor: $R = 0, D = 1$ (b) Tensor: $R = 0, D = 2$ (c) Tensor: $R = 0, D = 3$

Figure 1. A scalar (tensor of rank zero) in any space.

rank zero is called a "scalar." A tensor of rank one has as many scalar components as there are dimensions in the space. A tensor of rank one, which is called a vector, may be visualized as shown in Fig. 2. A tensor of

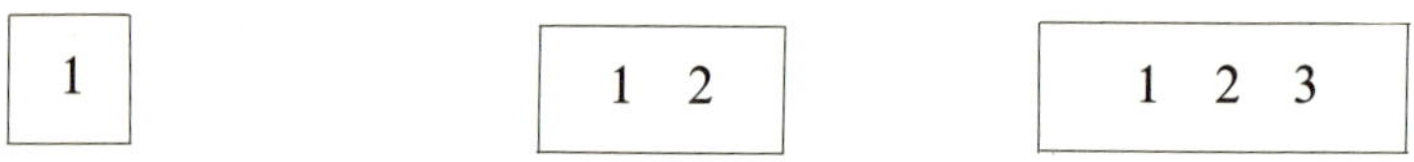

(a) Tensor: $R = 1, D = 1$ (b) Tensor: $R = 1, D = 2$ (c) Tensor: $R = 1, D = 3$

Figure 2. A vector (tensor of rank one) in any space.

rank two may be expressed as a square array of scalar components: symbols, letters, or numbers. A tensor of rank two, called a matrix, may be visualized in a space having any number of dimensions, as shown in Fig. 3.

(a) Tensor: $R = 2, D = 1$

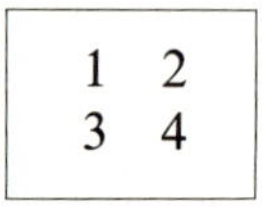

(b) Tensor: $R = 2, D = 2$

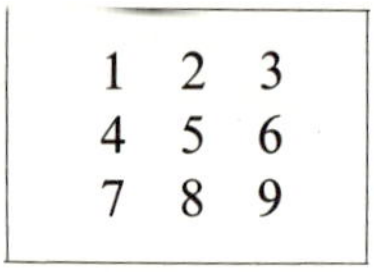

(c) Tensor: $R = 2, D = 3$

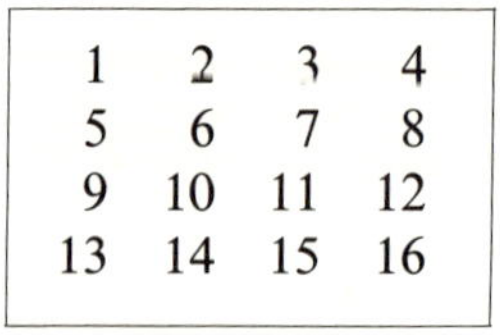

(d) Tensor: $R = 2, D = 4$

Figure 3. A matrix (tensor of rank two) in any space.

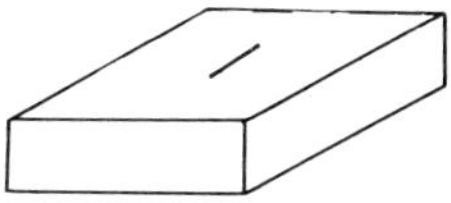

(*a*) Tensor: $R = 3, D = 1$

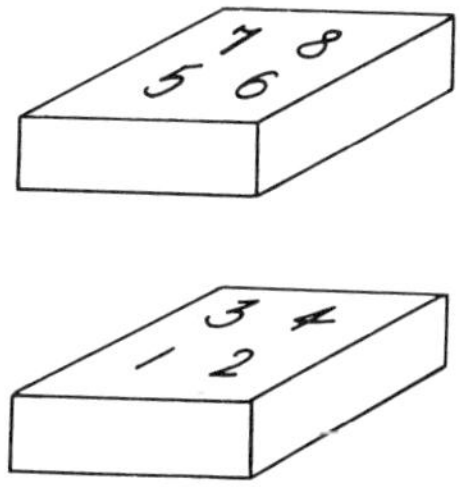

(*b*) Tensor: $R = 3, D = 2$

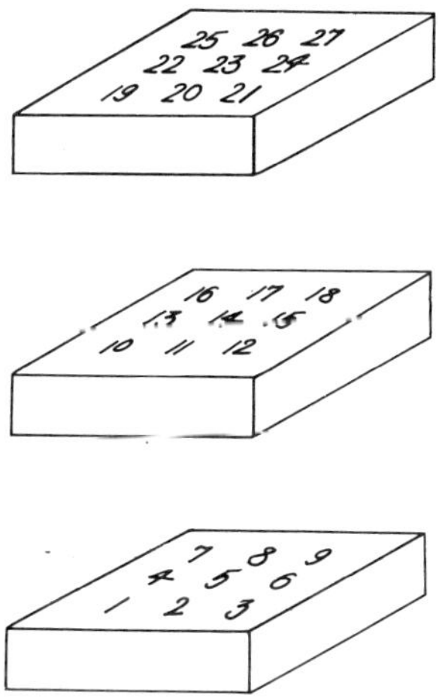

(*c*) Tensor: $R = 3, D = 3$

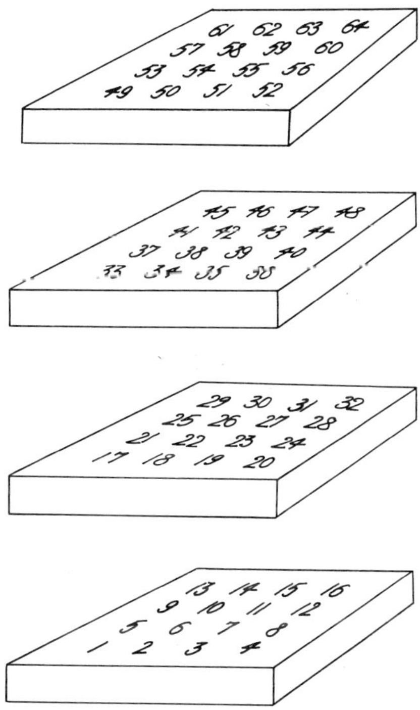

(*d*) Tensor: $R = 3, D = 4$

Figure 4. A tensor of rank three (true tensor) in any space.

A tensor of rank three may be expressed as a cubic array of symbols, letters, or numbers, as shown in Fig. 4. It is convenient to represent this three-dimensional array in perspective on a sheet of paper. So that the dimension of the tensor will be unlimited, for display purposes the roles of D (denoting "dimension") and R (denoting "rank") are interchanged in these figures.

B. Tensor Products

The associativity of the outer product is verified as follows for 2 by 2 matrices:

$$(L \cdot\times M) \cdot\times N = \begin{vmatrix} L_{11}M_{11} & L_{11}M_{12} & L_{12}M_{11} & L_{12}M_{12} \\ L_{11}M_{21} & L_{11}M_{22} & L_{12}M_{21} & L_{12}M_{22} \\ L_{21}M_{11} & L_{21}M_{12} & L_{22}M_{11} & L_{22}M_{12} \\ L_{21}M_{21} & L_{21}M_{22} & L_{22}M_{21} & L_{22}M_{22} \end{vmatrix} \cdot\times N$$

$$= \left|\begin{array}{c:c:c:c:c:c:c} L_{11}M_{11}N_{11} & L_{11}M_{11}N_{12} & L_{11}M_{12}N_{11} & L_{11}M_{12}N_{12} & & & \\ L_{11}M_{11}N_{21} & L_{11}M_{11}N_{22} & L_{11}M_{12}N_{21} & L_{11}M_{12}N_{22} & & & \\ L_{11}M_{21}N_{11} & L_{11}M_{21}N_{12} & L_{11}M_{22}N_{11} & L_{11}M_{22}N_{12} & & & \\ L_{11}M_{21}N_{21} & L_{11}M_{21}N_{22} & L_{11}M_{22}N_{22} & L_{11}M_{22}N_{22} & & & \\ \hdashline & & & & & & \\ & & & & & & \end{array}\right| \tag{2}$$

Expression (2), however, can be "factored" as follows:

$$\left|\begin{array}{cc} L_{11}\begin{vmatrix} M_{11}N_{11} & M_{11}N_{12} & M_{12}N_{11} & M_{12}N_{12} \\ M_{11}N_{21} & M_{11}N_{22} & M_{12}N_{21} & M_{12}N_{22} \\ M_{21}N_{11} & M_{21}N_{12} & M_{22}N_{11} & M_{22}N_{12} \\ M_{22}N_{22} & M_{21}N_{22} & M_{22}N_{21} & M_{22}N_{22} \end{vmatrix} & L_{12}\left(\quad\right) \\ L_{21}\left(\quad\right) & L_{22}\left(\quad\right) \end{array}\right|$$

$$= L \cdot\times \begin{vmatrix} M_{11}N_{11} & M_{11}N_{12} & M_{12}N_{11} & M_{12}N_{12} \\ M_{11}N_{21} & M_{11}N_{22} & M_{12}N_{12} & M_{12}N_{22} \\ M_{21}N_{11} & M_{23}N_{12} & M_{22}N_{11} & M_{22}N_{12} \\ M_{21}N_{21} & M_{21}N_{22} & M_{22}N_{21} & M_{22}N_{22} \end{vmatrix}. \tag{3}$$

Consequently (3) becomes

$$L \cdot\times (M \cdot\times N). \tag{4}$$

Thus

$$(L \cdot\times M) \cdot\times N = L \cdot\times (M \cdot\times N). \tag{5}$$

A similar verification for a general four-dimensional Lorentz transformation,

$$T' = L \cdot\times L \cdot\times LT = (4, 4) \cdot\times (4, 4) \cdot\times (4, 4)T, \tag{6}$$

would involve the following dimensionality equation:

$$T' = (64, 64)T = (4096, 1)T \tag{7}$$

Since T is a (4, 4, 4) tensor with 64 terms, applying the "Stretch Rule" would yield the following dimensionality equation:

$$(64, 1) = (64, 64) \cdot (64, 1). \tag{8}$$

EXERCISE 6

Prove the associativity of the outer product for n by n matrices.

C. Tensor Transformations

Assume that in some abstract way vectors $\mathbf{x}$, $\mathbf{x}'$, $\mathbf{y}$, $\mathbf{y}'$, operators A, A', and transformations Q, Q' exist and combine as follows:

$$\mathbf{y} = A\mathbf{x}, \quad \mathbf{x} = Q\mathbf{x}', \quad \text{and} \quad \mathbf{y} = Q\mathbf{y}'. \tag{1}$$

The relation between A and A' is determined as follows:

$$\begin{aligned} \mathbf{y} &= Q\mathbf{y}' = A\mathbf{x} = AQ\mathbf{x}', \\ Q^{-1}Q\mathbf{y}' &= Q^{-1}AQ\mathbf{x}', \\ \mathbf{y}' &= Q^{-1}AQ\mathbf{x}', \\ Q^{-1}Q &= I, \end{aligned} \tag{2}$$

or:

$$\mathbf{y}' = A'\mathbf{x}'.$$

Thus

$$A' = Q^{-1}AQ$$

or;

$$A = QA'Q^{-1}.$$

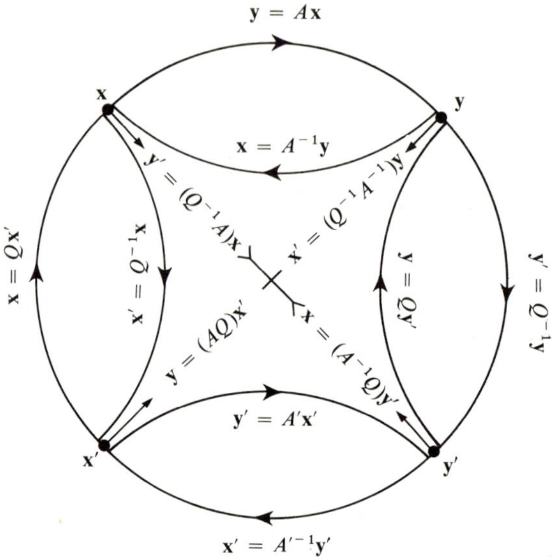

Figure 5. Direct graph showing connections between vectors, operators, and transformations.

A second way of determining this relationship is to start with

$$\mathbf{y} = A\mathbf{x},$$

then multiply by Q^{-1}:

$$Q^{-1}\mathbf{y} = Q^{-1}A\mathbf{x},$$

insert $QQ^{-1} = I$:

$$Q^{-1}\mathbf{y} = Q^{-1}AQQ^{-1}\mathbf{x}, \tag{3}$$

regroup terms:

$$(Q^{-1}\mathbf{y}) = (Q^{-1}AQ)(Q^{-1}\mathbf{x}),$$

and substitute:

$$\mathbf{y} = A'\mathbf{x}'.$$

Equations (2) are examples of a similarity transformation. Figure 5 illustrate the connections between these vectors, operators, and transformations.

Scalar quantities, measured in one coordinate systems as

$$\mathbf{x} \cdot \mathbf{y} = \mathbf{x} \cdot (A\mathbf{x}) \equiv \mathbf{x}A\mathbf{x}, \tag{4}$$

are measured as

$$\mathbf{x}' \cdot \mathbf{y}' = \mathbf{x}' \cdot (A'\mathbf{x}') = \tilde{\mathbf{x}}'A'\mathbf{x}' \tag{5}$$

in another coordinate system.

Physics requires that scalar measurements be the same for all observers, that is,

$$\tilde{\mathbf{x}}A\mathbf{x} = \tilde{\mathbf{x}}'A'\mathbf{x}'. \tag{6}$$

To verify this equality proceed as follows:

$$\begin{aligned} \tilde{\mathbf{x}}A\mathbf{x} &= (\widetilde{Q\mathbf{x}'})A(Q\mathbf{x}') \\ &= (\tilde{\mathbf{x}}'\tilde{Q})A(Q\mathbf{x}') \\ &= (\tilde{\mathbf{x}}'(\tilde{Q}AQ)\mathbf{x}' \\ &= \tilde{\mathbf{x}}'(Q^{-1}AQ)\mathbf{x}' \\ &= \tilde{\mathbf{x}}'A'\mathbf{x}'. \end{aligned} \tag{7}$$

In terms of components, the general transformation of a tensor T, which may be written as

$$T' = MT\tilde{N} \tag{8a}$$

$$T = (M \times\cdot N)T \tag{8b}$$

or

$$T' = mT, \tag{8c}$$

where m denotes the direct product. $M \times\cdot N$ becomes

$$T'_{ij} = M_{il}N_{jk}T_{lk} = m_{ijlk}T_{lk}. \tag{8d}$$

The formal similarity existing between equation (8c) and the vector equation

$$t' = Lt \tag{9}$$

can be exploited in the following manner. The transformation

$$T' = mT \tag{10a}$$

or

$$T'_{ij} = m_{ijkl}T_{kl} = M_{ik}N_{jl}T_{kl} \tag{10b}$$

for a tensor of rank two and dimension two may be written as follows:

$$\begin{pmatrix} T'_{11} & T'_{12} \\ T'_{21} & T'_{22} \end{pmatrix} = m_{ijkl}\begin{pmatrix} T_{11} & T_{12} \\ T_{21} & T_{22} \end{pmatrix}, \tag{10c}$$

where the matrix m_{ijkl}, formed by the law of dictionary order, has been derived earlier.

To find each individual element T'_{ij} appearing in equation (10c), the procedure of matrix elongation can be applied. That is, elements of the tensor T'_{ij} treated as the rows of a 2 by 2 matrix can be stretched into a column:

$$\begin{pmatrix} T'_{11} & T'_{12} \\ & \\ T'_{21} & T'_{22} \end{pmatrix} = \begin{pmatrix} T'_{11} \\ T'_{12} \\ T'_{21} \\ T'_{22} \end{pmatrix} = VT. \tag{11}$$

Thus the $T'_{ij} = (M \times N) T_{kl}$ becomes simply the following ordinary matrix product:

$$\begin{pmatrix} T'_{11} \\ T'_{12} \\ T'_{21} \\ T'_{22} \end{pmatrix} = \begin{pmatrix} M_{11}N_{11} & M_{11}N_{12} & M_{12}N_{11} & M_{12}N_{12} & T_{11} \\ M_{11}N_{21} & M_{11}N_{22} & M_{12}N_{21} & M_{12}N_{22} & T_{12} \\ M_{21}N_{11} & M_{21}N_{12} & M_{22}N_{11} & M_{22}N_{12} & T_{21} \\ M_{21}N_{21} & M_{21}N_{22} & M_{22}N_{21} & M_{22}N_{22} & T_{22} \end{pmatrix} \tag{12}$$

$$= T' = (M \times N) : T.$$

Therefore, to find any of the elements T'_{ij}, the regular row-by-column rule of ordinary matrix multiplication is employed:

$$T'_{i1} = M_{i1}N_{11}T_{11} + M_{i1}N_{12}T_{12} + M_{i2}N_{11}T_{21} + M_{i2}N_{12}T_{22}. \tag{13}$$

Similarly, the elements of the tensor T_{ij} treated as the rows of a 3 by 3 matrix would be stretched into a column as follows:

$$\begin{pmatrix} T'_{11} & T'_{12} & T'_{13} \\ & & \\ T'_{21} & T'_{22} & T'_{23} \\ & & \\ T'_{31} & T'_{32} & T'_{13} \end{pmatrix} = \begin{pmatrix} T'_{11} \\ T'_{12} \\ T'_{13} \\ T'_{21} \\ T'_{22} \\ T'_{23} \\ T'_{31} \\ T'_{32} \\ T'_{33} \end{pmatrix}. \tag{14}$$

The Stretch Rule for a three-dimensional tensor of rank two is

$$(T'_{ijk}) \Rightarrow \begin{pmatrix} T'_{111} \\ T'_{121} \\ T'_{211} \\ T'_{221} \\ T'_{112} \\ T'_{122} \\ T'_{212} \\ T'_{222} \end{pmatrix}. \tag{15}$$

Using the law of dictionary order, matrices of any size can be multiplied. Thus a matrix of dimensionality (4, 3) can be multiplied by a matrix of dimensionality (2, 2) to give a matrix of dimensionality (8, 6):

$$(4, 3) \times (2, 2) = (8, 6). \tag{16}$$

A third rank tensor will transform, as follows:

$$T' = (L \times M \times N)T, \tag{17}$$

or:

$$T'_{ijk} = L_{il} M_{jm} N_{kn} T_{lmn}. \tag{18}$$

Consider the following transformation of a two-dimensional tensor of second rank:

$$T'_{ij} = M_{ik} N_{jk} T_{kl} \qquad i, j, k, l = 1, 2. \tag{19}$$

Now expand the Einstein sums, as follows:

$$\begin{aligned} T'_{ij} &= M_{ik}(N_{j1} T_{k1} + N_{j2} T_{k2}) \\ &= M_{i1} N_{j1} T_{11} + M_{i1} N_{j2} T_{12} + M_{i2} N_{j1} T_{21} \\ &\quad + M_{i2} N_{j2} T_{22}. \end{aligned} \tag{20}$$

The element of the tensor indexed by $i - j = 1$ is

$$T'_{11} = M_{11} N_{11} T_{11} + M_{11} N_{12} T_{12} + M_{12} N_{11} T_{21} + M_{12} N_{12} T_{22}. \tag{21}$$

Now approach the problem in a different way:

$$T'_{ij} = N_{jl}(M_{ik} T_{kl}) = M_{ik} T_{kl} (\tilde{N})_{lj}, \tag{22}$$

where $\tilde{N}$ is the transpose of N

Thus

$$\begin{pmatrix} T'_{11} & T'_{12} \\ T'_{21} & T_{22} \end{pmatrix} = \begin{pmatrix} M_{11} & M_{12} \\ M_{21} & M_{22} \end{pmatrix} \begin{pmatrix} T_{11} & T_{12} \\ T_{21} & T_{22} \end{pmatrix} \begin{pmatrix} N_{11} & N_{21} \\ N_{12} & N_{22} \end{pmatrix}, \tag{23}$$

which becomes

$$\begin{pmatrix} T'_{11} & T'_{12} \\ T'_{21} & T'_{22} \end{pmatrix} = \begin{pmatrix} M_{11} & M_{12} \\ M_{21} & M_{22} \end{pmatrix} \begin{pmatrix} T_{11}N_{11} + T_{12}N_{12} & T_{11}N_{21} + T_{12}N_{22} \\ T_{21}N_{11} + T_{22}N_{12} & T_{21}N_{21} + T_{22}N_{22} \end{pmatrix}. \tag{24}$$

Multiply the right side to extract the T'_{11} term:

$$T'_{11} = (M_{11}T_{11}N_{11} + M_{11}T_{12}N_{12} + M_{12}T_{21}N_{11} + M_{12}T_{22}N_{12}), \tag{25}$$

which is the same as expression (21).

The general tensor transformation,

$$T' = MT\tilde{N}, \tag{26}$$

if repeated many times, requires less computational effort if handled in the equivalent form,

$$T' = (M \times\cdot N)T. \tag{27}$$

Let

$$T = \begin{pmatrix} T_{11} & T_{12} \\ T_{21} & T_{22} \end{pmatrix}, \quad M = \begin{pmatrix} M_{11} & M_{12} \\ M_{21} & M_{22} \end{pmatrix}, \quad \text{and} \quad N = \begin{pmatrix} N_{11} & N_{12} \\ N_{21} & N_{22} \end{pmatrix}. \tag{28}$$

Then equation (26) becomes

$$\begin{aligned} T' &= \begin{pmatrix} M_{11} & M_{12} \\ M_{21} & M_{22} \end{pmatrix} \begin{pmatrix} T_{11} & T_{12} \\ T_{21} & T_{22} \end{pmatrix} \widetilde{\begin{pmatrix} N_{11} & N_{12} \\ N_{21} & N_{22} \end{pmatrix}} \\ &= \begin{pmatrix} M_{11} & M_{12} \\ M_{21} & M_{22} \end{pmatrix} \begin{pmatrix} T_{11} & T_{12} \\ T_{21} & T_{22} \end{pmatrix} \begin{pmatrix} N_{11} & N_{21} \\ N_{12} & N_{22} \end{pmatrix} \\ &= \begin{pmatrix} M_{11} & M_{12} \\ M_{21} & M_{22} \end{pmatrix} \begin{pmatrix} T_{11}N_{11} + T_{12}N_{12} & T_{11}N_{21} + T_{12}N_{22} \\ T_{21}N_{11} + T_{21}N_{12} & T_{22}N_{21} + T_{22}N_{22} \end{pmatrix} \\ &= \left(\begin{matrix} M_{11}T_{11}N_{11} + M_{11}T_{12}N_{12} + M_{12}T_{21}N_{11} + M_{12}T_{21}N_{12} \\ M_{21}T_{11}N_{11} + M_{21}T_{12}N_{12} + M_{22}T_{21}N_{11} + M_{22}T_{21}N_{12} \end{matrix} \right. \\ &\qquad \left. \begin{matrix} M_{11}T_{11}N_{21} + M_{11}T_{12}N_{22} + M_{12}T_{22}N_{21} + M_{12}T_{22}N_{22} \\ M_{21}T_{11}N_{21} + M_{21}T_{12}N_{22} + M_{22}T_{22}N_{21} + M_{22}T_{22}N_{22} \end{matrix} \right). \end{aligned} \tag{29}$$

In the transformation of a two-dimensional tensor T,

$$T'_{ij} = M_{ik}N_{jl}T_{kl} = (M \times\cdot\, N):T_{kl}, \tag{30}$$

the quantity $M \times N$ is a right direct product of the transformation matrices, M and N. For two dimensions, transformation (30) becomes

$$\begin{pmatrix} T'_{11} & T'_{12} \\ T'_{21} & T'_{22} \end{pmatrix} = \begin{pmatrix} M_{11} & M_{12} \\ M_{21} & M_{22} \end{pmatrix} \times\cdot \begin{pmatrix} N_{11} & N_{12} \\ N_{21} & N_{22} \end{pmatrix} : \begin{pmatrix} T_{11} & T_{12} \\ T_{21} & T_{22} \end{pmatrix} \tag{31}$$

or

$$\begin{pmatrix} T'_{11} & T'_{12} \\ T'_{21} & T'_{22} \end{pmatrix} = \left(\begin{array}{cc|cc} M_{11}N_{11} & M_{11}N_{12} & M_{12}N_{11} & M_{12}N_{12} \\ M_{11}N_{21} & M_{11}N_{22} & M_{12}N_{21} & M_{12}N_{22} \\ \hline M_{21}N_{11} & M_{21}N_{12} & M_{22}N_{11} & M_{22}N_{12} \\ M_{21}N_{21} & M_{21}N_{22} & M_{22}N_{21} & M_{22}N_{22} \end{array}\right) : \begin{pmatrix} T_{11} & T_{12} \\ T_{21} & T_{22} \end{pmatrix}. \tag{32}$$

In order to show the mechanics of this operation, restate equation (32) as follows:

$$\begin{pmatrix} T'_{11} \\ T'_{12} \\ T'_{21} \\ T'_{22} \end{pmatrix} = \begin{pmatrix} M_{11}N_{11} & M_{11}N_{12} & M_{12}N_{11} & M_{12}N_{12} \\ M_{11}N_{21} & M_{11}N_{22} & M_{12}N_{21} & M_{12}N_{22} \\ M_{21}N_{11} & M_{21}N_{12} & M_{22}N_{11} & M_{22}N_{12} \\ M_{21}N_{21} & M_{21}N_{22} & M_{22}N_{21} & M_{22}N_{12} \end{pmatrix} \begin{pmatrix} T_{11} \\ T_{12} \\ T_{21} \\ T_{22} \end{pmatrix}. \tag{33}$$

Then, carrying out the ordinary matrix multiplication indicated by equation (33), equation (32) becomes

$$\begin{pmatrix} T'_{11} & T'_{12} \\ T'_{21} & T'_{22} \end{pmatrix}$$

$$= \left(\begin{array}{c|c} M_{11}N_{11}T_{11} + M_{11}N_{12}T_{12} + M_{12}N_{11}T_{21} + M_{12}N_{12}T_{22} & M_{11}N_{21}T_{11} + M_{11}N_{22}T_{12} + M_{12}N_{21}T_{21} + M_{12}N_{22}T_{22} \\ \hline M_{21}N_{11}T_{11} + M_{21}N_{12}T_{12} + M_{22}N_{11}T_{21} + M_{22}N_{12}T_{22} & M_{21}N_{21}T_{11} + M_{21}N_{22}T_{12} + M_{22}N_{21}T_{21} + M_{22}N_{22}T_{22} \end{array}\right) \tag{34}$$

As previously remarked, the computational advantage of this direct product is that the operator $M \times N$ has to be evaluated only once for each problem and it then can be applied to every tensor separately, whereas the product MTN has to be calculated for each tensor.

To make equations (33) and (34) more quantitative:

$$
M \cdot\!\times N = \begin{array}{c} \\ (11) \\ (12) \\ (21) \\ (22) \end{array}
\begin{array}{c}
\begin{array}{cccc} (11) \quad\quad & (12) \quad\quad & (21) \quad\quad & (22) \end{array} \\
\begin{pmatrix}
M_{11}N_{11} & M_{11}N_{12} & M_{12}N_{11} & M_{12}N_{12} \\
M_{11}N_{21} & M_{11}N_{22} & M_{12}N_{21} & M_{12}N_{22} \\
M_{21}N_{11} & M_{21}N_{12} & M_{22}N_{11} & M_{22}N_{12} \\
M_{21}N_{21} & M_{21}N_{22} & M_{22}N_{21} & M_{22}N_{22}
\end{pmatrix}
\end{array}. \tag{35}
$$

The proper element of T' is identified by the index couplet on the row and evaluated by a proper sum over the columns.

The following direct product tensor transformation of an antisymmetric tensor,

$$
\begin{pmatrix} T'_{11} & T'_{12} \\ T'_{21} & T'_{22} \end{pmatrix}
= \left(\begin{array}{cc:cc}
M_{11}N_{11} & M_{11}N_{12} & M_{12}N_{11} & M_{12}N_{12} \\
M_{11}N_{21} & M_{11}N_{22} & M_{12}N_{21} & M_{12}N_{22} \\
\hdashline
M_{21}N_{11} & M_{21}N_{12} & M_{22}N_{11} & M_{22}N_{12} \\
M_{21}N_{21} & M_{21}N_{22} & M_{22}N_{21} & M_{22}N_{22}
\end{array}\right) : \begin{pmatrix} 0 & T_{12} \\ -T_{12} & 0 \end{pmatrix}, \tag{36}
$$

results in

$$
\begin{pmatrix} T'_{11} & T'_{12} \\ T'_{21} & T'_{22} \end{pmatrix} = \begin{pmatrix} (M_{11}N_{12} - M_{12}N_{11})T_{12} & (M_{11}N_{22} - M_{12}N_{21})T_{12} \\ (M_{21}N_{12} - M_{22}N_{11})T_{12} & (M_{21}N_{22} - M_{22}N_{21})T_{22} \end{pmatrix}. \tag{37}
$$

Here T_{12}, the only tensor element is common to all terms, and therefore it comes out as a simple scalar factor:

$$
\begin{pmatrix} T'_{11} & T'_{12} \\ T'_{21} & T'_{22} \end{pmatrix} = \begin{pmatrix} (M_{11}N_{12} - M_{12}N_{11}) & (M_{11}N_{22} - M_{12}N_{21}) \\ (M_{21}N_{12} - M_{22}N_{11}) & (M_{21}N_{22} - M_{22}N_{21}) \end{pmatrix} T_{12}.
$$

Notice that T' cannot be antisymmetric unless $M = N$.

Then:

$$
\begin{aligned}
\begin{pmatrix} 0 & T'_{12} \\ T'_{21} & 0 \end{pmatrix} &= \begin{pmatrix} 0 & (M_{11}M_{22} - M_{12}M_{21}) \\ (M_{21}M_{12} - M_{22}M_{11}) & 0 \end{pmatrix} T_{12} \\
&= \begin{pmatrix} 0 & (M_{11}M_{22} - M_{12}M_{21}) \\ -(M_{11}M_{22} - M_{12}M_{21}) & 0 \end{pmatrix} \\
&= \begin{pmatrix} 0 & T'_{12} \\ -T'_{12} & 0 \end{pmatrix},
\end{aligned}
\tag{39}
$$

which leads to a simple relationship for the single element:

$$T'_{12} = (M_{11}M_{22} - M_{12}M_{21})T_{12} = (\det M)T_{12}. \tag{40}$$

If M has a determinant equal to one, then

$$T'_{12} = T_{12}. \tag{41}$$

Beginning anew, since T is antisymmetric, certain terms may be dropped from $M \times\cdot N$ without loss of accuracy as long as it is then only applied to antisymmetric tensors:

$$
M \times\cdot N = \left(\begin{array}{cc:cc}
0 & M_{11}N_{12} & M_{12}N_{11} & 0 \\
0 & M_{11}N_{22} & M_{12}N_{21} & 0 \\
\hdashline
0 & M_{21}N_{12} & M_{22}N_{11} & 0 \\
0 & M_{21}N_{12} & M_{22}N_{21} & 0
\end{array}\right). \tag{42}
$$

Further, if $M = N$, then

$$
M \times\cdot M = \left(\begin{array}{cc:cc}
0 & 0 & 0 & 0 \\
0 & M_{11}M_{22} & M_{12}M_{21} & 0 \\
\hdashline
0 & M_{21}M_{12} & M_{22}M_{11} & 0 \\
0 & 0 & 0 & 0
\end{array}\right), \tag{43}
$$

which agrees with the development (40), (41), and (42).

Let

$$
M = \begin{pmatrix} M_{11} & M_{12} & M_{13} \\ M_{21} & M_{22} & M_{23} \\ M_{31} & M_{32} & M_{33} \end{pmatrix} \quad \text{and} \quad N = \begin{pmatrix} N_{11} & N_{12} & N_{13} \\ N_{21} & N_{22} & N_{23} \\ N_{31} & N_{32} & N_{33} \end{pmatrix} \tag{44}
$$

be two 3 by 3 transformation matrices. Then the explicit form of the direct product $M \times\cdot N$ is as follows:

$$M \cdot\times N =$$

$$\begin{array}{c|ccc:ccc:ccc|}
 & (11) & (12) & (13) & (21) & (22) & (23) & (31) & (32) & (33) \\
(11) & M_{11}N_{11} & M_{11}N_{12} & M_{11}N_{13} & M_{12}N_{11} & M_{12}N_{12} & M_{12}N_{13} & M_{13}N_{11} & M_{13}N_{12} & M_{13}N_{13} \\
(12) & M_{11}N_{21} & M_{11}N_{22} & M_{11}N_{23} & M_{12}N_{21} & M_{12}N_{22} & M_{12}N_{23} & M_{13}N_{21} & M_{13}N_{22} & M_{13}N_{23} \\
(13) & M_{11}N_{31} & M_{11}N_{32} & M_{11}N_{33} & M_{12}N_{31} & M_{12}N_{32} & M_{12}N_{33} & M_{13}N_{31} & M_{13}N_{32} & M_{13}N_{33} \\
\hdashline
(21) & M_{21}N_{11} & M_{21}N_{12} & M_{21}N_{13} & M_{22}N_{11} & M_{22}N_{12} & M_{22}N_{13} & M_{23}N_{11} & M_{23}N_{12} & M_{23}N_{13} \\
(22) & M_{21}N_{21} & M_{21}N_{22} & M_{21}N_{23} & M_{22}N_{21} & M_{22}N_{22} & M_{22}N_{23} & M_{23}N_{21} & M_{23}N_{22} & M_{23}N_{23} \\
(23) & M_{21}N_{31} & M_{21}N_{32} & M_{21}N_{33} & M_{22}N_{31} & M_{22}N_{32} & M_{22}N_{33} & M_{23}N_{31} & M_{23}N_{32} & M_{23}N_{33} \\
\hdashline
(31) & M_{31}N_{11} & M_{31}N_{12} & M_{31}N_{13} & M_{32}N_{11} & M_{32}N_{12} & M_{32}N_{12} & M_{33}N_{11} & M_{33}N_{12} & M_{33}N_{13} \\
(32) & M_{31}N_{21} & M_{31}N_{22} & M_{31}N_{23} & M_{32}N_{21} & M_{32}N_{22} & M_{32}N_{23} & M_{33}N_{21} & M_{33}N_{22} & M_{33}N_{23} \\
(33) & M_{31}N_{31} & M_{31}N_{32} & M_{31}N_{33} & M_{32}N_{31} & M_{32}N_{32} & M_{32}N_{33} & M_{33}N_{31} & M_{33}N_{32} & M_{33}N_{33}
\end{array}. \tag{45}$$

For an antisymmetric 3 by 3 tensor T,

$$T_{ij} = -T_{ji} \quad \text{and} \quad T_{ii}. \tag{46}$$

Therefore

$$\begin{pmatrix} T'_{11} & T'_{12} & T'_{12} \\ T'_{21} & T'_{22} & T'_{23} \\ T'_{31} & T'_{32} & T'_{33} \end{pmatrix} = (M \times\cdot N) \begin{pmatrix} 0 & T_{12} & T_{13} \\ T_{21} & 0 & T_{23} \\ T_{31} & T_{32} & 0 \end{pmatrix}. \tag{47}$$

Since T has three zeros, the $M \times\cdot N$ direct product transformation matrix may contain as many as 27 zeros and yet apply correctly to any antisymmetric tensor. The general appearance of such a tensor is as follows:

$$M \cdot\times N =$$

$$\begin{array}{c|ccc:ccc:ccc|}
 & (11) & (12) & (13) & (21) & (22) & (23) & (31) & (32) & (33) \\
(11) & 0 & M_{11}N_{12} & M_{11}N_{13} & M_{12}N_{11} & 0 & M_{12}N_{13} & M_{13}N_{11} & M_{13}N_{12} & 0 \\
(12) & 0 & M_{11}N_{22} & M_{11}N_{23} & M_{12}N_{21} & 0 & M_{12}N_{23} & M_{13}N_{21} & M_{13}N_{22} & 0 \\
(13) & 0 & M_{11}N_{32} & M_{11}N_{33} & M_{12}N_{31} & 0 & M_{12}N_{33} & M_{13}N_{31} & M_{13}N_{32} & 0 \\
\hdashline
(21) & 0 & M_{21}N_{12} & M_{21}N_{13} & M_{22}N_{11} & 0 & M_{12}N_{13} & M_{23}N_{11} & M_{23}N_{12} & 0 \\
(22) & 0 & M_{21}N_{22} & M_{21}N_{23} & M_{22}N_{21} & 0 & M_{22}N_{23} & M_{23}N_{21} & M_{23}N_{22} & 0 \\
(23) & 0 & M_{21}N_{32} & M_{21}N_{33} & M_{22}N_{31} & 0 & M_{12}N_{33} & M_{23}N_{31} & M_{23}N_{32} & 0 \\
\hdashline
(31) & 0 & M_{31}N_{12} & M_{31}N_{13} & M_{32}N_{11} & 0 & M_{32}N_{13} & M_{33}N_{11} & M_{33}N_{12} & 0 \\
(32) & 0 & M_{31}N_{22} & M_{31}N_{23} & M_{32}N_{21} & 0 & M_{32}N_{23} & M_{33}N_{21} & M_{33}N_{22} & 0 \\
(33) & 0 & M_{31}N_{32} & M_{31}N_{33} & M_{32}N_{31} & 0 & M_{32}N_{33} & M_{33}N_{31} & M_{33}N_{32} & 0
\end{array}. \tag{48}$$

Actually, if T' is also antisymmetric, more elements of $M \times\cdot N$ are zero.

Any scalar quantity is an isotropic tensor of the zeroth order. First-order tensors, however, cannot be isotropic, because they are vectors, and vectors transform under a change of axes.

The Kronecker delta δ_{ij} can be considered to be a second-order symmetric tensor that is preserved under a change of coordinates. Write

$$u'_{jl} = l_{ij} l_{kl} \delta_{ik}, \tag{49}$$

and sum on i, k to obtain

$$\begin{aligned} u'_{jl} = l_{ij} l_{il} &= 1, \qquad \text{if} \quad j = l, \\ &= 0, \qquad \text{if} \quad j \neq l. \end{aligned} \tag{50}$$

Since u' and δ thus seem to obey the same rules, u' can be replaced by δ, so that

$$\delta_{ik} \quad \text{goes into} \quad \delta_{jl} \tag{51}$$

under a rotation of axes.

Similarly, the Levi-Civita Symbol ϵ_{ijk} can be considered to be a third-order antisymmetric tensor that is preserved under a change of coordinates. The transformation of this symbol is

$$\begin{aligned} \epsilon'_{jln} &= l_{ij} l_{kl} l_{mn} \epsilon_{ikm} \\ &= l_{1j} l_{2l} l_{3n} + l_{2j} l_{3l} l_{1n} + l_{3j} l_{1l} l_{2n} - l_{2j} l_{1l} l_{3n} - l_{3j} l_{2l} l_{1n} - l_{1j} l_{3l} l_{2n}. \end{aligned} \tag{52}$$

If $j = l$ or $j = n$ or $l = n$, all components cancel; if j, l, n are all different, the expression reduces to

$$\begin{pmatrix} l_{1j} & l_{1l} & l_{1n} \\ l_{2j} & l_{2l} & l_{2n} \\ l_{3j} & l_{3l} & l_{3n} \end{pmatrix}, \tag{53}$$

which is equal to 1 if j, l, n are in an even permutation and equal to -1 if they are in an odd permutation. Hence ϵ_{ikm} transforms into ϵ_{jln} under the rule for third-rank tensors.

D. Quadratic Forms and Diagonalization

In the theory of matrices and determinants it is learned that certain properties of a matrix are invariant when it is changed into its diagonal form. For a 2 by 2 matrix, these invariants are the trace and the determinant. As an example, consider the Pauli spin matrices:

$$\sigma_x = \begin{pmatrix} 0 & 1 \\ 1 & 0 \end{pmatrix}; \qquad \sigma_y = \begin{pmatrix} 0 & -i \\ i & 0 \end{pmatrix}; \qquad \sigma_z = \begin{pmatrix} 1 & 0 \\ 0 & -1 \end{pmatrix}. \tag{1}$$

All of these matrices have a trace of zero and a determinant of -1. The diagonal form of σ_x and σ_y is the same and is equal to σ_z, which is already in diagonal form.
The conditions that the trace and determinant of a 2 by 2 matrix remain invariant under diagonalization are sufficient to define the diagonalized form of the original matrix.
Consider a general matrix being transformed into its diagonalized form:

$$\begin{pmatrix} a & b \\ c & d \end{pmatrix} \rightarrow \begin{pmatrix} \lambda_1 & 0 \\ 0 & \lambda_2 \end{pmatrix}. \tag{2}$$

The conditions on the invariants in this case are

$$a + d = \lambda_1 + \lambda_2, \tag{3}$$

expressing the fact that the trace is preserved, and

$$(ad - cb) = \lambda_1 \lambda_2, \tag{4}$$

expressing the fact that the determinant is preserved.

These two simultaneous equations can be solved algebraically for the unknowns λ_1 and λ_2. From equation (3),

$$\lambda_2 = a + d - \lambda_1. \tag{5}$$

Substituting this into equation (4) yields a quadratic equation in λ_1:

$$(ad - bc) = (a + d - \lambda_1)\lambda_1, \tag{6}$$

or

$${\lambda_1}^2 - \lambda_1(a + d) + (ad - bc) = 0. \tag{7}$$

Equation (7) has the solution

$$\lambda_1 = \tfrac{1}{2}[(a + d) \pm \sqrt{(a + d)^2 - 4(ad - bc)}], \tag{8}$$

which can be written in the convenient form

$$\tfrac{1}{2}(\text{trace} \pm \sqrt{\text{trace} - 4 \times \text{determinant}}). \tag{9}$$

From equation (8), it can be seen that if the plus sign in equation (9) is chosen for λ_1, then λ_2 is found from equation (9) by using the minus sign and vice versa. Hence there are two different diagonal forms for a 2×2 matrix depending upon whether the + or − sign is chosen for λ_1 in equation (9). One diagonal form is merely the other with the diagonal elements interchanged.

An alternate approach to the diagonalization of a matrix, which when applied to a 2 by 2 matrix gives equation (9) directly, is derived from the consideration of the "secular determinant" or characteristic equation. Assume a 2 by 2 matrix M multiplies a 2-vector, $\begin{pmatrix} x \\ y \end{pmatrix}$, as follows:

$$M\begin{pmatrix} x \\ y \end{pmatrix} = \lambda\begin{pmatrix} x \\ y \end{pmatrix}, \tag{10}$$

where λ is the matrix M in diagonal form. Equation (10) can be written as a set of simultaneous linear equations, in matrix form:

$$\begin{pmatrix} a & b \\ c & d \end{pmatrix}\begin{pmatrix} x \\ y \end{pmatrix} = \begin{pmatrix} \lambda_1 & 0 \\ 0 & \lambda_2 \end{pmatrix}\begin{pmatrix} x \\ y \end{pmatrix}. \tag{11}$$

This set of equations has nontrivial solutions if and only if the secular determinant vanishes:

$$\begin{aligned} \left|\begin{pmatrix} a & b \\ c & d \end{pmatrix} - \begin{pmatrix} \lambda & 0 \\ 0 & \lambda \end{pmatrix}\right| &= 0, \\ \begin{vmatrix} a-\lambda & b \\ c & d-\lambda \end{vmatrix} &= 0, \\ (a-\lambda)(d-\lambda) - bc &= 0, \\ \lambda^2 - (a+d)\lambda + (ad-bc) &= 0, \end{aligned} \tag{12}$$

which is the same as equation (7) and hence yields equation (9).

For an n by n matrix there are n invariant relations, which allow the determination of the n eigenvalues or nonzero elements of the diagonalized matrix. These invariants are the sums of all the determinants of principal minors of a given dimension m ($1 \le m \le n$). A "principal minor" is an m by m submatrix of an n by n matrix whose diagonal elements coincide with diagonal elements of the n by n matrix. For each value of m, an invariant equation is formed.

For successive values of m these equations are:

$$\begin{aligned} \textstyle\sum \text{(1 by 1) principal minor determinant} &= \lambda_1 + \lambda_2 \cdots \lambda_n = a_1 \\ \textstyle\sum \text{(2 by 2) principal minor determinant} &= \lambda_1\lambda_2 + \lambda_2\lambda_3 + \cdots \lambda_n\lambda_1 \\ \textstyle\sum \text{(3 by 3) principal minor determinant} &= \lambda_1\lambda_2\lambda_3 + \cdots \\ &\vdots \\ \textstyle\sum (n \text{ by } n) \text{ principal minor determinant} &= \lambda_1\lambda_2\lambda_3 \cdots \lambda_n = a_n, \end{aligned} \tag{13}$$

where these symmetric functions in the λs are also the coefficients of the characteristic equation:

$$|M - \lambda I| = \pm \lambda^n \pm a_1 \lambda^{n-1} \pm \cdots \pm a_n = 0. \tag{14}$$

To illustrate, the invariant equations (13) are shown below for a general 4 by 4 matrix:

$$\begin{pmatrix} a_{11} & a_{12} & a_{13} & a_{14} \\ a_{21} & a_{22} & a_{23} & a_{24} \\ a_{31} & a_{32} & a_{33} & a_{34} \\ a_{41} & a_{42} & a_{43} & a_{44} \end{pmatrix} \rightarrow \begin{pmatrix} \lambda_1 & 0 & 0 & 0 \\ 0 & \lambda_2 & 0 & 0 \\ 0 & 0 & \lambda_3 & 0 \\ 0 & 0 & 0 & \lambda_4 \end{pmatrix}. \tag{15}$$

Sums of all the m by m Principal Minor Determinants

1. m 1:

$$a_{11} + a_{22} + a_{33} + a_{44} = \lambda_1 + \lambda_2 + \lambda_3 + \lambda_4. \tag{16}$$

2. m 2:

$$\begin{aligned} &\begin{vmatrix} a_{11} & a_{12} \\ a_{21} & a_{22} \end{vmatrix} + \begin{vmatrix} a_{11} & a_{13} \\ a_{31} & a_{33} \end{vmatrix} + \begin{vmatrix} a_{11} & a_{14} \\ a_{41} & a_{44} \end{vmatrix} \\ &+ \begin{vmatrix} a_{22} & a_{23} \\ a_{32} & a_{33} \end{vmatrix} + \begin{vmatrix} a_{22} & a_{24} \\ a_{42} & a_{44} \end{vmatrix} + \begin{vmatrix} a_{33} & a_{34} \\ a_{43} & a_{44} \end{vmatrix} \\ &= \lambda_1\lambda_2 + \lambda_1\lambda_3 + \lambda_1\lambda_4 + \lambda_2\lambda_3 \\ &\quad + \lambda_2\lambda_4 + \lambda_3\lambda_4. \end{aligned} \tag{17}$$

3. m 3:

$$\begin{aligned} &\begin{vmatrix} a_{11} & a_{12} & a_{13} \\ a_{21} & a_{22} & a_{23} \\ a_{31} & a_{32} & a_{33} \end{vmatrix} + \begin{vmatrix} a_{11} & a_{12} & a_{14} \\ a_{21} & a_{22} & a_{24} \\ a_{41} & a_{42} & a_{44} \end{vmatrix} \\ &+ \begin{vmatrix} a_{11} & a_{13} & a_{14} \\ a_{31} & a_{33} & a_{34} \\ a_{41} & a_{43} & a_{44} \end{vmatrix} + \begin{vmatrix} a_{22} & a_{23} & a_{24} \\ a_{32} & a_{33} & a_{34} \\ a_{42} & a_{43} & a_{44} \end{vmatrix} \\ &= \lambda_1\lambda_2\lambda_3 + \lambda_1\lambda_2\lambda_4 + \lambda_1\lambda_3\lambda_4 \\ &\quad + \lambda_2\lambda_3\lambda_4. \end{aligned} \tag{18}$$

4. m 4:

$$\begin{vmatrix} a_{11} & a_{12} & a_{13} & a_{14} \\ a_{21} & a_{22} & a_{23} & a_{24} \\ a_{31} & a_{32} & a_{33} & a_{34} \\ a_{41} & a_{42} & a_{43} & a_{44} \end{vmatrix} = \lambda_1 \lambda_2 \lambda_3 \lambda_4. \tag{19}$$

Equations (16) and (19) are special cases that express the invariance of the trace and determinant, respectively. Notice that the four equations, (16) to (19) in the four unknowns λ_i are not linear.

A quadratic form has many possible sets of conjugate diameters, one of which is the familiar set of orthogonal principal axes. The concept of conjugate diameters, which is necessary for the general problem of diagonalization of a quadratic form, almost always is required for the simultaneous diagonalization of two quadratic forms. Consider the elliptical quadratic form shown in Fig. 6. The orthogonal principal axes, a and b, are shown. Two chords, drawn parallel to the minor axis, are bisected by the major axis. Chords not shown, if parallel to the major axis, would be bisected by the minor axis. All diameters and axes pass through the center of the ellipse; chords do not.

Now consider the ellipse shown in Fig. 7. Its center is marked by an asterisk. Line segment P is drawn through the center. The dotted line segments indicate chords constructed parallel to P. A line q drawn through the points of bisection of these chords is the axis conjugate to P.

Line segments a and b in Fig. 6 are orthogonal principal axes and there is only one set of them. The ellipse is in diagonal form with respect to a and b.

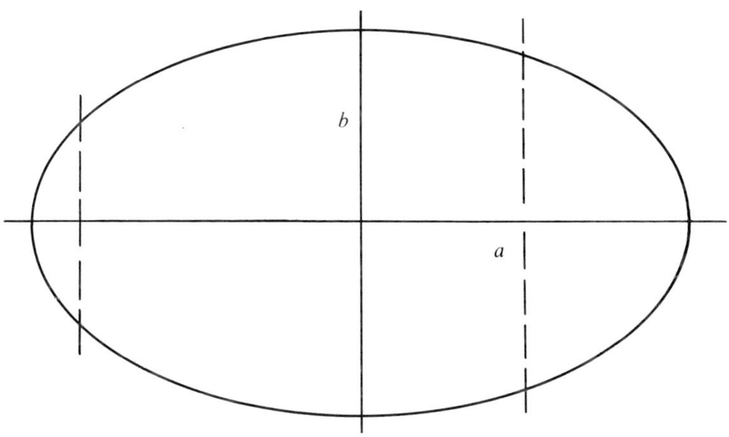

Figure 6. Major and minor axes of ellipse.

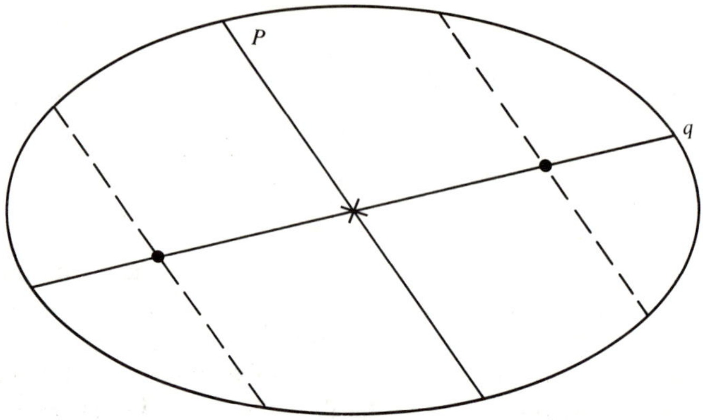

Figure 7. Conjugate diameters of an ellipse.

Line segments P and q in Fig. 7 are only two of an infinite number of pairs of nonorthogonal conjugate axes. The ellipse is in diagonal form with respect to any set of conjugate axes, such as P and q.

One final point—a parallelogram may be circumscribed about the ellipse using the axes P and q as shown in Fig. 8. For major and minor axes a and b (see Fig. 6) the parallelogram becomes a circumscribed rectangle, as shown in Fig. 9.

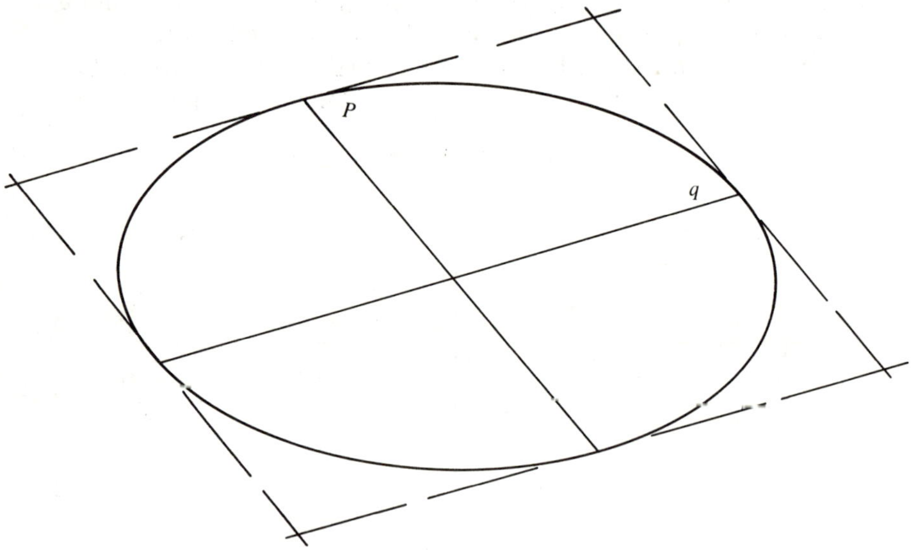

Figure 8. The circumscribed parallelogram.

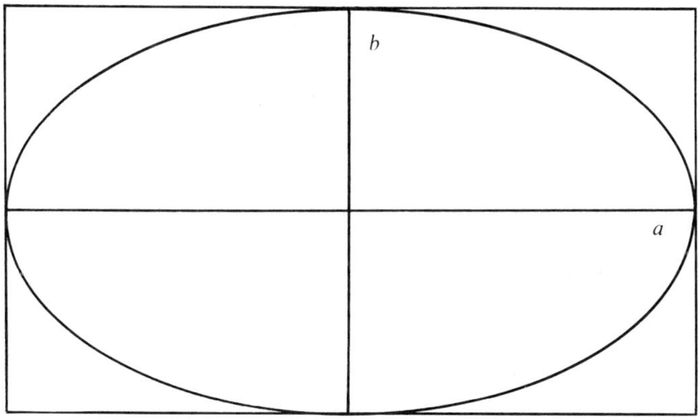

Figure 9. Rectangle tangent to ellipse at endpoints of orthogonal principal axes.

Consider the following quadratic expression:

$$F(x, y) = 2x^2 + 2\sqrt{2}\, xy + 3y^2. \tag{20}$$

The coefficients appearing in equation (20) were chosen merely to make the calculation easy. Since this expression is symmetric in the two coordinates, it may be written

$$F(x, y) = 2x^2 + \sqrt{2}\, xy + \sqrt{2}\, yx + 3y^2 \tag{21}$$

or

$$F(x, y) = \begin{Bmatrix} +2x^2 + \sqrt{2}\, xy \\ +\sqrt{2}\, yx + 3y^2 \end{Bmatrix}. \tag{22}$$

Treat x and y as the components of a two-dimensional vector $\mathbf{r}$ written in column form:

$$\mathbf{r} = \begin{pmatrix} x \\ y \end{pmatrix}, \quad \tilde{\mathbf{r}} \equiv \mathbf{r}^{\mathrm{Tr}} = (x, y). \tag{23}$$

Now F can be expressed in matrix form as follows:

$$\begin{aligned} F = \tilde{\mathbf{r}} A \mathbf{r} &= (x, y) \begin{pmatrix} 2 & \sqrt{2} \\ \sqrt{2} & 3 \end{pmatrix} \begin{pmatrix} x \\ y \end{pmatrix} \\ &= (x, y) \begin{pmatrix} 2x + \sqrt{2}\, y \\ \sqrt{2}\, x + 3y \end{pmatrix} \\ &= 2x^2 + \sqrt{2}\, xy + \sqrt{2}\, yx + 3y^2. \end{aligned}$$

Recovery of the same scalar quantity F from the product of two vectors with a tensor proves that the expression is correct. Somehow the matrix or tensor,

$$A = \begin{pmatrix} 2 & \sqrt{2} \\ \sqrt{2} & 3 \end{pmatrix}, \tag{25}$$

is related to the quadratic form F. It is useful to find the eigenvalues λ of this matrix. This will lead to an understanding of the special eigenvectors $\mathbf{r}$ of this matrix, defined by the equation

$$A\mathbf{r} = \lambda\mathbf{r}. \tag{26}$$

Eigenvectors are those vectors $\mathbf{r}$ that are unchanged in direction by the application of the operator A. The eigenvalues λ are the resultant changes in lengths of the eigenvectors as a result of the operation by A. For simplicity set F equal to unity. Then

$$2x^2 + 2\sqrt{2}\,xy + 3y^2 = 1 \tag{27}$$

represents some quadratic form (in this case an ellipse). The following table is needed to graph this ellipse:

x	y	Computations
0	+0.578	$y^2 = \frac{1}{3}$
0	−0.578	
+0·707	0	$x^2 = \frac{1}{2}$
−0·707	0	
+0·5	+0.235	$0{\cdot}5 + \sqrt{2}\,y + 3y^2 = 1$
+0·5	−0.707	
−0·5	−0.235	
−0·5	+0.707	

In Fig. 10, the ellipse is sketched with the aid of eight points.

The two concentric circles have radii a and b, respectively equal to the semi-major and semi-minor axes. The lengths of these axes are given by

$$\begin{aligned} a &= \frac{1}{\sqrt{\lambda_1}} = \frac{1}{\sqrt{1}} = 1 \\ b &= \frac{1}{\sqrt{\lambda_2}} = \frac{1}{\sqrt{4}} = 0.5. \end{aligned} \tag{28}$$

The eight points used to plot the ellipse are depicted by heavy dots. Numerical values are given for points that fall on the coordinate axes.

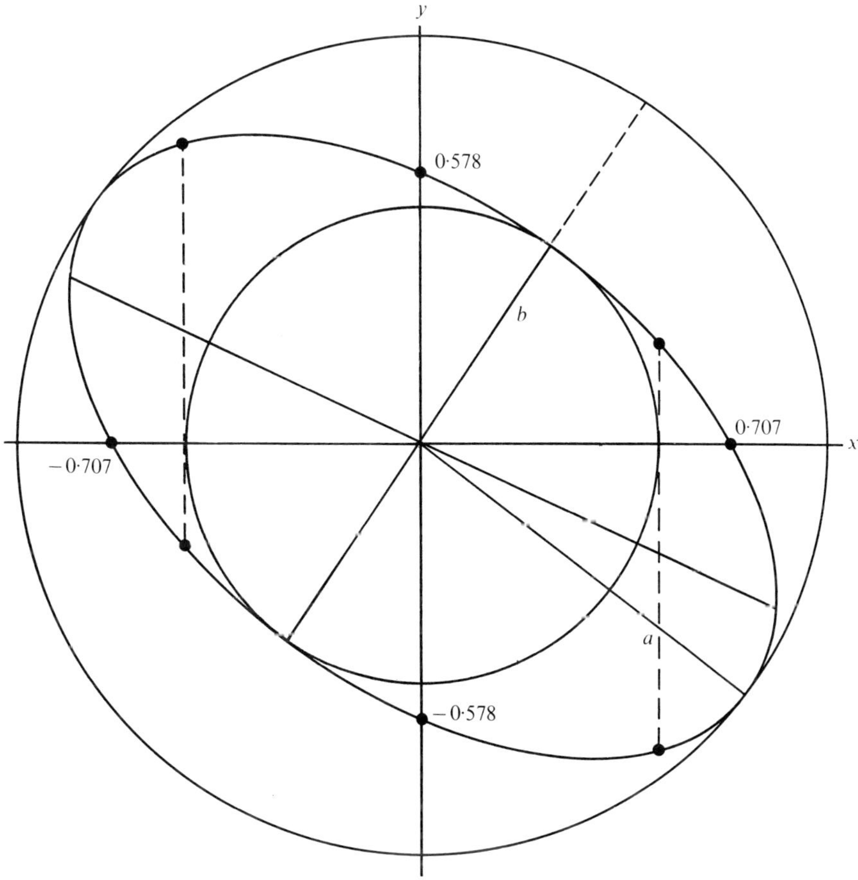

Figure 10. Graph of ellipse.

Returning to the eigenvalue problem,

$$\begin{pmatrix} 2 & \sqrt{2} \\ \sqrt{2} & 3 \end{pmatrix}\begin{pmatrix} x \\ y \end{pmatrix} = \begin{pmatrix} \lambda & 0 \\ 0 & \lambda \end{pmatrix}\begin{pmatrix} x \\ y \end{pmatrix}, \tag{29}$$

or

$$\begin{pmatrix} (2-\lambda) & \sqrt{2} \\ \sqrt{2} & (3-\lambda) \end{pmatrix}\begin{pmatrix} x \\ y \end{pmatrix} = 0, \tag{30}$$

or

$$\begin{cases} (2-\lambda)x + \sqrt{2}\,y = 0, \\ \sqrt{2}\,x + (3-\lambda)y = 0. \end{cases} \tag{31}$$

This set of simultaneous linear equations, being homogeneous, has a solution only if the determinant of the coefficient matrix is equal to zero, since by Cramer's rule the solutions are of the form

$$x, y = \frac{0}{\Delta}, \tag{32}$$

where Δ is the value of the determinant of the coefficient matrix. It is seen readily that if $\Delta \neq 0$, then expressions (31) for x, y are indeterminate. Conversely, if the set of simultaneous linear equations is nonhomogeneous, then it must be true that $\Delta \neq 0$.

$$\det A = \begin{vmatrix} 2 - \lambda & \sqrt{2} \\ \sqrt{2} & 3 - \lambda \end{vmatrix} = 0. \tag{33}$$

Then

$$\begin{aligned} (2 - \lambda)(3 - \lambda) - 2 &= 0, \\ 6 - 5\lambda + \lambda^2 - 2 &= 0, \\ \lambda^2 - 5\lambda + 4 &= 0, \end{aligned} \tag{37}$$

or

$$(\lambda - 4)(\lambda - 1) = 0. \tag{35}$$

Thus

$$\lambda_1 = 4, \qquad \lambda_2 = 1. \tag{36}$$

The importance of these eigenvalues stems from the fact that in terms of a new rotated coordinate system the quadratic form can be reduced to a sum of squares:

$$F(x', y') = 4x'^2 + 1y'^2, \tag{37}$$

or, in our special case

$$4x'^2 + 1y'^2 = 1. \tag{38}$$

The major and minor axes are 1 and 0.5, respectively. For that reason the concentric circles in Fig. 6 represent the inner and outer limits of the ellipse. Thus eight definite points and four points of tangency are known. Also known are the quadrants for the major and minor axes.

Now determine the axes or eigenvectors of the ellipse or tensor. Place λ_1 and λ_2 in expression (31):

$$\left.\begin{aligned} (2 - 1)x + \sqrt{2}\,y &= 0 \\ \sqrt{2}\,x + (3 - 1)y &= 0 \end{aligned}\right\} \quad \lambda_1 = 1 \tag{39}$$

or

$$\left.\begin{aligned} x + \sqrt{2}\,y &= 0 \\ \sqrt{2}\,x + 2y &= 0 \end{aligned}\right\}. \tag{40}$$

Since these equations are proportional and thus not independent, only the ratio of lengths can be determined:

$$\frac{y}{x} = -\frac{1}{\sqrt{2}} = -0.707. \tag{41}$$

The proper normalization is obtained as follows:

$$x^2 + y^2 = 1 = x^2 + (0.707x)^2 = 1{\cdot}5x^2 = 1 \tag{42}$$

or

$$x = \frac{1}{\sqrt{1.5}} = \sqrt{\tfrac{2}{3}}, \qquad y = -\frac{1}{\sqrt{3}}. \tag{43}$$

The second eigenvalue, $\lambda_2 = 4$, yields

$$\left\{\begin{aligned} (2-4)x + \sqrt{2}\,y &= 0 \\ \sqrt{2}\,x + (3-4)y &= 0 \end{aligned}\right. \tag{44}$$

or

$$\left\{\begin{aligned} -2x + \sqrt{2}\,y &= 0 \\ \sqrt{2}\,x - 1y &= 0. \end{aligned}\right. \tag{45}$$

Again these are proportional and give

$$\frac{y}{x} = \sqrt{2}. \tag{46}$$

The proper normalization yields

$$x^2 + y^2 = 1 = x^2 + (1.414x)^2 = 3x^2 = 1 \tag{47}$$

or

$$x = \frac{1}{\sqrt{3}}, \qquad y = \sqrt{\tfrac{2}{3}}. \tag{48}$$

Now form a similarity transformation matrix from the normalized and orthogonal eigenvectors:

$$Q = \begin{pmatrix} \sqrt{\frac{2}{3}} & -\sqrt{\frac{1}{3}} \\ \sqrt{\frac{1}{3}} & \sqrt{\frac{2}{3}} \end{pmatrix}, \qquad Q^{-1} = \tilde{Q} = Q^{\mathrm{Tr}} = \begin{pmatrix} \sqrt{\frac{2}{3}} & \sqrt{\frac{1}{3}} \\ -\sqrt{\frac{1}{3}} & \sqrt{\frac{2}{3}} \end{pmatrix}. \tag{49}$$

The two orthogonal vectors drawn on the graph define the perpendicular principal axes of the ellipse.

Now diagonalize the matrix, or tensor, A:

$$
\begin{aligned}
A' &= QAQ^{-1} \\
&= \begin{pmatrix} \sqrt{\frac{2}{3}} & -\frac{1}{\sqrt{3}} \\ \frac{1}{\sqrt{3}} & \sqrt{\frac{2}{3}} \end{pmatrix} \begin{pmatrix} 2 & \sqrt{2} \\ \sqrt{2} & 3 \end{pmatrix} \begin{pmatrix} \sqrt{\frac{2}{3}} & \frac{1}{\sqrt{3}} \\ -\frac{1}{\sqrt{3}} & \sqrt{\frac{2}{3}} \end{pmatrix} \\
&= \begin{pmatrix} \sqrt{\frac{2}{3}} & -\frac{1}{\sqrt{3}} \\ \frac{1}{\sqrt{3}} & \sqrt{\frac{2}{3}} \end{pmatrix} \begin{pmatrix} \sqrt{\frac{2}{3}} & \frac{4}{\sqrt{3}} \\ -\frac{1}{\sqrt{3}} & 4\sqrt{\frac{2}{3}} \end{pmatrix} = \begin{pmatrix} 1 & 0 \\ 0 & 4 \end{pmatrix},
\end{aligned} \tag{50}
$$

where the two diagonal terms are the eigenvalues. Notice the trace of the matrix is left unchanged by the similarity transformation and hence gives the sum of the eigenvalues:

$$2 + 3 = 1 + 4 = \text{Trace } A = \text{Trace } A'. \tag{51}$$

The orthogonality of the eigenvectors is also obvious:

$$\mathbf{v}_1 = \left[\sqrt{\tfrac{2}{3}}\,\hat{\mathbf{i}} - \frac{1}{\sqrt{3}}\,\hat{\mathbf{j}}\right], \tag{52}$$

$$\mathbf{v}_2 = \left[\frac{1}{\sqrt{3}}\,\hat{\mathbf{i}} + \sqrt{\tfrac{2}{3}}\,\hat{\mathbf{j}}\right]. \tag{53}$$

Thus

$$\mathbf{v}_1 \cdot \mathbf{v}_2 = \frac{\sqrt{2}}{3} - \frac{\sqrt{2}}{3} = 0. \tag{54}$$

Reconsider the original quadratic expression:

$$2x^2 + 2\sqrt{2}\,xy + 3y^2 = 1. \tag{55}$$

Complete the square by the usual method:

$$[2x^2 + 2\sqrt{2}\,xy] + 3y^2 = 1, \tag{56}$$

$$2\left[x^2 + \frac{2\sqrt{2}}{2}\,xy + \left(\frac{\sqrt{2}}{2}\,y\right)^2\right] + [3y^2 - y^2] = 1, \tag{57}$$

or

$$2\left[x + \left(\frac{1}{\sqrt{2}}\right)y\right]^2 + 2y^2 = 1. \tag{58}$$

The transformation

$$\begin{aligned} y' &= y, & y &= y', \\ x' &= \left(x + \frac{1}{\sqrt{2}}y\right), & x &= \left(x' - \frac{1}{\sqrt{2}}y'\right), \end{aligned} \tag{59}$$

leads to

$$2x'^2 + 2y'^2 = 1, \tag{60}$$

which again is the sum of perfect squares. But the axes are no longer orthogonal. The transformation matrix is triangular:

$$\mathbf{r} = Q\mathbf{r}', \qquad \mathbf{r}' = Q^{-1}\mathbf{r}, \tag{61}$$

with

$$Q = \begin{pmatrix} 1 & -\frac{1}{\sqrt{2}} \\ 0 & 1 \end{pmatrix}, \qquad Q^{-1} = \begin{pmatrix} 1 & \frac{1}{\sqrt{2}} \\ 0 & 1 \end{pmatrix}, \qquad \tilde{Q} = \begin{pmatrix} 1 & 0 \\ -\frac{1}{\sqrt{2}} & 1 \end{pmatrix}. \tag{62}$$

For an orthogonal matrix,

$$Q^{-1} = \tilde{Q}, \qquad \det Q = 1. \tag{63}$$

Although this does not apply to triangular Q, $\det Q$ does equal 1. For an orthogonal Q the similarity transformation diagonalizes the matrix and preserves the trace. Here

$$\begin{aligned} \tilde{Q}AQ &= \begin{pmatrix} 1 & 0 \\ -\frac{1}{\sqrt{2}} & 1 \end{pmatrix} \begin{pmatrix} 2 & \sqrt{2} \\ \sqrt{2} & 3 \end{pmatrix} \begin{pmatrix} 1 & -\frac{1}{\sqrt{2}} \\ 0 & 1 \end{pmatrix} \\ &= \begin{pmatrix} 1 & 0 \\ -\frac{1}{\sqrt{2}} & 1 \end{pmatrix} \begin{pmatrix} 2 & 0 \\ \sqrt{2} & 2 \end{pmatrix} = \begin{pmatrix} 2 & 0 \\ 0 & 2 \end{pmatrix}. \end{aligned} \tag{64}$$

The matrix A' is diagonal, but the trace of A is not preserved in A'. The transformation QAQ^{-1} would preserve the trace but not diagonalize the matrix. Given the transformation equations (59) how are new primed axes located? Take

$$x' = x + \frac{1}{\sqrt{2}}y, \qquad y' = y. \tag{65}$$

When $x' = 0$, $y = \sqrt{2}x$. This is an equation of a straight line with slope equal to $-\sqrt{2}$ in the old coordinate system. Since x' is zero everywhere on this line, this line must be the new y' axis.

The first equation in (65) can also be written

$$x' = x + \frac{1}{\sqrt{2}}y'. \tag{66}$$

Since $y' = 0$ everywhere along the x' axis, setting $y' = 0$ in equation (66) shows that the new x' axis is the same as the old x axis. To show the reciprocal nature of the choice of one of the original axes as one of the final conjugate axes, start again with the quadratic form as stated in equation (55). Complete the square in a second way:

$$2x^2 + [2\sqrt{2}\,xy + 3y^2] = 1, \tag{67}$$

$$\left[2x^2 - \frac{\sqrt{2}}{3}x^2\right] + 3\left[y^2 + \tfrac{2}{3}\sqrt{2}\,xy + \left(\frac{\sqrt{2}}{3}x\right)^2\right], \tag{68}$$

or

$$\left(\frac{2-\sqrt{2}}{3}\right)x^2 + 3\left(y + \frac{\sqrt{2}}{3}x\right)^2 = 1. \tag{69}$$

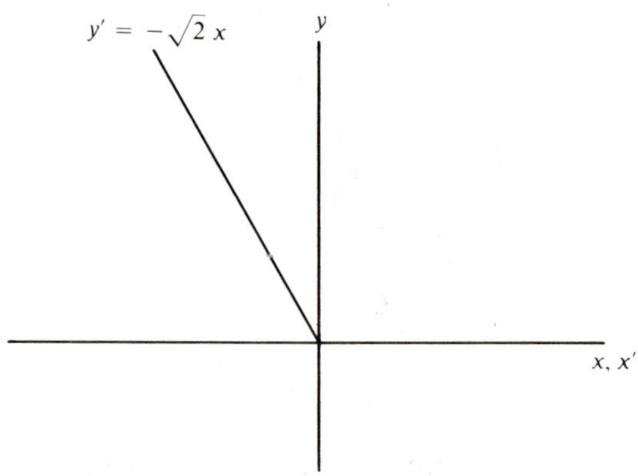

Figure 11

The transformation

$$\left.\begin{aligned} x'' &= x, & x &= x'', \\ y'' &= \left(y + \frac{\sqrt{2}}{3}x\right), & y &= \left(y'' - \frac{\sqrt{2}}{3}x''\right) \end{aligned}\right\} \tag{70}$$

leads to

$$0.195x''^2 + 3y''^2 = 1. \tag{71}$$

Coordinate transformations have a fascination that can be increased only by learning more about their proper interpretation. Usually coordinate transformations are limited to the well-known orthogonal systems and represent simple rotations. It becomes almost instinctive to write cartesian coordinate systems r, r' then rotate r' through an angle θ:

$$r' = Mr \quad \begin{cases} x' = x\cos\theta + y\sin\theta \\ y' = -x\sin\theta + y\cos\theta \end{cases} \tag{72}$$

It should be emphasized that the coordinate transformation shown in Fig. 12 is not intended simply to represent two axis crosses that are tilted with respect to each other. Their primary purpose is to allow the numerical evaluation of a function $F(\mathbf{r})$ or of even such a simple thing as the position of point P. If a function has a value in the $\mathbf{r}$ system, it should have the same numerical value in the $\mathbf{r}'$ system even if its functional form is changed. This fact can be illustrated conveniently by reference to the simple but nonroutine example of conjugate axes of a quadratic form (ellipse) in two dimensions.

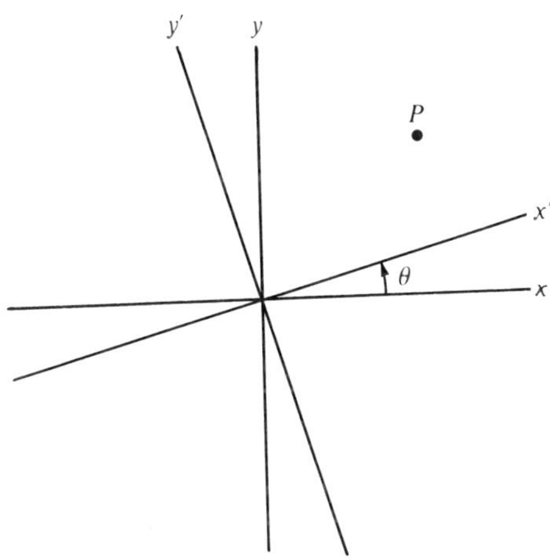

Figure 12. Simple rotation of axes.

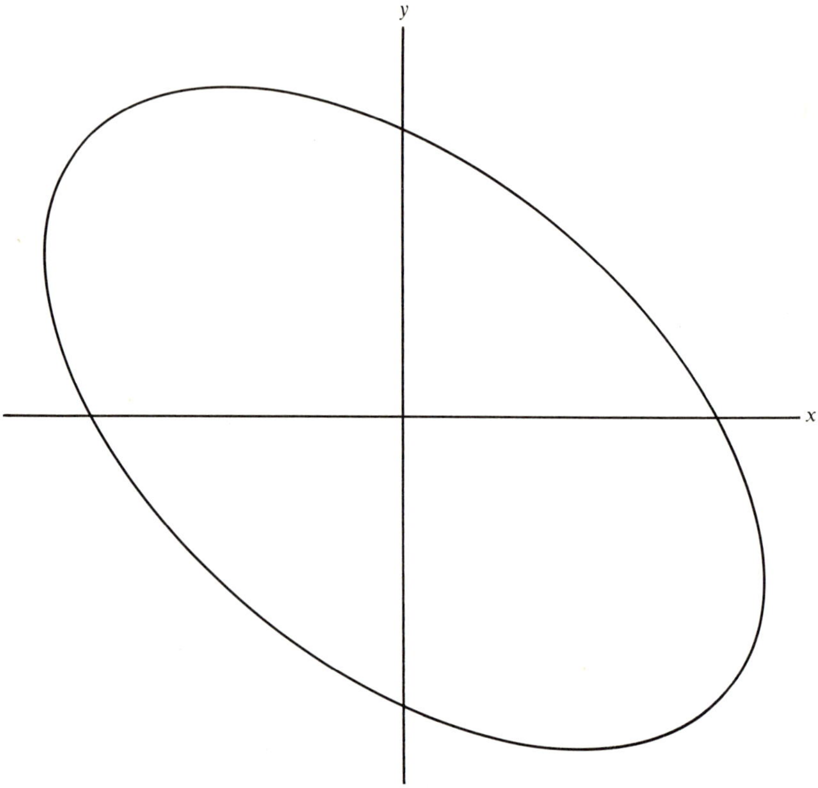

Figure 13. The nondiagonal representation of an ellipse in simple orthogonal coordinates.

Consider the ellipse:

$$2x^2 + 2\sqrt{2}\,xy + 3y^2 = 1, \tag{73}$$

shown in Fig. 13. To diagonalize the ellipse we may complete the square in two simple but distinct ways:

$$\begin{aligned} F_1 &= 2x^2 + 2\sqrt{2} + 3y^2 = 1 \rightarrow 2x^2 + 2\sqrt{2}\,xy + y^2 + 2y^2 \\ &= 2\left(x + \frac{1}{\sqrt{2}}y\right)^2 + 2y^2 = 1 = (\sqrt{2}\,x + y)^2 + 2y^2 \\ &\qquad\qquad \downarrow \qquad\qquad\qquad\qquad\qquad \downarrow \\ &\underbrace{2x'^2 + 2y'^2 = 1}_{\text{a circle}} \qquad\qquad \underbrace{x''^2 + 2y''^2 = 1}_{\text{an ellipse}} \end{aligned} \tag{74}$$

The corresponding transformations are

$$x' = x + \frac{1}{\sqrt{2}}y, \qquad x'' = \sqrt{2}\,x + y,$$
$$y' = y \qquad \text{or} \qquad y'' = y. \tag{75}$$

It is clear that

$$F_1 = F_1' = F_1'' = 1, \tag{76}$$

and that they would graph the same, but the coordinate axes would be changed. However, the statement that all three x axes are identical,

$$x = x' = x'', \tag{77}$$

may seem strange and incompatible with equation (75). This can be shown by graphing the ellipse in the three coordinate systems. First, find the

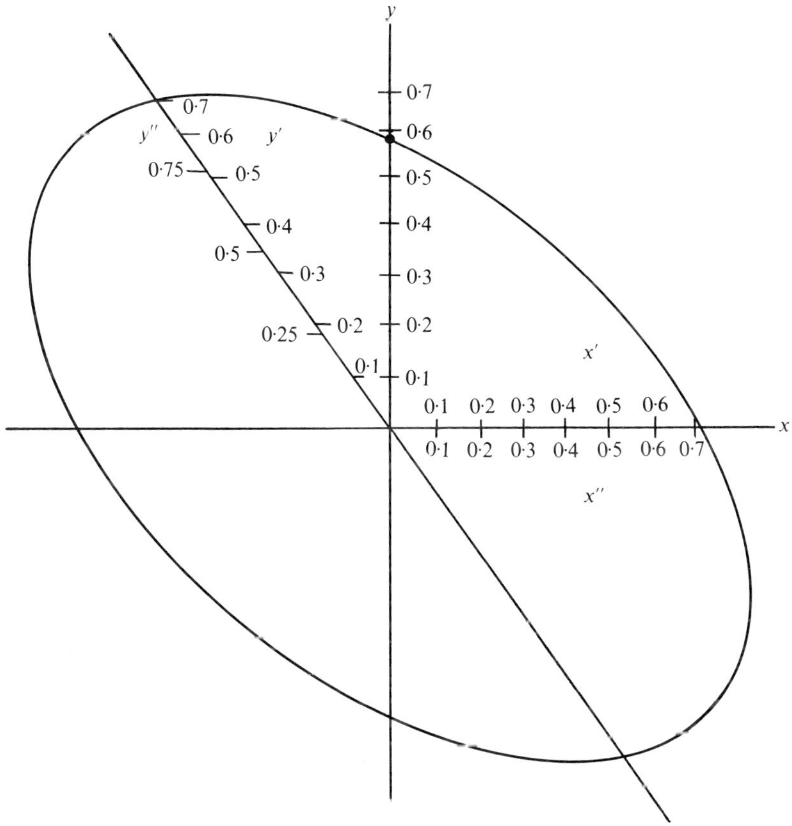

Figure 14. Ellipse shown in all three coordinate systems.

direction of the new x', x'' and y' y'' axes. To find the x' axis set

$$y = y' = 0, \tag{78}$$

then

$$x' = x, \tag{79}$$

which can be layed out in the original x, y system. To find the y' axis set

$$x' = \left(x + \frac{1}{\sqrt{2}}y\right) = 0, \tag{80}$$

or

$$y' = y = -\sqrt{2}\,x. \tag{81}$$

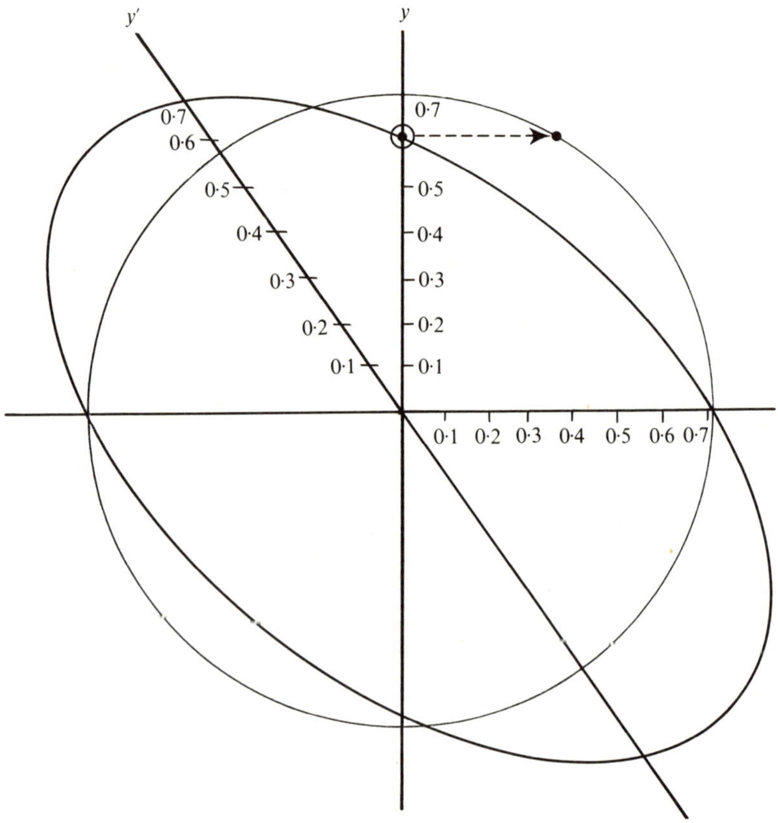

Figure 15. Equivalent ghost circle.

This is the equation of the y' axis in the x, y coordinate system and can be graphed. It turns out that $x'y'$ are conjugate diameters. The ellipse now may be graphed as in Fig. 14. Notice that the scales along the $\mathbf{x}$ and $\mathbf{x}'$ axes are identical. Also notice that the scales along the y and y' axes are numerically the same for equal distances north of the x, x' axes, but the units are of different absolute length.

The point marked on the intersection of the ellipse with the y axes in Fig. 14 has coordinates $x = 0$, $y = 0.585$, but in the new system it has coordinates $X' = 0.4$, $y' = 0.585$. However, notice the equivalent ghost circle shown in Fig. 15.
The point on the x axis has moved to coordinate $x = 4$. Thus the ghost circle in the x, y coordinate system is identical to the ellipse in the non-orthogonal $x'y'$ system. In the language of the theory of elasticity, the circle suffers a shear along the x direction to become the ellipse.

Finally, it should be remembered that there are an infinite variety of ways in which the square of F_1 could be accomplished, each leading to a new set of conjugate axes.

EXERCISES

7. Determine the scales along the x'' and y'' axes.

8. What is the deformation of the circle when transforming to x'', y''?

Assume that two quadratic forms

$$F_1 = x^2 + 4y^2 = 1 \tag{82}$$

and

$$F_2 = \tfrac{5}{2}x^2 - 3xy + \tfrac{5}{2}y^2 = 1 \tag{83}$$

are given. To diagonalize F_1 and F_2 simultaneously, let

$$x' = x \tag{84}$$

and

$$y' = 2y. \tag{85}$$

Then F_1 becomes

$$F_1' = x'^2 + y'^2 = 1 \tag{86}$$

and F_2 becomes

$$F_2' = \tfrac{5}{2}x'^2 - \tfrac{3}{2}x'y' + \tfrac{5}{8}y'^2 = 1. \tag{87}$$

Since F'_1 is symmetric about the z' axis, the x' and y' axes can be rotated until the cross-terms of F'_2 vanish. The amount of rotation, θ, can be determined graphically or by writing F'_2 in matrix form:

$$F'_2 = (x', y')\begin{pmatrix} \frac{5}{2} & -\frac{3}{4} \\ -\frac{3}{4} & \frac{5}{8} \end{pmatrix}\begin{pmatrix} x' \\ y' \end{pmatrix} = (x', y')M\begin{pmatrix} x' \\ y' \end{pmatrix}, \tag{88}$$

and then diagonalizing M by finding its eigenvalues. This is equivalent to finding λ, such that

$$\begin{pmatrix} \frac{5}{2} - \lambda & -\frac{3}{4} \\ -\frac{3}{4} & \frac{5}{8} - \lambda \end{pmatrix} = 0 = (\tfrac{5}{2} - \lambda)(\tfrac{5}{8} - \lambda) - \tfrac{9}{16}. \tag{89}$$

This implies that

$$\lambda = \frac{25 \pm 19.209}{16}, \tag{90}$$

or

$$\lambda_1 = \frac{44.2}{16} = \frac{22.1}{8}, \tag{91}$$

and

$$\lambda_2 = \frac{2.895}{8} \tag{92}$$

Since rotation is an orthogonal transformation in two dimensions, only one of the eigenvectors associated with the λs needs to be calculated. The eigenvectors associated with λ_2 will lie along the major axis of F'_2 and thus permits θ to be found easily.

Setting

$$\begin{pmatrix} \frac{5}{2} & -\frac{3}{4} \\ -\frac{3}{4} & \frac{5}{8} \end{pmatrix}\hat{\mathbf{r}}_2 = \lambda_2\hat{\mathbf{r}}_2 = \begin{pmatrix} \frac{5}{2} & -\frac{3}{4} \\ -\frac{3}{4} & \frac{5}{8} \end{pmatrix}\begin{pmatrix} r_2{}^1 \\ r_2{}^2 \end{pmatrix} = \frac{2.9}{8}\begin{pmatrix} r_2{}^1 \\ r_2{}^2 \end{pmatrix} \tag{93}$$

gives

$$\tfrac{5}{2}r_2{}^1 - \tfrac{3}{4}r_2{}^2 = \frac{2.9}{8}r_2{}^1 \quad \text{and} \quad -\tfrac{3}{4}r_2{}^1 + \tfrac{5}{8}r_2{}^2 = \frac{2.9}{8}r_2{}^2 \tag{94}$$

or

$$r_2{}^2 = \frac{17.1}{6}r_2{}^1 \tag{95}$$

Normalizing shows that $r_2' = 1/3.02$, but to find θ requires only $r_2{}^2/r_2{}^1$. In the x', y' coordinate system $r_2{}^2/r_2{}^1 = \tan\theta$, since $r_2{}^1$ is the component of r_2 along x^1 and $r_2{}^2$ is the component of r_2 along y^1. Therefore

$$\tan\theta \cong 2.85 = \frac{17.1}{6}, \tag{96}$$

or

$$\theta = 70.66°. \tag{97}$$

Now let the x'' axis be along P_2 and the y'' axis be orthogonal to it, as shown in Fig. 16. Also, let

$$\begin{aligned} x'' &= \cos\theta x' + \sin\theta y', \\ y'' &= -\sin\theta x' + \cos\theta y', \end{aligned} \tag{98}$$

and

$$\begin{aligned} x' &= \cos\theta x'' - \sin\theta\, y'' \\ y' &= \sin\theta\, x'' + \cos\theta\, y''. \end{aligned} \tag{99}$$

therefore

$$F_1' \to F_1'' = x''^2 + y''^2 = 1 \tag{100}$$

and

$$\begin{aligned} F_2' \to F_2'' &= \tfrac{5}{2}(\cos\theta\, x'' - \sin\theta\, y'')^2 \\ &\quad - \tfrac{3}{2}(\cos\theta\, x'' - \sin\theta\, y'')(\sin\theta\, x'' + \cos\theta\, y'') \\ &\quad + \tfrac{5}{8}(\sin\theta\, x'' + \cos\theta\, y'')^2 \\ &= 0.362x''^2 - 0.0016x''y'' + 2.863y''^2 = 1. \end{aligned} \tag{101}$$

As a result, F_1 and F_2 are diagonalized almost simultaneously in the doubly primed coordinate system.

By expressing the x'' and y'' axes in the x, y coordinate system the two conjugate axes p and q are found, such that p is conjugate to q and q is conjugate to p. Also, $\tan\phi = \frac{1}{2}\tan\theta$ and $\tan\beta = \frac{1}{2}\tan\alpha$ (see the diagram in Fig. 16).

Some final remarks can be made concerning a buffer matrix corresponding to the transformation to conjugate axes. For the given quadratic forms:

$$\begin{aligned} F_1 &= (x \quad y)\begin{pmatrix} 1 & 0 \\ 0 & 4 \end{pmatrix}\begin{pmatrix} x \\ y \end{pmatrix} = \tilde{\mathbf{r}}\begin{pmatrix} 1 & 0 \\ 0 & 4 \end{pmatrix}\mathbf{r}, \\ F_2 &= (x \quad y)\begin{pmatrix} 2.5 & -1.5 \\ -1.5 & 2.5 \end{pmatrix}\begin{pmatrix} x \\ y \end{pmatrix} = \tilde{\mathbf{r}}\begin{pmatrix} 2.5 & -1.5 \\ -1.5 & 2.5 \end{pmatrix}\mathbf{r}. \end{aligned} \tag{102}$$

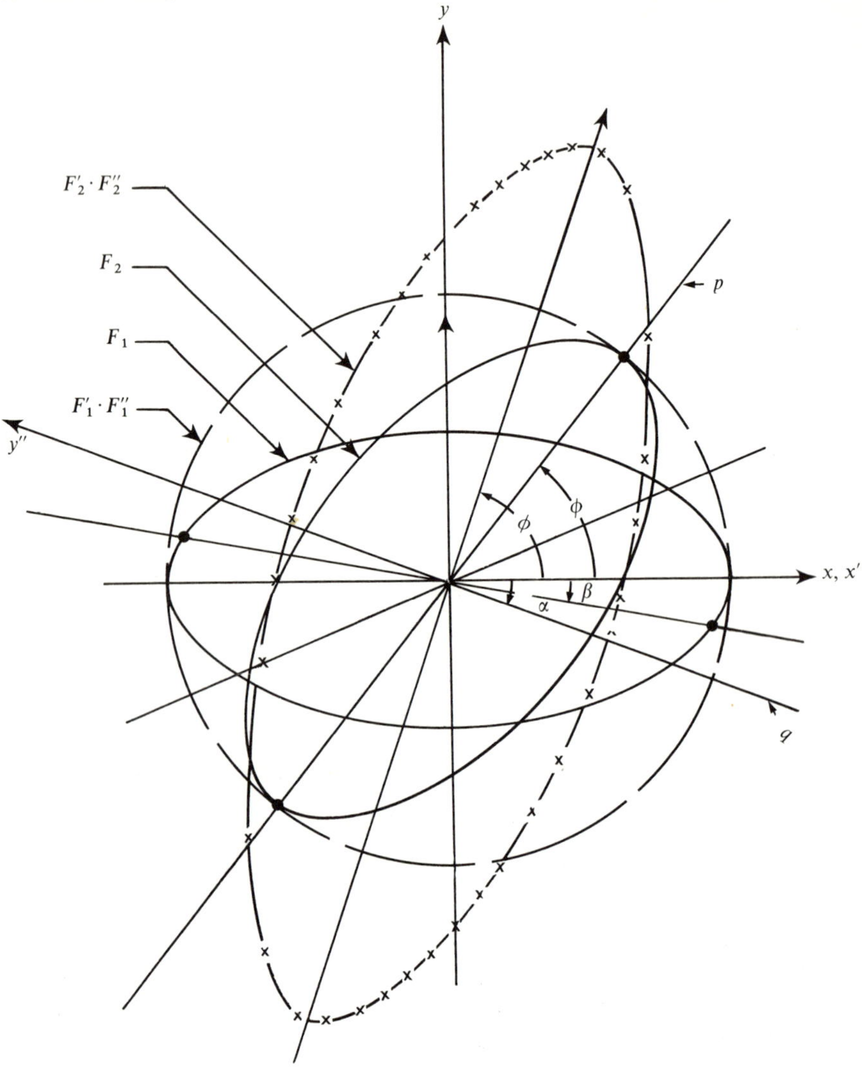

Figure 16. Diagonalization of two quadratic forms.

Two coordinate transformations have been performed:

$$\mathbf{r} = M\mathbf{r}' = \begin{pmatrix} 1 & 0 \\ 0 & 0{\cdot}5 \end{pmatrix}\mathbf{r}', \qquad \mathbf{r}' = N\mathbf{r}'' = \begin{pmatrix} \cos\theta & -\sin\theta \\ \sin\theta & \cos\theta \end{pmatrix}\mathbf{r}'' \quad (103)$$

or

$$\mathbf{r} = MN\mathbf{r}'' = Q\mathbf{r}'' = \begin{pmatrix} 1 & 0 \\ 0 & 0.5 \end{pmatrix} \begin{pmatrix} \cos\theta & -\sin\theta \\ \sin\theta & \cos\theta \end{pmatrix} \mathbf{r}''$$
$$= \begin{pmatrix} \cos\theta & -\sin\theta \\ 0.5\sin\theta & 0.5\cos\theta \end{pmatrix} \mathbf{r}''. \tag{104}$$

Q and $\tilde{Q}$ are

$$Q = \begin{pmatrix} \cos\theta & -\sin\theta \\ 0.5\sin\theta & 0.5\cos\theta \end{pmatrix}, \qquad \tilde{Q} = \begin{pmatrix} \cos\theta & 0.5\sin\theta \\ -\sin\theta & 0.5\cos\theta \end{pmatrix}, \tag{105}$$

with

$$\det Q = \tfrac{1}{2}. \tag{106}$$

The commutation buffer matrix B is

$$Q\tilde{Q} = \begin{pmatrix} \cos\theta & -\sin\theta \\ 0.5\sin\theta & 0.5\cos\theta \end{pmatrix} \begin{pmatrix} \cos\theta & 0.5\sin\theta \\ -\sin\theta & 0.5\cos\theta \end{pmatrix} = \begin{pmatrix} 1 & 0 \\ 0 & 0.25 \end{pmatrix}. \tag{107}$$

The original matrices representing quadratic forms F_1 and F_2 do not commute, that is,

$$F_1 F_2 \neq F_2 F_1$$
$$\begin{pmatrix} 1 & 0 \\ 0 & 4 \end{pmatrix} \begin{pmatrix} 2.5 & -1.5 \\ -1.5 & 2.5 \end{pmatrix} = \begin{pmatrix} 2.5 & -1.5 \\ 6 & 10 \end{pmatrix} \neq \begin{pmatrix} 2.5 & -1.5 \\ -1.5 & 2.5 \end{pmatrix} \begin{pmatrix} 1 & 0 \\ 0 & 4 \end{pmatrix} \tag{108}$$
$$= \begin{pmatrix} 2.5 & 6 \\ -1.5 & 10 \end{pmatrix}.$$

When the buffer B matrix is used, however,

$$F_1 B F_2 = F_2 B F_1$$
$$\begin{pmatrix} 1 & 0 \\ 0 & 4 \end{pmatrix} \begin{pmatrix} 1 & 0 \\ 0 & \frac{1}{4} \end{pmatrix} \begin{pmatrix} 2.5 & -1.5 \\ -1.5 & 2.5 \end{pmatrix} = I \begin{pmatrix} 2.5 & -1.5 \\ -1.5 & 2.5 \end{pmatrix}$$
$$= \begin{pmatrix} 2.5 & 1.5 \\ -1.5 & 2.5 \end{pmatrix} \begin{pmatrix} 1 & 0 \\ 0 & \frac{1}{4} \end{pmatrix} \begin{pmatrix} 1 & 0 \\ 0 & 4 \end{pmatrix} \tag{109}$$
$$= \begin{pmatrix} 2.5 & -1.5 \\ -1.5 & 2.5 \end{pmatrix} I.$$

The buffer matrix has a particularly simple form as the reciprocal of one of the quadratic forms F_1. It is slightly easier to use than Q in the following:

$$\tilde{Q}F_1Q = \tilde{Q}\begin{pmatrix}1 & 0\\ 0 & 1\end{pmatrix}Q = \begin{pmatrix}\cos\theta & 2\sin\theta\\ -\sin\theta & 2\cos\theta\end{pmatrix}\begin{pmatrix}\cos\theta & -\sin\theta\\ 0.5\sin\theta & 0.5\cos\theta\end{pmatrix} = \begin{pmatrix}1 & 0\\ 0 & 1\end{pmatrix}, \tag{110}$$

which is diagonal, and

$$\begin{aligned}\tilde{Q}F_2Q &= \tilde{Q}\begin{pmatrix}2.5 & -1.5\\ -1.5 & 2.5\end{pmatrix}Q\\ &= \begin{pmatrix}(2.5\cos\theta - 0.75\sin\theta) & (-1.5\cos\theta + 1.25\sin\theta)\\ (-2.5\sin\theta - 0.75\cos\theta) & (1.5\sin\theta + 1.25\cos\theta)\end{pmatrix}\begin{pmatrix}\cos\theta & -\sin\theta\\ +0.5\sin\theta & 0.5\cos\theta\end{pmatrix}\\ &= \begin{pmatrix}(+2.5\cos^2\theta - 1.5\cos\theta\sin\theta + \frac{5}{8}\sin^2\theta) & (-0.75\cos^2\theta - \frac{15}{8}\cos\theta\sin\theta + 0.75\sin^2\theta)\\ (-0.75\cos^2\theta - \frac{15}{8}\cos\theta\sin\theta + 0.75\sin^2\theta) & (2.5\sin^2\theta + 1.5\cos\theta\sin\theta + \frac{5}{8}\cos^2\theta)\end{pmatrix},\end{aligned} \tag{111}$$

which is also diagonal when the proper value of $\theta \sim 70°$ is specified.

Two symmetric quadratic forms can be diagonalized simultaneously if at least one of them is positive definite, that is, ellipsoidal. For the resultant nonorthogonal coordinates, tensors or matrices also commute. Given

$$\tilde{Q}AQ = D_A \tag{112}$$

and

$$\tilde{Q}BQ = D_B, \tag{113}$$

then

$$\tilde{Q}AQ\tilde{Q}BQ = \tilde{Q}BQ\tilde{Q}AQ, \tag{114}$$

or, provided Q has an inverse,

$$A(Q\tilde{Q})B = B(Q\tilde{Q})A. \tag{115}$$

Thus

$$AMB = BMA, \tag{116}$$

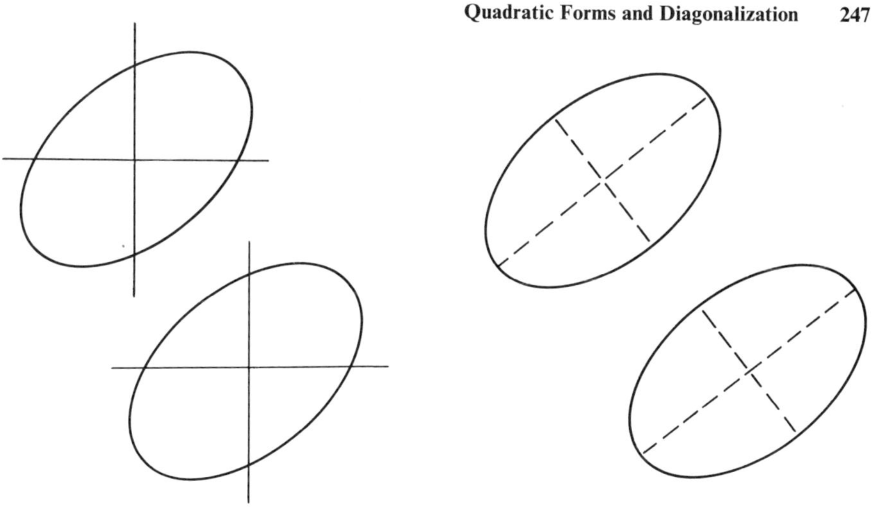

(*a*) Two quadratic forms before diagonalization

(*b*) Same quadratic forms after diagonalization

Figure 17

where M is a sort of buffer matrix for diagonalization and commutation. Sometimes it will be possible to find a matrix M such that $Q\tilde{Q} = I$, for orthogonal transformations Q, since $\tilde{Q} = Q^{-1}$.

In quantum mechanics the problem of the simultaneous diagonalization of quadratic operators is of paramount importance in a determination of

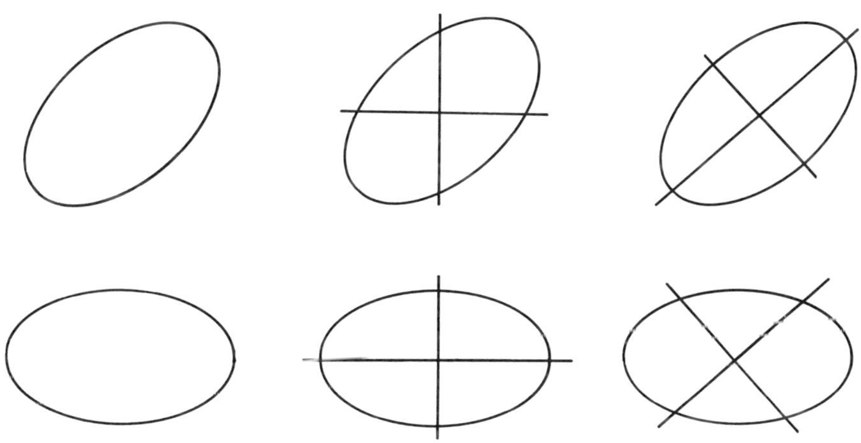

(*a*) Quadratic forms before diagonalization

(*b*) Unsuccessful attempts to diagonalize simultaneously

Figure 18

those observables, represented by the quantum mechanical operators, that may be simultaneously measured. Simultaneous diagonalization is guaranteed if the operators commute. In the literature, discussions of simultaneous diagonalization are usually limited to orthogonal coordinates. The situation is easily understood by referring to the illustrations in Fig. 17. Since the

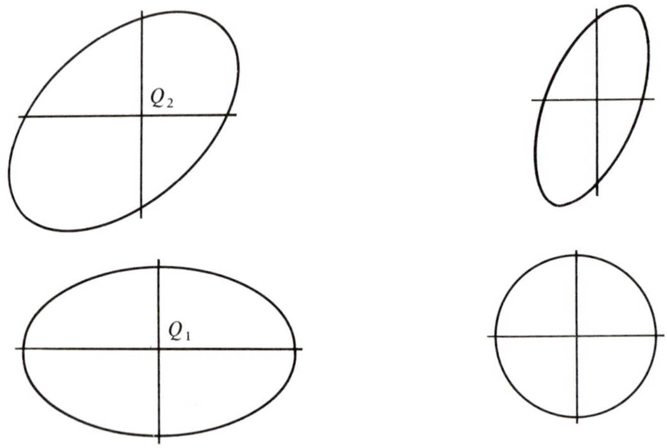

(a) Nonparallel quadratic forms

(b) Effect of a coordinate shrinkage in one direction

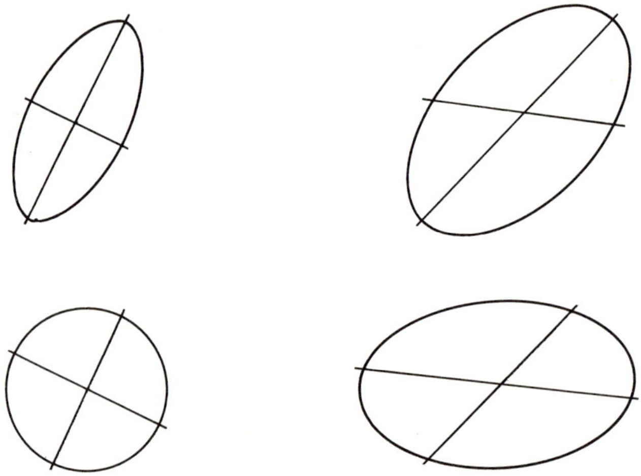

(c) Effect of a simple rotation chosen to diagonalize Q_2

(d) Original shape of quadratic forms reestablished (but original axes have become nonorthogonal conjugate diameters)

Figure 19

quadratic forms in Fig. 17 are essentially parallel to each other, they are diagonalized easily by the same single rotation of coordinates.

Two quadratic forms that cannot be diagonalized simultaneously by a simple rotation of coordinates are shown in Fig. 18. The problem illustrated in Fig. 18 arises because the quadratic forms initially are not placed parallel.

By considering the theory of conjugate diameters as described elsewhere, it is possible to diagonalize nonparallel quadratic forms simultaneously by using nonorthogonal coordinates that are simultaneously conjugate axes for both quadratic forms.

The logical steps for finding the diagonal representation of the two nonparallel quadratic forms are shown in Fig. 19. The first step is to diagonalize one of the quadratic forms by a simple rotation of coordinates if necessary, as shown in Fig. 19*a*. Then shrink the diagonalized form to a circle as shown in Fig. 19*b*. The coordinates can now be rotated in the circle, without loosing diagonalization, such as to diagonalize the second quadratic form as shown in Fig. 19*c*. By reversing the shrinkage of Fig. 19*b*, the two diagonalized matrices are obtained as shown in Fig. 19*d*.

A necessary and sufficient condition that conjugate axes may be found simultaneously for two quadratic forms is that at least one of the quadratic forms is positive definite. Figure 20 illustrates two different instances in which this minimum requirement is satisfied. The ellipse is the primary quadratic form that is reduced to a circle.

The essential point in the success of nonorthogonal conjugate diameters in the diagonalization of even a single quadratic form is most easily understood by recalling that an ellipse is a conic section and therefore provides a

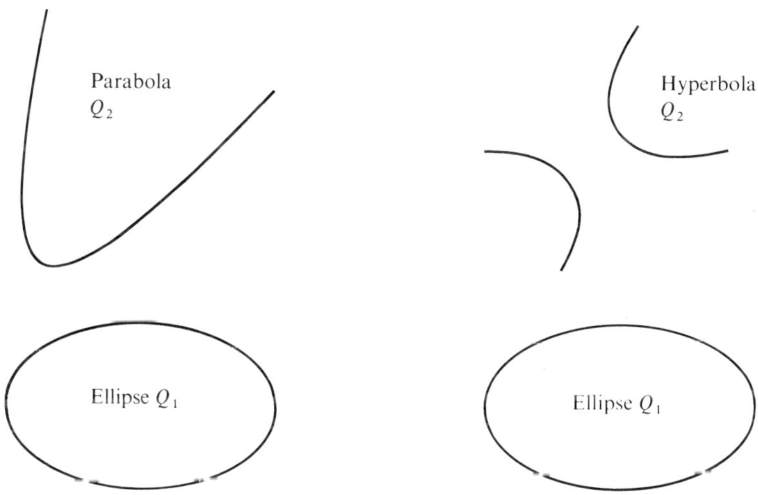

Figure 20. Two possible minimum cases of simultaneous diagonalization.

perspective view of a circle or another ellipse. The reader may satisfy himself on this point by drawing a circle or ellipse with a set of perpendicular principal axes on a sheet of stiff paper, and then viewing it at various angles and rotations. The perpendicular axes become any of an infinite variety of correct conjugate axes. This principle is used in constructing plastic ellipse templates, which are projections of circles. The special angle of 33°16′ produces the isometric projection and a set of nonorthogonal conjugate axes are shown on isometric templates as small tick marks on the ellipse.

EXERCISES

9. A "quadratic form" is defined in matrix notation as $\mathbf{x}^T A\mathbf{x}$. Such a form called "positive definite" if and only if $\mathbf{x}^T A\mathbf{x} \geq 0$. Let $\mathbf{x}^T A\mathbf{x} \geq \mathbf{x}^T B\mathbf{x} \geq 0$; prove that:

(*a*) $\det A \geq \det B$.

(*b*) The roots of $|A - \lambda B|$ are real and greater than or equal to one.

(*c*) Any principal minor of A is not less than the corresponding principal minor of C.

10. Determine a change of variables $x \to y$ under which the quadratic form $-x_1{}^2 + 2x_1x_2 + x_2{}^2$ becomes $\sum y_i{}^2$.

11. For amorphous materials, the dielectric constant and conductivity are simple scalar quantities. For crystals these quantities become tensors, which are in general not proportional element-for-element. What physical phenomenon does the fact that the axes are no longer parallel represent? How can the concept of principal axis transformation be applied here? What minimum symmetry is necessary to assure congruence of axes?

E. Tensor Calculus

The derivatives

$$\frac{\partial \mathbf{a}_i}{\partial x^j} = \mathbf{a}_{i,j} = \frac{\partial \mathbf{r}}{\partial x^i \partial x^j} \tag{1}$$

of the set of base vectors

$$\mathbf{a}_i = \frac{\partial \mathbf{r}}{\partial x^i} \tag{2}$$

often arise in tensor analysis work.

These derivatives, which are still vectors, are symmetric in the indices i and j. Thus it is possible to express them as a linear combination of the base vectors $\mathbf{a}_i$:

$$\frac{\partial \mathbf{a}_i}{\partial x^j} = \Gamma_{ij}^k \mathbf{a}_k. \tag{3}$$

A group theoretician would say that the elements Γ^k_{ij} are reminiscent of the *structure constants* of a group, but in the present work, Γ^k_{ij} represents what is called a "Christoffel three-index symbol" and emphatically not a mixed tensor of rank three.

Remember the definition:

$$g_{ik} = \mathbf{a}_i \cdot \mathbf{a}_k = \mathbf{a}_k \cdot \mathbf{a}_i = g_{ki}. \tag{4}$$

Evidently g is also symmetric in the indices i, k.

Now differentiate g_{ik} to get

$$\frac{\partial g_{ik}}{\partial x^j} = \frac{\partial}{\partial x^j}(\mathbf{a}_i \cdot \mathbf{a}_k) = \frac{\partial \mathbf{a}_i}{\partial x^j} \cdot \mathbf{a}_k + \mathbf{a}_i \cdot \frac{\partial \mathbf{a}_k}{\partial x^j}, \tag{5}$$

and put expression (3) into equation (5):

$$\frac{\partial g_{ik}}{\partial x^j} = (\Gamma^k_{ij}\mathbf{a}_k) \cdot \mathbf{a}_k + \mathbf{a}_i \cdot (\Gamma^{ij}_k \mathbf{a}_i). \tag{6}$$

That the symmetry between i, k and i, j also exists between j, k suggests that $\Gamma^\sigma_{\mu\nu}$ may be solved by a proper linear combination of derivatives (6) with the indices properly permuted. Cyclic permutation—always fashionable—requires $i \to j, j \to k, k \to i$, and yields

$$\frac{\partial g_{ji}}{\partial x^k} = (\Gamma^i_{jk}\mathbf{a}_i) \cdot \mathbf{a}_i + \mathbf{a}_j \cdot (\Gamma^j_{ik}\mathbf{a}_j). \tag{7}$$

Reapplication of $i \to j, j \to k, k \to i$ in equation (7) then yields

$$\frac{\partial g_{kj}}{\partial x^i} = (\Gamma^j_{ki}\mathbf{a}_j) \cdot \mathbf{a}_j + \mathbf{a}_k \cdot (\Gamma^k_{ji}\mathbf{a}_k). \tag{8}$$

Now adding equation (7) to equation (8) and subtracting equation (6) produces

$$\begin{aligned}\left[\frac{\partial g_{ji}}{\partial x^k} + \frac{\partial g_{kj}}{\partial x^i} - \frac{\partial g_{ik}}{\partial x^j}\right] &= [(\Gamma^j_{ki}\mathbf{a}_j) \cdot \mathbf{a}_j + \mathbf{a}_k \cdot (\Gamma^k_{ji}\mathbf{a}_k) \\ &+ (\Gamma^i_{jk}\mathbf{a}_i) \cdot \mathbf{a}_i + \mathbf{a}_j \cdot (\Gamma^j_{ik}\mathbf{a}_j) - (\Gamma^k_{ij}\mathbf{a}_k) \cdot \mathbf{a}_k - \mathbf{a}_i \cdot (\Gamma^i_{kj}\mathbf{a}_i)],\end{aligned} \tag{9}$$

and using symmetry properties:

$$\left[\frac{\partial g_{ji}}{\partial x^k} + \frac{\partial g_{kj}}{\partial x^i} - \frac{\partial g_{ik}}{\partial x^j}\right] = 2\mathbf{a}_j \cdot (\Gamma^P_{ik}\mathbf{a}_P) = 2\mathbf{a}_j \cdot \mathbf{a}_P \Gamma^P_{ik} - 2g_{jP}\Gamma^P_{ik}. \tag{10}$$

So

$$g_{jP}\Gamma^P_{ik} = \tfrac{1}{2}\left[\frac{\partial g_{ji}}{\partial x^k} + \frac{\partial g_{kj}}{\partial x^i} - \frac{\partial g_{ik}}{\partial x^j}\right], \tag{11}$$

where P is a dummy variable. Now multiply by g^{jm}:

$$g^{jm}g_{jP}\Gamma^P_{ik} = \partial^m_P\Gamma^P_{ik} = \Gamma^m_{ik} = \tfrac{1}{2}g^{jm}\left[\frac{\partial g_{ji}}{\partial x^k} + \frac{\partial g_{ki}}{\partial x^i} - \frac{\partial g_{ik}}{\partial x^j}\right]. \tag{12}$$

Finally,

$$\Gamma^m_{ik} = \tfrac{1}{2}g^{jm}\left[\frac{\partial g_{ji}}{\partial x^k} + \frac{\partial g_{kj}}{\partial x^i} - \frac{\partial g_{ik}}{\partial x^j}\right]. \tag{13}$$

There are other Christoffel symbols and other ways of writing them. For instance, definition (12) may be written:

$$\Gamma^m_{ik} = \tfrac{1}{2}g^{jm}\Gamma_{j,\,ik}, \tag{14}$$

where

$$\Gamma_{j,\,ik} = \tfrac{1}{2}\left[\frac{\partial g_{ji}}{\partial x^k} + \frac{\partial g_{kj}}{\partial x^i} - \frac{\partial g_{ik}}{\partial x^j}\right]. \tag{15}$$

Also

$$\Gamma^m_{ik} \equiv \begin{Bmatrix} m \\ ik \end{Bmatrix} \qquad \text{and} \quad \Gamma_{j,\,ik} \equiv [ik, j]. \tag{16}$$

Starting with new indices, the first kind of Christoffel symbol is defined to be

$$[mn, r] = \tfrac{1}{2}\left[\frac{\partial g_{rm}}{\partial x^n} + \frac{\partial g_{nr}}{\partial x^m} - \frac{\partial g_{mn}}{\partial x^r}\right]. \tag{17}$$

The metric tensor g_{mn} transforms as follows:

$$g_{mn} = \bar{g}_{ij}\frac{\partial \bar{x}^i}{\partial x^m}\frac{\partial \bar{x}^j}{\partial x^n}. \tag{18}$$

To determine how the Christoffel symbol transforms, differentiate equation (18) as follows:

$$\frac{\partial g_{mn}}{\partial x^r} = \frac{\partial \bar{x}^k}{\partial x^r}\frac{\partial \bar{x}^j}{\partial x^n}\frac{\partial \bar{x}^i}{\partial x^m}\left(\frac{\partial \bar{g}_{ij}}{\partial \bar{x}^k}\right) + \bar{g}_{ij}\left[\frac{\partial^2 \bar{x}^i}{\partial x^r \partial x^m}\frac{\partial \bar{x}^j}{\partial x^n} + \frac{\partial \bar{x}^i}{\partial x^m}\frac{\partial^2 \bar{x}^j}{\partial x^n \partial x^r}\right], \tag{19}$$

and substitute the result into equation (17).

First, change the indexes in a cyclic manner:

$$\frac{\partial g_{nr}}{\partial x^m} = \frac{\partial \bar{x}^i}{\partial x^m}\frac{\partial \bar{x}^j}{\partial x^n}\frac{\partial \bar{x}^k}{\partial x^r}\left(\frac{\partial \bar{g}_{jk}}{\partial \bar{x}^i}\right) + \bar{g}_{ij}\left[\frac{\partial^2 \bar{x}^i}{\partial x^m \partial x^n}\frac{\partial \bar{x}^j}{\partial x^r} + \frac{\partial \bar{x}^i}{\partial x^n}\frac{\partial^2 \bar{x}^j}{\partial x^r \partial x^m}\right].$$

Another cyclic permutation gives

$$\frac{\partial g_{rm}}{\partial x^n} = \frac{\partial \bar{x}^i}{\partial x^m}\frac{\partial \bar{x}^j}{\partial x^n}\frac{\partial \bar{x}^k}{\partial x^r}\left(\frac{\partial \bar{g}_{ki}}{\partial \bar{x}^j}\right) + \bar{g}_{ij}\left[\frac{\partial^2 \bar{x}^i}{\partial x^n \partial x^r}\frac{\partial \bar{x}^i}{\partial x^m} + \frac{\partial \bar{x}^i}{\partial x^r}\frac{\partial^2 \bar{x}^j}{\partial x^m \partial x^n}\right]. \tag{21}$$

Now subtract expression (19) from the sum of expressions (20) and (21).

$$\begin{aligned}\left[\frac{\partial g_{rm}}{\partial x^n} + \frac{\partial g_{nr}}{\partial x^m} - \frac{\partial g_{mn}}{\partial x^r}\right]& \\ = \left[\frac{\partial \bar{g}_{ki}}{\partial \bar{x}^j} + \frac{\partial \bar{g}_{jk}}{\partial \bar{x}^i} - \frac{\partial \bar{g}_{ij}}{\partial \bar{x}^k}\right]&\frac{\partial \bar{x}^i}{\partial x^m}\frac{\partial \bar{x}^j}{\partial x^n}\frac{\partial \bar{x}^k}{\partial x^s} + \partial \bar{g}_{ij}\frac{\partial \bar{x}^j}{\partial x^n}\frac{\partial^2 \bar{x}^i}{\partial x^m \partial x^n}.\end{aligned} \tag{22}$$

Divide by 2 and introduce the Christoffel symbol:

$$[mn, r] = [ij, k]\frac{\partial \bar{x}^i}{\partial x^m}\frac{\partial \bar{x}^j}{\partial x^n}\frac{\partial \bar{x}^k}{\partial x^r} + \bar{g}_{ij}\frac{\partial^2 \bar{x}^i}{\partial x^m \partial x^n}\frac{\partial \bar{x}^j}{\partial x^r}. \tag{23}$$

If the coordinate transformation were affine or linear, this Christoffel symbol of the first kind would transform as a third rank covariant tensor.

The second kind of Christoffel symbol is defined to be

$$\begin{Bmatrix} s \\ mn \end{Bmatrix} = g^{sr}[mn, r] = \tfrac{1}{2}g^{sr}\left[\frac{\partial g_{rm}}{\partial x^n} + \frac{\partial g_{nr}}{\partial x^m} \quad \frac{\partial g_{mn}}{\partial x^r}\right]. \tag{24}$$

From equation (23):

$$g^{sr}[mn, r] = g^{sr}[\overline{ij, k}]\frac{\partial \bar{x}^i}{\partial x^n}\frac{\partial \bar{x}j}{\partial x^n}\frac{\partial \bar{x}^k}{\partial x^r} + g^{sr}\bar{g}_{ij}\frac{\partial^2 \bar{x}^i}{\partial x^m \partial x^n}\frac{\partial \bar{x}^j}{\partial x^r}. \tag{25}$$

Thus

$$\begin{Bmatrix} s \\ mn \end{Bmatrix} = g^{sr}[\overline{ij, k}]\frac{\partial \bar{x}^i}{\partial x^m}\frac{\partial \bar{x}^j}{\partial x^n}\frac{\partial \bar{x}^k}{\partial x^r} + g^{sr}\bar{g}_{ij}\frac{\partial^2 \bar{x}^i}{\partial x^m \partial x^n}\frac{\partial \bar{x}^j}{\partial x^r}. \tag{26}$$

Now introduce

$$g^{sr} = \bar{g}^{tk}\frac{\partial x^s}{\partial \bar{x}^t}\frac{\partial x^r}{\partial \bar{x}^k}. \tag{27}$$

Then

$$\begin{Bmatrix} s \\ mn \end{Bmatrix} = \begin{Bmatrix} t \\ ij \end{Bmatrix}\frac{\partial \bar{x}^i}{\partial x^m}\frac{\partial \bar{x}^j}{\partial x^n}\frac{\partial x^s}{\partial \bar{x}^t} + \bar{g}^{tj}\bar{g}_{ij}\frac{\partial^2 \bar{x}^i}{\partial x^m \partial x^n}\frac{\partial \bar{x}^j}{\partial x^r}\frac{\partial x^s}{\partial \bar{x}^t}\frac{\partial x^r}{\partial \bar{x}^k}, \tag{28}$$

which reduces to

$$\begin{Bmatrix} s \\ mn \end{Bmatrix} = \begin{Bmatrix} \bar{t} \\ ij \end{Bmatrix}\frac{\partial \bar{x}^i}{\partial x^m}\frac{\partial \bar{x}^j}{\partial x^n}\frac{\partial x^s}{\partial \bar{x}^t} + \bar{g}^{tj}\bar{g}^{ij}\frac{\partial^2 \bar{x}^i}{\partial x^m \partial x^n}\frac{\partial \bar{x}^j}{\partial \bar{x}^k}\frac{\partial x^s}{\partial \bar{x}^t}. \tag{29}$$

Recognizing a $\delta_k{}^j$ function in the second term,

$$\begin{Bmatrix} s \\ mn \end{Bmatrix} \quad \begin{Bmatrix} \bar{t} \\ ij \end{Bmatrix} \frac{\partial \bar{x}^i}{\partial x^m} \frac{\partial \bar{x}^j}{\partial x^n} \frac{\partial x^s}{\partial \bar{x}^t} + \bar{g}^{tj} \bar{g}_{ij} \frac{\partial^2 \bar{x}^i}{\partial x^m \partial x^n} \frac{\partial x^s}{\partial \bar{x}^t}. \tag{30}$$

Recognizing a second $\delta_i{}^T$ function,

$$\begin{Bmatrix} s \\ mn \end{Bmatrix} \quad \begin{Bmatrix} \bar{t} \\ ij \end{Bmatrix} \frac{\partial \bar{x}^i}{\partial x^m} \frac{\partial \bar{x}^j}{\partial x^n} \frac{\partial x^s}{\partial \bar{x}^t} + \frac{\partial^2 \bar{x}^t}{\partial x^m \partial x^n} \frac{\partial x^s}{\partial \bar{x}^T}. \tag{31}$$

If the coordinate transformation is affine or linear, the Christoffel symbol of the second kind would transform as a mixed tensor of rank three. Now multiply equation (31) by $\partial \bar{x}^p / \partial x^s$:

$$\begin{Bmatrix} s \\ mn \end{Bmatrix} \frac{\partial \bar{x}^p}{\partial x^s} = \begin{Bmatrix} \bar{t} \\ ij \end{Bmatrix} \frac{\partial \bar{x}^p}{\partial x^s} \frac{\partial x^s}{\partial \bar{x}^T} \frac{\partial \bar{x}^i}{\partial x^m} \frac{\partial \bar{x}^j}{\partial x^n} + \frac{\partial^2 \bar{x}^T}{\partial x^m \partial x^n} \frac{\partial \bar{x}^P}{\partial \bar{x}^T}. \tag{32}$$

The delta function then gives

$$\begin{Bmatrix} s \\ mn \end{Bmatrix} \frac{\partial \bar{x}^p}{\partial x^i} = \begin{Bmatrix} \bar{p} \\ ij \end{Bmatrix} \frac{\partial \bar{x}^i}{\partial x^m} \frac{\partial \bar{x}^j}{\partial x^n} + \frac{\partial^2 \bar{x}^P}{\partial x^m \partial x^n}. \tag{33}$$

Consider a position vector $\mathbf{r}$ and its differential $d\mathbf{r}$ under the coordinate transformation:

$$x'^{\mu} = x'^{\mu}(x^{\sigma}), \tag{34a}$$

or its inverse

$$x^{\sigma} = x^{\sigma}(x'^{\mu}). \tag{34b}$$

Write

$$d\mathbf{r} = \frac{\partial \mathbf{r}}{\partial x^{\sigma}} dx^{\sigma} = \frac{\partial \mathbf{r}}{\partial x'^{\mu}} dx'^{\mu} = \mathbf{a}_{\sigma} dx^{\sigma} = \mathbf{a}'_{\mu} dx'^{\mu}. \tag{35}$$

It is possible to express $\mathbf{r}$ in terms of its cartesian components x, y, z and then relate x^{σ} and x'^{μ} to x, y, z. Figure 21 suggests the symmetry and equality of the coordinate systems. This symmetry is also evident in the following expression for $d\mathbf{r}$:

$$\begin{aligned} d\mathbf{r} &= \hat{\imath} dx + \jmath dy + \hat{\mathbf{k}} dz \doteq \mathbf{a}_1 dx' + \mathbf{a}_2 dx^2 + \mathbf{a}_3 dx^3 \\ &= \mathbf{a}'_1 dx^{1'} + \mathbf{a}'_2 dx^{2'} + \mathbf{a}'_3 dx^{3'}. \end{aligned} \tag{36}$$

Form the square of the differential position vectors, which is called the "positive definite differential arc length square" or "interval square." The value

$$d\mathbf{r}^2 \equiv ds^2 = \left(\frac{\partial \mathbf{r}}{\partial x^{\sigma}} dx^{\sigma} \right)^2 = \left(\frac{\partial \mathbf{r}}{\partial x^{n'}} dx^{n'} \right)^2 \tag{37}$$

is termed the "metric."

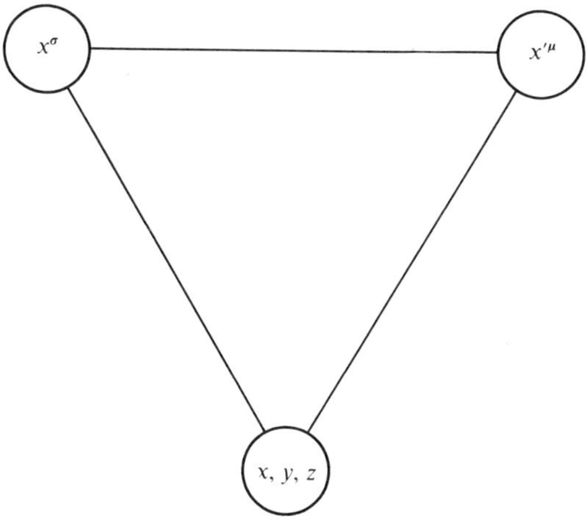

Figure 21. Three-way connection between components and coordinate systems.

To keep all the products straight, rotate indices as follows:

$$\begin{aligned} d\mathbf{r}^2 &\equiv ds^2 = dx^2 + dy^2 + dz^2 \\ &= \left(\frac{\partial \mathbf{r}}{\partial x^m} \cdot \frac{\partial \mathbf{r}}{\partial x^n}\right) dx^m dx^n = \left(\frac{\partial \mathbf{r}}{\partial x^{p'}} \cdot \frac{\partial \mathbf{r}}{\partial x^{q'}}\right) dx^{p'} dx^{q'}. \end{aligned} \tag{38}$$

The scalar products (in parentheses) are given a shorthand notation, as follows:

$$d\mathbf{r}^2 \equiv ds^2 = g_{mn} dx^m dx^n = g'_{pq} dx^{p'} dx^{q'}, \tag{39}$$

where

$$g_{mn} = \frac{\partial \mathbf{r}}{\partial x^n} \cdot \frac{\partial \mathbf{r}}{\partial x^n} = \mathbf{a}_m \cdot \mathbf{a}_n = \left[\frac{\partial r_x}{\partial x^m}\frac{\partial r_x}{\partial x^n} + \frac{\partial r_y}{\partial x^m}\frac{\partial r_y}{\partial x^n} + \frac{\partial r_z}{\partial x^m}\frac{\partial r_z}{\partial x^n}\right] \tag{40}$$

and

$$g'_{pq} = \frac{\partial \mathbf{r}}{\partial x^{p'}} \cdot \frac{\partial \mathbf{r}}{\partial x^{q'}} = \mathbf{a}'_p \cdot \mathbf{a}'_q = \left[\frac{\partial r_x}{\partial x^{p'}}\frac{\partial r_x}{\partial x^{q'}} + \frac{\partial r_y}{\partial x^{p'}}\frac{\partial r_y}{\partial x^{q'}} + \frac{\partial r_z}{\partial x^{p'}}\frac{\partial r_z}{\partial x^{q'}}\right]. \tag{41}$$

Here g is called the "metric tensor," as it defines the geometric properties of the space.

The generalized coordinate expressions

$$g_{mn}\, dx^m\, dx^n = g'_{pq} = dx^{p'}\, dx^{q'} \tag{42}$$

may be changed by using transformation (34), as follows:

$$dx^{p'} = \frac{\partial x^{p'}}{\partial x^m}\, dx^n \tag{43}$$

and

$$dx^{q'} = \frac{\partial x^{q'}}{\partial x^n} dx^n. \tag{43}$$

Moreover,

$$g_{mn} dx^m dx^n = g'_{pq} \left[\frac{\partial x^{p'}}{\partial x^m} \frac{\partial x^{q'}}{\partial x^n} \right] dx^m dx^n, \tag{44}$$

and thus

$$g_{mn} = g'_{pq} \left[\frac{\partial x^{p'}}{\partial x^m} \frac{\partial x^{q'}}{\partial x^n} \right] \tag{45}$$

or

$$g_{mn} = M_m{}^p M_n{}^q g'_{pq}. \tag{46}$$

Consequently, g_{mn} acts like a covariant tensor. To relate these to the definitions used earlier, interchange the primed and unprimed coordinate systems. Then

$$\begin{aligned} dx^m &= \frac{\partial x^n}{\partial x^{p'}} dx^{p'}, \qquad & dx^n &= \frac{\partial x^n}{\partial x^{q'}} dx^{q'}, \\ dx^m &= M_p{}^n dx^{p'}, \qquad & dx^n &= M_q{}^n dx^{q'}, \end{aligned} \tag{47}$$

and

$$g'_{pq} dx^{p'} dx^{q'} = g_{mn} \left[\frac{\partial x^m}{\partial x_p{}'} \frac{\partial x^n}{\partial x_q{}'} \right] dx^{p'} dx^{q'}, \tag{48}$$

or, manipulating as before,

$$g'_{pq} = M_p{}^m M_q{}^n g_{mn}. \tag{49}$$

Thus, once again, the saying "co-lo-prime-below" applies.

Since the metric tensor for cartesian coordinates is:

$$g_{\mu\nu} = \begin{pmatrix} 1 & 0 & 0 \\ 0 & 1 & 0 \\ 0 & 0 & 1 \end{pmatrix}, \tag{50}$$

it follows that:

$$\begin{aligned} ds^2 &= 1\,dx^2 + 0\,dx\,dy + 0\,dx\,dz, \\ &\quad + 0\,dy\,dx + 1\,dy^2 + 0\,dy\,dz, \\ &\quad + 0\,dz\,dx + 0\,dz\,dy + 1\,dz^2, \\ &= dx^2 + dy^2 + dz^2. \end{aligned} \tag{51}$$

The metric tensor prescribes the local scale factors of a coordinate system. In addition, it is a symmetric tensor.

Recall the unit vectors $\hat{\mathbf{i}}$, $\hat{\mathbf{j}}$ and $\hat{\mathbf{e}}_r$, $\hat{\mathbf{e}}_\theta$, and the unitary vectors $\mathbf{a}_r$, $\mathbf{a}_\theta$ and $\mathbf{a}^r$, $\mathbf{a}^\theta$. Also, changing r, θ to 1, 2, recall that

$$\begin{gathered} \mathbf{a}_i \cdot \mathbf{a}^j = \delta_i^{\,j} \\ \text{and} \\ \mathbf{a}^j \cdot \mathbf{a}_i = \delta_i^{\,j}. \end{gathered} \tag{52}$$

But what can be said about the values of

$$\begin{gathered} \mathbf{a}_i \cdot \mathbf{a}_j, \\ \mathbf{a}^i \cdot \mathbf{a}^j? \end{gathered} \tag{53}$$

Carrying out individual investigations,

$$\begin{aligned} \mathbf{a}_1 \cdot \mathbf{a}_1 &= 1, & \mathbf{a}_1 \cdot \mathbf{a}_2 &= 0, \\ \mathbf{a}_2 \cdot \mathbf{a}_1 &= 0, & \mathbf{a}_2 \cdot \mathbf{a}_2 &= r^2, \\ \mathbf{a}^1 \cdot \mathbf{a}^1 &= 1, & \mathbf{a}^1 \cdot \mathbf{a}^2 &= 0, \\ \mathbf{a}^2 \cdot \mathbf{a}^1 &= 0, & \mathbf{a}^2 \cdot \mathbf{a}^2 &= \frac{1}{r^2}. \end{aligned} \tag{54}$$

So define

$$g_{ij} = \begin{pmatrix} 1 & 0 \\ 0 & r^2 \end{pmatrix} \qquad \text{and} \quad g^{ij} = \begin{pmatrix} 1 & 0 \\ 0 & \frac{1}{r^2} \end{pmatrix}. \tag{55}$$

Their matrix product is

$$(g_{ij})(g^{ij}) = \begin{pmatrix} 1 & 0 \\ 0 & r^2 \end{pmatrix} \begin{pmatrix} 1 & 0 \\ 0 & \frac{1}{r^2} \end{pmatrix} = \begin{pmatrix} 1 & 0 \\ 0 & 1 \end{pmatrix} = I. \tag{56}$$

Their determinants are

$$\begin{gathered} g = \det g_{ij} = r^2 \\ \text{and} \\ g = \det g^{ij} = \frac{1}{r^2}. \end{gathered} \tag{57}$$

Finally, define quantities h_i as follows:

$$h_i = \sqrt{g_{ii}}, \tag{58}$$

so that

$$h_1 = \sqrt{1} = 1, \qquad h_2 = \sqrt{r^2} = r,$$
$$\text{and} \tag{59}$$
$$h_1 h_2 = \sqrt{g} = r.$$

In vector and tensor analysis a quantity called the "Levi-Civita symbol" or "Levi-Civita density" sometimes is used to simplify calculations and equations. This symbol may be denoted in several ways: ϵ_{ijk}, E_{ijk}, e_{ijk}, or, since it is a glorified Krönecker delta, δ_{ijk}, all of which are equivalent:

$$\epsilon_{ijk} = E_{ijk} = e_{ijk} = \delta_{ijk}. \tag{60}$$

This book will use ϵ_{ijk}. The Levi-Civita symbol is defined as a two-valued function of the three indices ijk that appear as subscripts. These indices are independent and may each take on the value 1, 2, or 3, resulting in a total of $3 \times 3 \times 3 = 27$ different combinations, as follows:

ϵ_{111}	ϵ_{112}	ϵ_{113}
ϵ_{121}	ϵ_{122}	ϵ_{123}
ϵ_{131}	ϵ_{132}	ϵ_{133}

ϵ_{211}	ϵ_{212}	ϵ_{213}
ϵ_{221}	ϵ_{222}	ϵ_{223}
ϵ_{231}	ϵ_{232}	ϵ_{233}

ϵ_{311}	ϵ_{312}	ϵ_{313}
ϵ_{321}	ϵ_{322}	ϵ_{323}
ϵ_{331}	ϵ_{332}	ϵ_{333}

By definition, the Levi-Civita symbol is zero if any two or three indices are the same. Thus the original 27 possibilities, shown above, become:

0	0	0
0	0	ϵ_{123}
0	ϵ_{132}	0

0	0	ϵ_{213}
0	0	0
ϵ_{231}	0	0

0	ϵ_{312}	0
ϵ_{321}	0	0
0	0	0

Furthermore, the remaining six symbols are defined to be ± 1 depending on whether the indices are an odd or even permutation of the three numbers 1, 2, 3. So for even permutations

$$123 = [123], \qquad 231 = \begin{bmatrix} 231 \\ 213 \\ 123 \end{bmatrix}, \qquad 312 = \begin{bmatrix} 312 \\ 132 \\ 123 \end{bmatrix}, \qquad +1 \text{ is used}, \tag{61}$$

and for odd permutations

$$132 = \begin{bmatrix} 132 \\ 123 \end{bmatrix}, \qquad 213 = \begin{bmatrix} 213 \\ 123 \end{bmatrix}, \qquad 321 = \begin{bmatrix} 321 \\ 312 \\ 132 \\ 123 \end{bmatrix}, \qquad -1 \text{ is used.} \tag{62}$$

Thus

$$\epsilon_{ijk} = \begin{array}{|c|c|c|} \hline 0 & 0 & 0 \\ \hline 0 & 0 & +1 \\ \hline 0 & -1 & 0 \\ \hline \end{array} \quad \begin{array}{|c|c|c|} \hline 0 & 0 & -1 \\ \hline 0 & 0 & 0 \\ \hline +1 & 0 & 0 \\ \hline \end{array} \quad \begin{array}{|c|c|c|} \hline 0 & +1 & 0 \\ \hline -1 & 0 & 0 \\ \hline 0 & 0 & 0 \\ \hline \end{array} \tag{63}$$

Figure 22 shows these 27 values in a three-dimensional array. From Fig. 22 it is easy to see that every plane has both a positive and a negative symbol.

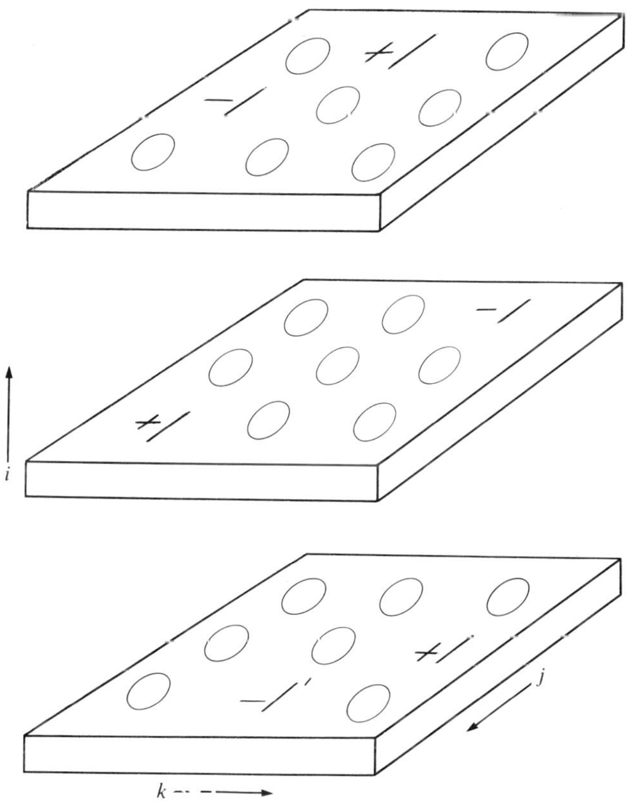

Figure 22. The Levi-Civita symbol as a tensor $R = 3$, $D = 3$.

Moreover, it is obvious that the Levi-Civita symbol is an antisymmetric tensor of rank three.

To summarize,

$$\epsilon_{123} = \epsilon_{231} = \epsilon_{312} = +1 \tag{64}$$

if ijk are found among (12312), and

$$\epsilon_{213} = \epsilon_{132} = \epsilon_{321} = -1 \tag{65}$$

if ijk are found among (21321). All other values are zero.

In many vector and tensor calculations, the Levi-Civita symbol ϵ_{ijk} appears in combination with itself. To illustrate, for convenience, change the dummy indices and form the product

$$\epsilon_{ijm}\epsilon_{klm}, \tag{66}$$

to be evaluated. Since it is understood that the Einstein summation convention applies to the index m,

$$\epsilon_{ijm}\epsilon_{klm} \equiv \sum_{m=1}^{3} \epsilon_{ijm}\epsilon_{klm} \tag{67}$$

with four variables, i, j, k, l, each of which is independent and can range over 1, 2, 3. Thus summation (67) really represents $3 \times 3 \times 3 \times 3 = 81$ separate terms. Fortunately, however, the majority of the Levi-Civita symbols in these terms are zero. To evaluate all 81 possible terms would not tax a student's reasoning power but would prove his patience and bookkeeping ability. Furthermore, it would lead only to a sequence of 81 numbers, all either 0 or ± 1, and would lead to no useful and interesting formulas or relationships. That is not the way a theoretical physicist would do things.

Instead, the theoretical physicist probably would notice that, in every term of summation (67), at least two of the four indices must be the same. If, for the sake of argument,

$$i = 1, \qquad j = 2, \qquad k = 3, \tag{68}$$

for instance, then l also must duplicate either 1, 2, or 3, and then summation (67) becomes

$$\epsilon_{121}\epsilon_{3l1} + \epsilon_{122}\epsilon_{3l2} + \epsilon_{133}\epsilon_{3l3} \equiv 0. \tag{69}$$

This is analogous to saying that any four three-dimensional vectors i, j, k, l are linearly dependent, because three of them can be chosen independently, but not the fourth.

Consider combination (66) again; obviously,

$$\epsilon_{ijm}\epsilon_{klm} = 0, \qquad \text{for} \quad i = j; \tag{70}$$

also,

$$\epsilon_{ijm}\epsilon_{klm} = 0, \qquad \text{for} \quad k = l. \tag{71}$$

If both $i = k$ and $j = l$

$$\epsilon_{ijm}\epsilon_{klm} = (+1)(+1) = (-1)(-1) = +1. \tag{72}$$

In addition, if both $i = l$ and $j = k$,

$$\epsilon_{ijm}\epsilon_{klm} = (+1)(-1) = (-1)(+1) = -1. \tag{73}$$

The last two cases may be combined to yield a final condition, which is illustrated by equations (68) and (69):

$$\begin{aligned} \epsilon_{ijm}\epsilon_{klm} = 0, \qquad &\text{for} \quad i \neq k \quad \text{or} \quad l. \\ &\text{or} \quad j \neq k \quad \text{or} \quad l. \end{aligned} \tag{74}$$

Equation (72) is true if $i = k$ and $j = l$. Hence the product $\epsilon_{ijm}\epsilon_{klm}$ can be replaced by a different product of Kronecker deltas, as follows:

$$\epsilon_{ijm}\epsilon_{klm} = \delta_{ik}\delta_{jl} = 1, \qquad \text{for} \quad i = k \qquad \text{and} \quad j = l. \tag{75}$$

Equation (73) is true if $i = l$ and $j = k$. Hence the product $\epsilon_{ijm}\epsilon_{klm}$ can be replaced by a different product of Kronecker deltas, as follows:

$$\epsilon_{ijm}\epsilon_{klm} = -\delta_{il}\delta_{jk} = -1, \qquad \text{for} \quad i = l \qquad \text{and} \quad j = k. \tag{76}$$

Now equations (72) and (73) may both be taken care of by a combination of equations (75) and (76), reminiscent of the rule for linear differential equations that the sum of two solutions is again a solution:

$$\begin{aligned} \epsilon_{ijm}\epsilon_{klm} = \delta_{ik}\delta_{jl} - \delta_{il}\delta_{jk} &= +1, \qquad \text{for} \quad i = k \qquad \text{and} \quad j = l, \\ &= -1, \qquad \text{for} \quad i = l \qquad \text{and} \quad j = k. \end{aligned} \tag{77}$$

Now notice that if $k = 1$, as it does in equation (71), then

$$\delta_{ik}\delta_{jk} - \delta_{ik}\delta_{jk} = 0, \qquad \text{for} \quad k = l. \tag{78}$$

Thus equations (71), (72), and (73) are all satisfied. Finally, since equation (77) is a restatement of equations (72) and (73), it also is satisfied, yielding the identity

$$\epsilon_{ijm}\epsilon_{klm} = \delta_{ik}\delta_{jl} - \delta_{il}\delta_{jk}. \tag{79}$$

This result can be used to evaluate $\epsilon_{ijm}\epsilon_{kjm}$. Set $l = j$ and invoke the Einstein convention to sum over the repeated index, as follows:

$$\begin{aligned} \epsilon_{ijm}\epsilon_{kjm} &= \delta_{ik}\delta_{jj} - \delta_{ij}\delta_{jk} \\ &= \delta_{ik}(\delta_{11} + \delta_{22} + \delta_{33}) - (\delta_{i1}\delta_{1k} + \delta_{i2}\delta_{2k} + \delta_{i3}\delta_{3k}). \end{aligned} \tag{80}$$

If $i \neq k$, then all the terms are zero, otherwise one of the last three terms will be nonzero and the first bracket also will be nonzero. So

$$\begin{aligned} 3\delta_{ik} - (\delta_{i1}\delta_{1k} + \delta_{i2}\delta_{2k} + \delta_{i3}\delta_{3k}) &= 3 - 1 = 2 \qquad \text{for} \quad i = k, \\ &= 0 - 0 = 0 \qquad \text{for} \quad i \neq k. \end{aligned} \tag{81}$$

This statement may be combined simply as

$$\epsilon_{ijm}\epsilon_{kjm} = 2\delta_{ik}. \tag{82}$$

Second, the result (82) can be employed to evaluate $\epsilon_{ijm}\epsilon_{ijm}$, summing over all the indices. Set $i = k$, then

$$\epsilon_{ijm}\epsilon_{ijm} = 2\delta_{ii} = 2(\delta_{11} + \delta_{22} + \delta_{33}) = 2(3) = 6, \tag{83}$$

in other words, an integer:

$$\epsilon_{ijm}\epsilon_{ijm} = 6. \tag{84}$$

It is now reasonable to ask: What is the value of this symbol and how is it used?

The symbol ϵ_{ijk} in conjunction with other quantities like vector components produces a short-hand notation for the vector or cross-product. Consider the two vectors **A** and **B** with components

$$\mathbf{A} = (A_x, A_y, A_z) \equiv (A_1, A_2, A_3) \equiv (A_j) \tag{85}$$

and

$$B = (B_x, B_y, B_z) \equiv (B_1, B_2, B_3) \equiv (B_k). \tag{86}$$

Their vector or cross-product will be designated

$$\mathbf{C} = \mathbf{A} \times \mathbf{B}, \tag{87}$$

with components

$$C = (C_x, C_y, C_z) = (C_1, C_2, C_3) = (C_i). \tag{88}$$

The letters i, j, k are dummy variables in the sense that they may take on all the values allowed 1, 2, and 3. Now form the product using the Levi-Civita symbol:

$$C_i = \epsilon_{ijk} A_j B_k, \tag{89}$$

and evaluate this product by using the Einstein summation convention, that is, by summing over a repeated index. It is permissible either to include the summation symbol $\sum$ or to omit it, as preferred:

$$C_i = \sum_{j=1}^{3} \sum_{k=1}^{3} \epsilon_{ijk} A_j B_k = \epsilon_{ijk} A_j B_k. \tag{90}$$

Take a test case, $i = 1$. Then summation (90) becomes:

$$C_1 = \sum_j \sum_k \epsilon_{ijk} A_j B_k = \sum_k (\epsilon_{12k} A_2 B_k + \epsilon_{13k} A_3 B_k) \tag{91}$$
$$= \epsilon_{123} A_2 B_3 + \epsilon_{132} A_3 B_2 = A_2 B_3 - A_3 B_2,$$

or, in familiar terms,

$$C_x = A_y B_z - A_z B_y. \tag{92}$$

The other two components are also easy to obtain. To show that this new method gives the correct answer for the cross-product of a vector with itself, consider **A** × **A**. Then

$$C_i = \epsilon_{ijk} A_j B_k \tag{93}$$

or

$$C_1 = \epsilon_{123} A_2 A_3 + \epsilon_{132} A_3 A_2 = A_2 A_3 - A_3 A_2 = 0. \tag{94}$$

Now proceed to the evaluation of the triple scalar product:

$$\mathbf{A} \cdot (\mathbf{B} \times\cdot \mathbf{C}) = \text{volume.} \tag{95}$$

Introduce the Levi-Civita symbol for the cross-product **B** × **C**:

$$(\mathbf{B} \times \mathbf{C})_i = \epsilon_{ijk} B_j C_k = \sum_j \sum_k \epsilon_{ijk} B_j C_k, \tag{96}$$

where the limits on the summation signs are implied. Now product (95) becomes

$$\begin{aligned} \text{volume} &= \mathbf{A} \cdot (\mathbf{B} \times\cdot \mathbf{C}) = A_i(\mathbf{B} \times\cdot \mathbf{C})_i = A_i \epsilon_{ijk} B_j C_k \\ &= \sum_i \epsilon_{ijk} A_i B_j C_k = \epsilon_{ijk} A_i B_j C_k = \sum_i \sum_j \sum_k \epsilon_{ijk} A_i B_j C_k. \end{aligned} \tag{97}$$

For reducing expression (97), the Einstein summation convention is used for all three repeated indices. Remember that of the 27 products in expression (97) only six are not zero. Thus

$$\begin{aligned} \text{volume} &= \epsilon_{123} A_1 B_2 C_3 + \epsilon_{132} A_1 B_3 C_2 + \epsilon_{231} A_2 B_3 C_1 + \epsilon_{213} A_2 B_1 C_3 \\ &\quad + \epsilon_{312} A_3 B_1 C_2 + \epsilon_{321} A_3 B_2 C_1 = A_1 B_2 C_3 - A_1 B_3 C_2 \\ &\quad + A_2 B_3 C_1 - A_2 B_1 C_3 + A_3 B_1 C_2 - A_3 B_2 C_1 \\ &= A_1 (B_2 C_3 - B_3 C_2) + A_2 (B_3 C_1 - B_1 C_3) \\ &\quad + A_3 (B_1 C_2 - B_2 C_1) = \begin{vmatrix} A_1 & A_2 & A_3 \\ B_1 & B_2 & B_3 \\ C_1 & C_2 & C_3 \end{vmatrix} = \text{determinant.} \end{aligned} \tag{98}$$

The last result is familiar and acts as a good check on this application of ϵ_{ijk}.

To evaluate the triple cross-product,

$$\mathbf{A} \times [\mathbf{B} \times \mathbf{C}], \tag{99}$$

by using the result just obtained for the combined Levi-Civita symbol $\epsilon_{ijm}\epsilon_{klm}$, consider the ith component:

$$\{\mathbf{A} \times [\mathbf{B} \times \mathbf{C}]\}_i = \epsilon_{ijm} A_j [\mathbf{B} \times \mathbf{C}]_m. \tag{100}$$

Now insert the mth component of the cross-product inside the square brackets:

$$\begin{aligned} \{\mathbf{A} \times [\mathbf{B} \times \mathbf{C}]\}_i &= \epsilon_{ijm} A_j \epsilon_{mkl} B_k C_l \\ &= \epsilon_{ijm}\epsilon_{mkl} A_j B_k C_l = \epsilon_{ijm}\epsilon_{klm} A_j B_k C_l. \end{aligned} \tag{101}$$

The last change occurs because of the cyclic nature of the indices, that is,

$$\epsilon_{mkl} = \epsilon_{lmk} = \epsilon_{klm}. \tag{102}$$

Now insert

$$\epsilon_{ijm}\epsilon_{klm} = \delta_{ik}\delta_{jl} - \delta_{il}\delta_{jk} \tag{103}$$

into equation (101) to get

$$\begin{aligned} \{\mathbf{A} \times [\mathbf{B} \times \mathbf{C}]\}_i &= \epsilon_{ijn}\epsilon_{klm} A_j B_k C_l \\ &= [\delta_{ik}\delta_{jl} - \delta_{il}\delta_{jk}] A_j B_k C_l. \end{aligned} \tag{104}$$

Hold the index i temporarily fixed in expression (104). That is, when there is a choice $i = k$ or $i = l$, change k and l into i, not the other way around. This procedure gives

$$\{\mathbf{A} \times [\mathbf{B} \times \mathbf{C}]\}_i = \delta_{ik}\delta_{jl} A_j B_k C_l - \delta_{il}\delta_{jk} A_j B_k C_l. \tag{105}$$

In searching for and adding all nonzero contributions to expression (105), j, k, l are allowed to take on all values. So:

$$\begin{aligned} \{\mathbf{A} \times [\mathbf{B} \times \mathbf{C}]\}_i &= \delta_{ii}\delta_{jl} A_j B_i C_l - \delta_{ii}\delta_{jk} A_j B_k C_i \\ &= \delta_{jl} A_j B_i C_l - \delta_{jk} A_j B_k C_i. \end{aligned} \tag{106}$$

Now proceed to let $j = l$ and $j = k$. It does not matter which is the final variable since the repeated index will disappear under the Einstein summation convention. Thus

$$\begin{aligned} \{\mathbf{A} \times [\mathbf{B} \times \mathbf{C}]\}_i &= \delta_{jj} A_j B_i C_j - \delta_{jj} A_i B_j C_i \\ &= B_i A_j C_j - C_i A_j B_j \end{aligned} \tag{107}$$

and

$$\{\mathbf{A} \times [\mathbf{B} \times \mathbf{C}]\}_i = B_i(\mathbf{A} \cdot \mathbf{C}) - C_i(\mathbf{A} \cdot \mathbf{B}). \tag{108}$$

This completes the evaluation of the triple cross-product using combined Levi-Civita symbols, yielding

$$\{\mathbf{A} \times [\mathbf{B} \times \mathbf{C}]\} = \mathbf{B}(\mathbf{A} \cdot \mathbf{C}) - \mathbf{C}(\mathbf{A} \cdot \mathbf{B}). \tag{109}$$

Part III

Applications to Relativity and Quantum Mechanics

5 Applications to Relativity

A. Maxwell's Equations and Electromagnetic Theory

A popular pastime of a few physicists is manipulating standard sets of equations upon which certain well-known theories are based, in an attempt to wrench new information from them. This intellectual activity may result in a trivial reexpression, or it may produce a simpler formulation that is easier to remember than the one before, or it may lead to a Nobel prize! Examples of the last include two great achievements: Dirac's equation in symmetric, four-dimensional notation, and Dirac's equation in two-component form that does not conserve parity.

Before showing (at the second level of attainment above) that Maxwell's equation permits concise statement in symmetric four-dimensional notation, a few tricks pulled from the complex variable magician's hat will be applied to simplify the form of Maxwell's equations in free space. But first, recall that for free space $\mathbf{E} = \mathbf{D}$, if gaussian units are used.

Start with

$$\nabla \times \mathbf{H} = \frac{1}{c}\frac{\partial \mathbf{D}}{\partial t} = \frac{1}{c}\frac{\partial \mathbf{E}}{\partial t} \tag{1}$$

$$\nabla \times \mathbf{E} = -\frac{1}{c}\frac{\partial \mathbf{B}}{\partial t} = -\frac{1}{c}\frac{\partial \mathbf{E}}{\partial t} \tag{2}$$

$$\nabla \cdot \mathbf{D} = \nabla \cdot \mathbf{E} = 0 \tag{3}$$

$$\nabla \cdot \mathbf{B} = \nabla \cdot \mathbf{H} = 0. \tag{4}$$

Now introduce a linear combination of E and H:

$$\mathbf{K} = \mathbf{H} + i\mathbf{E}. \tag{5}$$

Replace equation (2) by

$$\nabla \times i\mathbf{E} = -\frac{i}{c}\frac{\partial \mathbf{H}}{\partial t} \tag{6}$$

and equation (3) by

$$\nabla \cdot i\mathbf{E} = 0. \tag{7}$$

Adding equations (1) and (6) gives

$$\nabla \times \mathbf{H} + \nabla \times i\mathbf{E} = \nabla \times (\mathbf{H} + i\mathbf{E}) = \nabla \times \mathbf{K} = \frac{1}{c}\frac{\partial \mathbf{E}}{\partial t} - \frac{i}{c}\frac{\partial \mathbf{H}}{\partial t}$$

$$= -\frac{i}{c}\frac{\partial}{\partial t}(\mathbf{H} + i\mathbf{E}) = -\frac{i}{c}\frac{\partial \mathbf{K}}{\partial t} \tag{8}$$

$$\text{or} \quad \nabla \times \mathbf{K} = -\frac{i}{c}\dot{K}.$$

Next, adding equations (4) and (7) gives

$$\nabla \cdot \mathbf{H} + \nabla \cdot i\mathbf{E} = \nabla \cdot (\mathbf{H} + i\mathbf{E}) = \nabla \cdot \mathbf{K} = 0. \tag{9}$$

Thus the number of Maxwell equations has been reduced to three, equations (5), (8), and (9).

Now start again to manipulate equations (1) through (4) in a different way to writing them in component form:

$$(\nabla \times \mathbf{H})_x - \frac{1}{c}\frac{\partial E_x}{\partial t} = \frac{\partial H_z}{\partial y} - \frac{\partial H_y}{\partial z} - \frac{1}{c}\frac{\partial E_x}{\partial t} = 0, \tag{10a}$$

$$(\nabla \times \mathbf{H})_y - \frac{1}{c}\frac{\partial E_y}{\partial t} = \frac{\partial H_x}{\partial t} - \frac{\partial H_z}{\partial x} - \frac{1}{c}\frac{\partial E_y}{\partial t} = 0, \tag{10b}$$

$$(\nabla \times \mathbf{H})_z - \frac{1}{c}\frac{\partial E_z}{\partial t} = \frac{\partial H_y}{\partial x} - \frac{\partial H_x}{\partial y} - \frac{1}{c}\frac{\partial E_z}{\partial t} = 0, \tag{10c}$$

$$(\nabla \times \mathbf{E})_x + \frac{1}{c}\frac{\partial H_x}{\partial t} = \frac{\partial E_z}{\partial y} - \frac{\partial E_y}{\partial z} + \frac{1}{c}\frac{\partial H_x}{\partial t} = 0, \tag{10d}$$

$$(\nabla \times \mathbf{E})_y + \frac{1}{c}\frac{\partial H_y}{\partial t} = \frac{\partial E_x}{\partial z} - \frac{\partial E_z}{\partial x} + \frac{1}{c}\frac{\partial H_y}{\partial t} = 0, \tag{10e}$$

$$(\nabla \times \mathbf{E})_z + \frac{1}{c}\frac{\partial H_z}{\partial t} = \frac{\partial E_y}{\partial x} - \frac{\partial E_x}{\partial y} + \frac{1}{c}\frac{\partial H_z}{\partial t} = 0, \tag{10f}$$

$$\nabla \cdot \mathbf{E} = \frac{\partial E_x}{\partial x} + \frac{\partial E_y}{\partial y} + \frac{\partial E_z}{\partial z} = 0, \tag{10g}$$

$$\nabla \cdot \mathbf{H} = \frac{\partial H_x}{\partial x} + \frac{\partial H_y}{\partial y} + \frac{\partial H_z}{\partial z} = 0. \tag{10h}$$

Next introduce four new symbols:

$$\psi_1 = H_x + iH_y \tag{11a}$$

$$\psi_2 = H_z \tag{11b}$$

$$\psi_3 = E_x + iE_y \tag{11c}$$

$$\psi_4 = E_z \tag{11d}$$

Next multiply equation (10*c*) by *i* and add the result to equation (10*h*):

$$\begin{aligned}\frac{\partial iH_y}{\partial x} - i\frac{\partial}{\partial y}H_x + \frac{1}{c}\frac{\partial}{\partial y}(-iE_z) + \frac{\partial H_z}{\partial z} + \frac{\partial H_y}{\partial y} + \frac{\partial H_z}{\partial z}\\ = \frac{\partial}{\partial x}(H_x + iH_y) - i\frac{\partial}{\partial y}(H_x + iH_y) + \frac{\partial H_z}{\partial z} + \frac{1}{c}\frac{\partial}{\partial t}(-iE_z) = 0\end{aligned} \tag{12a}$$

or

$$\left(\frac{\partial}{\partial x} - i\frac{\partial}{\partial y}\right)\psi_1 + \frac{\partial \psi_2}{\partial z} + \frac{1}{ic}\frac{\partial \psi_4}{\partial t} = 0. \tag{12b}$$

Now multiply equation (10*g*) by $-i$ and add this result to (10*f*):

$$\begin{aligned}-i\frac{\partial}{\partial x}E_x - \frac{\partial iE_y}{\partial y} + \frac{\partial(-iE_z)}{\partial z} + \frac{\partial E_y}{\partial x} - \frac{\partial E_x}{\partial y} + \frac{1}{c}\frac{\partial H_z}{\partial t}\\ = \frac{\partial}{\partial x}(-i[E_x + iE_y]) - i\frac{\partial}{\partial y}(-i[E_x + iE_y]) + \frac{\partial}{\partial z}(-iE_z) + \frac{1}{c}\frac{\partial H_z}{\partial t}\\ = \left(\frac{\partial}{\partial x} - i\frac{\partial}{\partial y}\right)(-i[E_x + iE_y]) + \frac{\partial}{\partial z}(-iE_z) + \frac{1}{c}\frac{\partial H_z}{\partial t} = 0\end{aligned} \tag{13a}$$

or

$$\left(\frac{\partial}{\partial x} - i\frac{\partial}{\partial y}\right)\psi_3 + \frac{\partial}{\partial z}\psi_4 - \frac{1}{ic}\frac{\partial}{\partial t}\psi_2 = 0. \tag{13b}$$

Next multiply equation (10*a*) by *i* and then subtract (10*b*) from it:

$$\begin{aligned}i\frac{\partial}{\partial y}H_z - \frac{\partial}{\partial z}iH_y + \frac{1}{c}\frac{\partial}{\partial t}(-iE_z) - \left(\frac{\partial H_x}{\partial z} - \frac{\partial H_z}{\partial x} - \frac{1}{c}\frac{\partial}{\partial t}E_y\right)\\ = \left(\frac{\partial}{\partial x} + i\frac{\partial}{\partial y}\right)H_z - \frac{\partial}{\partial z}(H_x + iH_y) + \frac{1}{ic}\frac{\partial}{\partial t}(E_x + iE_y) = 0\end{aligned} \tag{14a}$$

or

$$\left(\frac{\partial}{\partial x} + i\frac{\partial}{\partial y}\right)\psi_2 - \frac{\partial}{\partial z}\psi_1 + \frac{1}{ic}\frac{\partial}{\partial t}\psi_3 = 0. \tag{14b}$$

Next multiply equation (10*e*) by *i* and add the result to equation (10*d*):

$$i\frac{\partial E_x}{\partial z} - i\frac{\partial}{\partial x}E_z + \frac{\partial}{c\partial t}iH_y + \left(\frac{\partial E_z}{\partial y} - \frac{\partial E_y}{\partial z} + \frac{1}{c}\frac{\partial H_x}{\partial t}\right)$$
$$= -i\left(\frac{\partial}{\partial x} + i\frac{\partial}{\partial y}\right)E_z - (-i)\frac{\partial}{\partial z}(E_x + iE_y) + \frac{1}{c}\frac{\partial}{\partial t}(H_x + iH_y) = 0 \tag{15a}$$

or

$$\cdot\left(\frac{\partial}{\partial x} + i\frac{\partial}{\partial y}\right)\psi_4 - \frac{\partial}{\partial z}\psi_3 - \frac{1}{ic}\frac{\partial}{\partial t}\psi_1 = 0. \tag{15b}$$

Equations (12) through (15) are now written as a system of linear algebraic equations:

$$\left\{\begin{aligned} \frac{\partial\psi_1}{\partial z} - \left(\frac{\partial}{\partial x} + i\frac{\partial}{\partial y}\right)\psi_2 - \frac{1}{ic}\frac{\partial}{\partial t}\psi_3 + 0\cdot\psi_4 &= 0\\ -\left(\frac{\partial}{\partial x} - i\frac{\partial}{\partial y}\right)\psi_1 - \frac{\partial}{\partial z}\psi_2 + 0\cdot\psi_3 - \frac{1}{ic}\frac{\partial}{\partial t}\psi_4 &= 0\\ -\frac{1}{ic}\frac{\partial\psi_1}{\partial t} + 0\cdot\psi_2 - \frac{\partial}{\partial t}\psi_3 + \left(\frac{\partial}{\partial x} + i\frac{\partial}{\partial y}\right)\psi_4 &= 0\\ 0\cdot\psi_1 - \frac{1}{ic}\frac{\partial}{\partial t}\psi_2 + \left(\frac{\partial}{\partial x} - i\frac{\partial}{\partial y}\right)\psi_3 + \frac{\partial}{\partial z}\psi_4 &= 0 \end{aligned}\right. \tag{16}$$

or, in matrix form

$$\begin{pmatrix} \frac{\partial}{\partial z} & -\left(\frac{\partial}{\partial x} + i\frac{\partial}{\partial y}\right) & -\frac{1}{ic}\frac{\partial}{\partial t} & 0\\ -\left(\frac{\partial}{\partial x} - i\frac{\partial}{\partial y}\right) & -\frac{\partial}{\partial t} & 0 & -\frac{1}{ic}\frac{\partial}{\partial t}\\ -\frac{1}{ic}\frac{\partial}{\partial t} & 0 & -\frac{\partial}{\partial z} & \left(\frac{\partial}{\partial x} + i\frac{\partial}{\partial y}\right)\\ 0 & -\frac{1}{ic}\frac{\partial}{\partial t} & \left(\frac{\partial}{\partial x} - i\frac{\partial}{\partial y}\right) & \frac{\partial}{\partial z} \end{pmatrix}\begin{pmatrix}\psi_1\\ \psi_2\\ \psi_3\\ \psi_4\end{pmatrix} = 0. \tag{17}$$

Now substitute $x = x_1$, $y = x_2$, $z = x_3$, $ict = x_4$, and $\Psi = \begin{pmatrix}\psi_1\\ \psi_2\\ \psi_3\\ \psi_4\end{pmatrix}$.

Then expression (17) becomes

$$\begin{pmatrix} \frac{\partial}{\partial x_3} & -\left(\frac{\partial}{\partial x_1} + i\frac{\partial}{\partial x_2}\right) & -\frac{\partial}{\partial x_4} & 0 \\ \left(-\frac{\partial}{\partial x_1} - i\frac{\partial}{\partial x_2}\right) & -\frac{\partial}{\partial x_3} & 0 & -\frac{\partial}{\partial x_4} \\ -\frac{\partial}{\partial x_4} & 0 & -\frac{\partial}{\partial x_3} & \left(\frac{\partial}{\partial x_1} + i\frac{\partial}{\partial x_2}\right) \\ 0 & -\frac{\partial}{\partial x_4} & \left(\frac{\partial}{\partial x_1} - i\frac{\partial}{\partial x_2}\right) & \frac{\partial}{\partial x_3} \end{pmatrix}\Psi$$

$$= \left\{\begin{pmatrix} 0 & -1 & 0 & 0 \\ -1 & 0 & 0 & 0 \\ 0 & 0 & 0 & 1 \\ 0 & 0 & 1 & 0 \end{pmatrix}\frac{\partial}{\partial x_1} + \begin{pmatrix} 0 & -i & 0 & 0 \\ i & 0 & 0 & 0 \\ 0 & 0 & 0 & i \\ 0 & 0 & -i & 0 \end{pmatrix}\frac{\partial}{\partial x_2}\right.$$

$$\left. + \begin{pmatrix} 1 & 0 & 0 & 0 \\ 0 & -1 & 0 & 0 \\ 0 & 0 & -1 & 0 \\ 0 & 0 & 0 & 1 \end{pmatrix}\frac{\partial}{\partial x_3} + \begin{pmatrix} 0 & 0 & -1 & 0 \\ 0 & 0 & 0 & -1 \\ -1 & 0 & 0 & 0 \\ 0 & -1 & 0 & 0 \end{pmatrix}\frac{\partial}{\partial x_4}\right\}\Psi = \tag{18}$$

$$= \left\{\gamma_1\frac{\partial}{\partial x_1} + \gamma_2\frac{\partial}{\partial x_2} + \gamma_3\frac{\partial}{\partial x_3} + \gamma_4\frac{\partial}{\partial x_4}\right\}\Psi = \gamma_\mu\frac{\partial}{\partial x_\mu}\Psi = 0,$$

where

$$\gamma_1 = \begin{pmatrix} 0 & -1 & 0 & 0 \\ -1 & 0 & 0 & 0 \\ 0 & 0 & 0 & 1 \\ 0 & 0 & 1 & 0 \end{pmatrix} = \begin{pmatrix} -\sigma_x & 0 \\ 0 & \sigma_x \end{pmatrix}, \tag{19}$$

$$\gamma_2 = \begin{pmatrix} 0 & -i & 0 & 0 \\ i & 0 & 0 & 0 \\ 0 & 0 & 0 & i \\ 0 & 0 & -i & 0 \end{pmatrix} = \begin{pmatrix} \sigma_y & 0 \\ 0 & -\sigma_y \end{pmatrix}, \tag{10}$$

$$\gamma_3 = \begin{pmatrix} 1 & 0 & 0 & 0 \\ 0 & -1 & 0 & 0 \\ 0 & 0 & -1 & 0 \\ 0 & 0 & 0 & 1 \end{pmatrix} = \begin{pmatrix} \sigma_z & 0 \\ 0 & -\sigma_z \end{pmatrix}, \tag{21}$$

and

$$\gamma_4 = \begin{pmatrix} 0 & 0 & -1 & 0 \\ 0 & 0 & 0 & -1 \\ -1 & 0 & 0 & 0 \\ 0 & -1 & 0 & 0 \end{pmatrix} = \begin{pmatrix} 0 & -I \\ -I & 0 \end{pmatrix}. \tag{22}$$

It is now possible to write Maxwell's equation in a concise form. Multiply equation (18) by $\hbar/i$ ($\partial/\partial x_\mu \to \hbar/i\ \partial/\partial x_\mu = p_\mu$ is the operator representing momentum in quantum mechanics). Therefore,

$$\gamma_\mu \frac{\partial}{\partial x_\mu}\Psi = 0 \Rightarrow \gamma_\mu p_\mu \Psi = \not{p}\Psi = 0. \tag{23}$$

Note that equation (23) is the relativistic wave equation for the photon. To see that this is true, add minus (12) and (13) to get:

$$\left(\frac{\partial}{\partial x} - i\frac{\partial}{\partial y}\right)(\psi_1 + \psi_3) + \left(\frac{\partial}{\partial z} + \frac{1}{c}\frac{\partial}{\partial t}\right)(\psi_2 + \psi_4) = 0. \tag{24}$$

Adding equations (14) and (15) gives

$$-\left(\frac{\partial}{\partial z} - \frac{1}{c}\frac{\partial}{\partial t}\right)(\psi_1 + \psi_3) + \left(\frac{\partial}{\partial x} + i\frac{\partial}{\partial y}\right)(\psi_2 + \psi_4) = 0. \tag{25}$$

Now let

$$(\psi_1 + \psi_3) = x_1 \tag{26}$$

and

$$(\psi_2 + \psi_4) = x_2. \tag{27}$$

Then with the definitions

$$\left(\frac{\partial}{\partial z} \pm \frac{1}{c}\frac{\partial}{\partial t}\right) \equiv \frac{\partial}{\partial t_\pm} \quad \text{and} \quad \left(\frac{\partial}{\partial x} \pm i\frac{\partial}{\partial y}\right) \equiv \frac{\partial}{\partial x_\pm}, \tag{28}$$

equations (24) and (25) become:

$$\left[\frac{\partial}{\partial x_-}x_1 + \frac{\partial}{\partial t_+}x_2\right] = 0 \quad \text{and} \quad \left[-\frac{\partial}{\partial t_-}x_1 + \frac{\partial}{\partial x_+}x_2\right] = 0. \tag{29}$$

By differentiating once more, these become:

$$\frac{\partial}{\partial t_-}\left[\frac{\partial}{\partial x_-}x_1 + \frac{\partial}{\partial t_+}x_2\right] + \frac{\partial}{\partial x_-}\left[-\frac{\partial}{\partial t_-}x_1 + \frac{\partial}{\partial x_+}x_2\right]$$
$$= \left[\frac{\partial^2}{\partial x^2} + \frac{\partial^2}{\partial y^2} + \frac{\partial^2}{\partial z^2} - \frac{1}{c^2}\frac{\partial^2}{\partial t^2}\right]x_2 = \square^2 x_2 = 0. \tag{30}$$

which is recognized as the classical wave equation.

Maxwell's equations for the six-component electromagnetic tensor field have been reduced to a Dirac-like equation for a four-component electromagnetic spinor field with zero mass. The Dirac γ-matrices, in this case

$$\gamma_1 = \begin{pmatrix} -\sigma_x & 0 \\ 0 & \sigma_x \end{pmatrix}, \quad \gamma_2 = \begin{pmatrix} \sigma_y & 0 \\ 0 & -\sigma_y \end{pmatrix},$$
$$\gamma_3 = \begin{pmatrix} \sigma_z & 0 \\ 0 & -\sigma_z \end{pmatrix}, \quad \gamma_4 = \begin{pmatrix} 0 & -1 \\ -1 & 0 \end{pmatrix}, \tag{31}$$

can be transformed into a different set that is suitable to a two-component form of the Dirac equation. In this connection see Part IV of *Modern Quantum Mechanics.**

$$\gamma_1' = \begin{pmatrix} 0 & -i\sigma_x \\ i\sigma_x & 0 \end{pmatrix}, \quad \gamma_2' = \begin{pmatrix} 0 & -i\sigma_y \\ i\sigma_y & 0 \end{pmatrix},$$
$$\gamma_3' = \begin{pmatrix} 0 & -i\sigma_z \\ i\sigma_z & 0 \end{pmatrix}, \quad \gamma_4' = \begin{pmatrix} 0 & -1 \\ -1 & 0 \end{pmatrix}. \tag{32}$$

The similarity transformation matrix that accomplishes this may be written

$$S = \begin{pmatrix} a & b & c & d \\ e & f & g & h \\ j & k & l & m \\ n & p & q & r \end{pmatrix}. \tag{33}$$

Then the first relationship is

$$S\gamma_1 S^{-1} = \gamma_1' \tag{34}$$

or

$$S\gamma_1 = \gamma_1' S, \tag{35}$$

* John A. Eisele, *Modern Quantum Mechanics with Applications to Elementary Particle Physics* (New York: Wiley-Interscience, 1969).

and becomes

$$\begin{pmatrix} a & b & c & d \\ e & f & g & h \\ j & k & l & m \\ n & p & q & r \end{pmatrix}\begin{pmatrix} 0 & -1 & 0 & 0 \\ -1 & 0 & 0 & 0 \\ 0 & 0 & 0 & 1 \\ 0 & 0 & 1 & 0 \end{pmatrix} = \begin{pmatrix} 0 & 0 & 0 & -i \\ 0 & 0 & -i & 0 \\ 0 & i & 0 & 0 \\ i & 0 & 0 & 0 \end{pmatrix}\begin{pmatrix} a & b & c & d \\ e & f & g & h \\ j & k & l & m \\ n & p & q & r \end{pmatrix},$$

which yields, upon equating elements,

$$\begin{aligned} j &= -if, \\ k &= -ie, \\ l &= ih, \\ m &= ig, \\ n &= -ib, \\ p &= -ia, \\ q &= id, \\ r &= ic. \end{aligned} \tag{27}$$

Thus S becomes

$$S = \begin{pmatrix} a & b & c & d \\ e & f & g & h \\ -if & -ie & ih & ig \\ -ib & -ia & id & ic \end{pmatrix}. \tag{38}$$

The second relationship is

$$S\gamma_3 = \gamma'_3 S \tag{39}$$

or

$$\begin{aligned} &\begin{pmatrix} a & b & c & d \\ e & f & g & h \\ -if & -ie & ih & ig \\ -ib & -ia & id & ic \end{pmatrix}\begin{pmatrix} 1 & 0 & 0 & 0 \\ 0 & -1 & 0 & 0 \\ 0 & 0 & -1 & 0 \\ 0 & 0 & 0 & 1 \end{pmatrix} \\ &\qquad = \begin{pmatrix} 0 & 0 & -i & 0 \\ 0 & 0 & 0 & i \\ i & 0 & 0 & 0 \\ 0 & -i & 0 & 0 \end{pmatrix}\begin{pmatrix} a & b & c & d \\ e & f & g & h \\ -if & -ie & ih & ig \\ -ib & -ia & id & ic \end{pmatrix}. \end{aligned} \tag{40}$$

Upon equating elements this yields

$$\begin{aligned} e &= b, \\ f &= -a, \\ g &= d, \\ h &= -c. \end{aligned} \tag{41}$$

Then S becomes:

$$S = \begin{pmatrix} a & b & c & d \\ b & -a & d & -c \\ ia & -ib & -ic & id \\ -ib & -ia & id & ic \end{pmatrix}. \tag{42}$$

The third relationship is

$$S\gamma_2 = \gamma_2' S \tag{43}$$

or

$$\begin{pmatrix} a & b & c & d \\ b & -a & d & -c \\ ia & -ib & -ic & id \\ -ib & -ia & id & ic \end{pmatrix} \begin{pmatrix} 0 & -i & 0 & 0 \\ i & 0 & 0 & 0 \\ 0 & 0 & 0 & i \\ 0 & 0 & -i & 0 \end{pmatrix} = \begin{pmatrix} 0 & 0 & 0 & -1 \\ 0 & 0 & 1 & 0 \\ 0 & 1 & 0 & 0 \\ -1 & 0 & 0 & 0 \end{pmatrix} \begin{pmatrix} a & b & c & d \\ b & -a & d & -c \\ ia & -ib & -ic & id \\ -ib & -ia & id & ic \end{pmatrix}. \tag{44}$$

Equating elements yields

$$a = 0, \qquad c = 0, \tag{45}$$

so the S matrix becomes

$$S = \begin{pmatrix} 0 & b & 0 & d \\ b & 0 & d & 0 \\ 0 & -ib & 0 & id \\ -ib & 0 & id & 0 \end{pmatrix}. \tag{46}$$

The fourth relationship is

$$S\begin{pmatrix} 0 & -1 \\ -1 & 0 \end{pmatrix} = \begin{pmatrix} 0 & -1 \\ -1 & 0 \end{pmatrix} S \tag{47}$$

or

$$\begin{pmatrix} 0 & b & 0 & d \\ b & 0 & d & 0 \\ 0 & -ib & 0 & id \\ -ib & 0 & id & 0 \end{pmatrix}\begin{pmatrix} 0 & -1 \\ -1 & 0 \end{pmatrix} = \begin{pmatrix} 0 & -1 \\ -1 & 0 \end{pmatrix}\begin{pmatrix} 0 & b & 0 & d \\ b & 0 & d & 0 \\ 0 & -ib & 0 & id \\ -ib & 0 & id & 0 \end{pmatrix}. \tag{48}$$

Equating elements yields

$$d = -ib. \tag{49}$$

Thus the S matrix becomes

$$S = \begin{pmatrix} 0 & b & 0 & -ib \\ b & 0 & -ib & 0 \\ 0 & -ib & 0 & b \\ -ib & 0 & b & 0 \end{pmatrix}. \tag{50}$$

The unitary requirement yields the final relationship:

$$SS^\dagger = I \tag{51}$$

or

$$\begin{pmatrix} 0 & b & 0 & -ib \\ b & 0 & -ib & 0 \\ 0 & -ib & 0 & b \\ -ib & 0 & b & 0 \end{pmatrix}\begin{pmatrix} 0 & b & 0 & ib \\ b & 0 & ib & 0 \\ 0 & ib & 0 & b \\ ib & 0 & b & 0 \end{pmatrix} = \begin{pmatrix} 1 & 0 & 0 & 0 \\ 0 & 1 & 0 & 0 \\ 0 & 0 & 1 & 0 \\ 0 & 0 & 0 & 1 \end{pmatrix}. \tag{52}$$

Equating elements yields

$$2b^2 = 1 \tag{53}$$

or

$$b = \frac{1}{\sqrt{2}}. \tag{54}$$

Thus the S matrix is

$$S = \frac{1}{\sqrt{2}}\begin{pmatrix} 0 & 1 & 0 & -i \\ 1 & 0 & -i & 0 \\ 0 & -i & 0 & 1 \\ -i & 0 & 1 & 0 \end{pmatrix} = \frac{1}{\sqrt{2}}\begin{pmatrix} \sigma_x & -i\sigma_x \\ -i\sigma_x & \sigma_x \end{pmatrix}. \tag{55}$$

Now apply this S matrix to a four-component spinor:

$$\frac{1}{\sqrt{2}}\begin{matrix} \sigma_x & -i\sigma_x \\ -i\sigma_x & \sigma_x \end{matrix}\begin{pmatrix} \psi_1 \\ \psi_2 \\ \psi_3 \\ \psi_4 \end{pmatrix} = \frac{1}{\sqrt{2}}\begin{pmatrix} (\psi_2 - i\psi_4) \\ (\psi_1 - i\psi_3) \\ -i(\psi_2 + i\psi_4) \\ -i(\psi_1 + i\psi_3) \end{pmatrix} = \begin{pmatrix} \psi'_1 \\ \psi'_2 \\ \psi'_3 \\ \psi'_4 \end{pmatrix}. \tag{56}$$

The four-component Dirac equation becomes

$$\begin{pmatrix} 0 & -i\boldsymbol{\sigma} \\ i\boldsymbol{\sigma} & 0 \end{pmatrix} \cdot \mathbf{p} + \begin{pmatrix} 0 & -I \\ -I & 0 \end{pmatrix} p_4 \Big\} \psi' = 0, \tag{57}$$

or two decoupled two-component equations:

$$[-i\boldsymbol{\sigma} \cdot \mathbf{p} + Ip_4]\begin{pmatrix} (\psi_2 - i\psi_4) \\ (\psi_1 - i\psi_3) \end{pmatrix} = 0$$

and (58)

$$[+i\boldsymbol{\sigma} \cdot \mathbf{p} + Ip_4]\begin{pmatrix} (\psi_2 + i\psi_4) \\ (\psi_1 + i\psi_3) \end{pmatrix} = 0,$$

which simplify to one equation:

$$[Ip_4 \pm i\boldsymbol{\sigma} \cdot \mathbf{p}]\psi'_{\pm} = 0. \tag{59}$$

For Maxwell's electromagnetic field the two-component spinor form becomes

$$\psi'_{+} = \frac{1}{\sqrt{2}}\begin{pmatrix} (H_z - iE_z) \\ (H_x - iE_x) + i(H_y - iE_y) \end{pmatrix} = \begin{pmatrix} \psi'_1 \\ \psi'_2 \end{pmatrix} \tag{60}$$

for the positive sign in equation (59).

EXERCISE

Find the S matrix that transforms

$$\gamma = \begin{pmatrix} \boldsymbol{\sigma} & 0 \\ 0 & -\boldsymbol{\sigma} \end{pmatrix}, \qquad \gamma_4 = \begin{pmatrix} 0 & -I \\ -I & 0 \end{pmatrix}$$

into

$$\gamma = \begin{pmatrix} 0 & -i\boldsymbol{\sigma} \\ i\boldsymbol{\sigma} & 0 \end{pmatrix}, \qquad \gamma_4 = \begin{pmatrix} 0 & -I \\ -I & 0 \end{pmatrix}.$$

How does this differ from the S matrix just derived in the text? What does this mean physically? Interpret the appearance of σ_x in S.

B. Lorentz Transformations and Special Relativity

For the special Lorentz transformation L (for motion in the x-direction, as illustrated in Fig. 23):

$$\begin{aligned} \mathbf{r} &= (x, y, z, ct) \qquad \text{or} \quad (x, y, z, ict), \\ \mathbf{r}' &= (x', y', z', ct') \qquad \text{or} \quad (x', y', z', ict'), \end{aligned} \tag{1}$$

and

$$\mathbf{r}' = L\mathbf{r}. \tag{2}$$

Now let the unprimed and primed systems be transformed by three-dimensional time invariant space transformations, m and n, respectively.

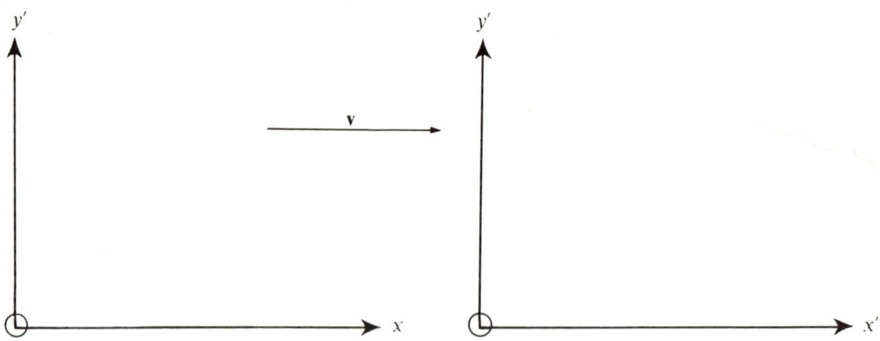

Figure 23. Original unprimed and primed coordinate systems.

These transformations can be obtained as the following specialized instances of four-dimensional transformations M and N:

$$\rho = Mr, \qquad M = \begin{pmatrix} & & & 0 \\ & m & & 0 \\ & & & 0 \\ 0 & 0 & 0 & 1 \end{pmatrix}, \tag{3}$$

and

$$\rho' = Nr', \qquad N = \begin{pmatrix} & & & 0 \\ & n & & 0 \\ & & & 0 \\ 0 & 0 & 0 & 1 \end{pmatrix}, \tag{4}$$

where M is not necessarily equal to N.

Now:

$$\begin{aligned} r' &= Lr, \\ Nr' &= NLr, \\ &\text{and} \\ Nr' &= NLM^{-1}Mr. \end{aligned} \tag{5}$$

Accordingly, from equations (3) and (4),

$$\rho' = \Lambda\,\rho, \tag{6}$$

where

$$\Lambda = NLM^{-1} \tag{7}$$

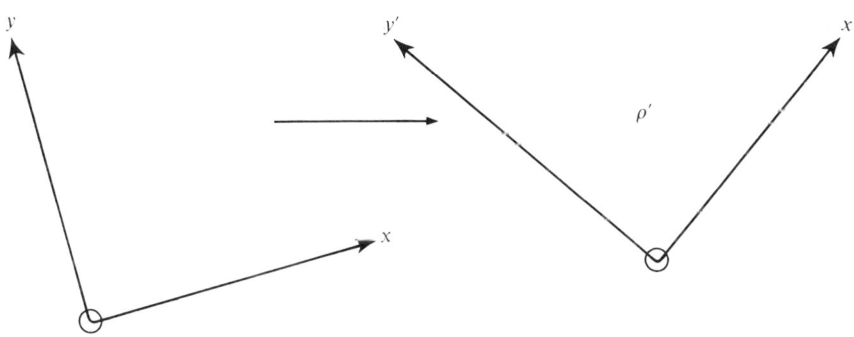

Figure 24. Tilted unprimed and primed coordinate systems.

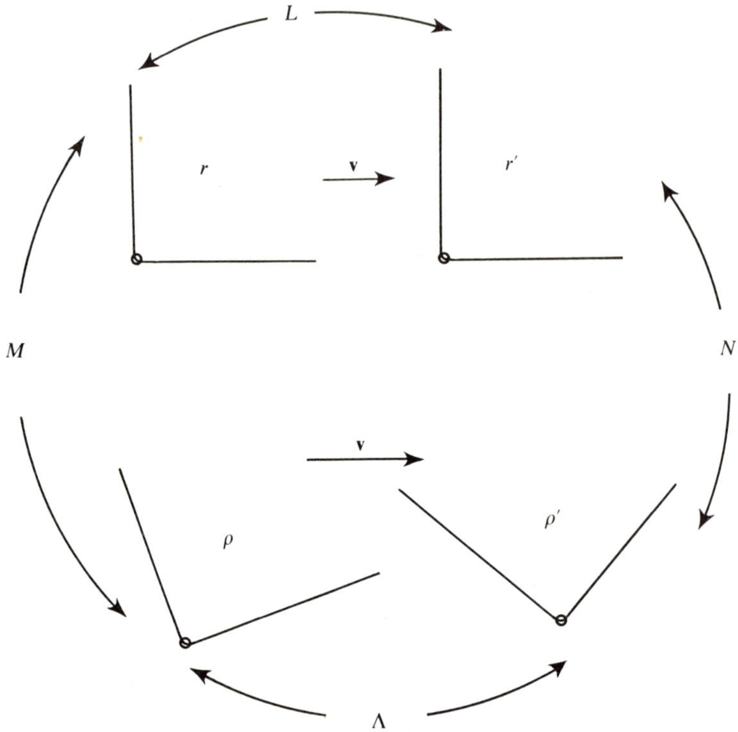

Figure 25. Diagram of the relationships arising in the transformation of the Lorentz Transformation L.

denotes a congruence transformation. Figure 25 helps visualize the relationships among the position vectors r, r', ρ, ρ', and the transformations L, M, N, and Λ.

If M is orthogonal, the special 4 by 4 transformation matrix:

$$M = (M_{ij}) = \begin{pmatrix} A & 0 & 0 & B \\ 0 & 1 & 0 & 0 \\ 0 & 0 & 1 & 0 \\ -B & 0 & 0 & A \end{pmatrix}, \tag{8}$$

where

$$A = \frac{1}{\sqrt{1-\beta^2}} \quad \text{and} \quad B = \frac{i\beta}{\sqrt{1-\beta^2}} \tag{9}$$

is a Lorentz transformation with $v = v_x$ and $v_y = v_z = 0$. The direct product of M with itself,

$$M \times M = (M_{ij}) \times \cdot (M_{kl}) = \begin{pmatrix} A & 0 & 0 & B \\ 0 & 1 & 0 & 0 \\ 0 & 0 & 1 & 0 \\ -B & 0 & 0 & A \end{pmatrix} \times \cdot \begin{pmatrix} A & 0 & 0 & B \\ 0 & 1 & 0 & 0 \\ 0 & 0 & 1 & 0 \\ -B & 0 & 0 & A \end{pmatrix}, \tag{10}$$

is needed for tensor transformations. The result of the indicated multiplication is as follows:

$$M \times \cdot M = \begin{array}{c|cccc:cccc:cccc:cccc|} & (11) & (12) & (13) & (14) & (21) & (22) & (23) & (24) & (31) & (32) & (33) & (34) & (41) & (42) & (43) & (44) \\ (11) & A^2 & 0 & 0 & AB & & & & & & & & & BA & 0 & 0 & B^2 \\ (12) & 0 & A & 0 & 0 & & \bigcirc & & & & \bigcirc & & & 0 & B & 0 & 0 \\ (13) & 0 & 0 & A & 0 & & & & & & & & & 0 & 0 & B & 0 \\ (14) & -AB & 0 & 0 & A^2 & & & & & & & & & -B^2 & 0 & 0 & BA \\ \hdashline (21) & & & & & A & 0 & 0 & B & & & & & & & & \\ (22) & & \bigcirc & & & 0 & 1 & 0 & 0 & & \bigcirc & & & & \bigcirc & & \\ (23) & & & & & 0 & 0 & 1 & 0 & & & & & & & & \\ (24) & & & & & -B & 0 & 0 & A & & & & & & & & \\ \hdashline (31) & & & & & & & & & A & 0 & 0 & B & & & & \\ (32) & & \bigcirc & & & & \bigcirc & & & 0 & 1 & 0 & 0 & & \bigcirc & & \\ (33) & & & & & & & & & 0 & 0 & 1 & 0 & & & & \\ (34) & & & & & & & & & -B & 0 & 0 & A & & & & \\ \hdashline (41) & -BA & 0 & 0 & -B^2 & & & & & & & & & A^2 & 0 & 0 & BA \\ (42) & 0 & -B & 0 & 0 & & \bigcirc & & & & \bigcirc & & & 0 & A & 0 & 0 \\ (43) & 0 & 0 & -B & 0 & & & & & & & & & 0 & 0 & A & 0 \\ (44) & B^2 & 0 & 0 & -BA & & & & & & & & & -AB & 0 & 0 & A^2 \end{array} \tag{11}$$

where the rows and columns are labeled with two-digit numbers, called "index couplets," in preparation for arranging the elements individually in dictionary order.

Replace the A and B just used in forming the direct product of a transformation matrix M by

$$A = \gamma \qquad \text{and} \qquad B = i\beta\gamma, \tag{12}$$

where

$$\gamma = \frac{1}{\sqrt{1 - \beta^2}}, \tag{13}$$

and call the resulting transformation L instead of M. Then

$$
L \times \cdot L = \left|\begin{array}{c:c:c:c}
\begin{matrix} \gamma^2 & 0 & 0 & i\beta\gamma^2 \\ 0 & \gamma & 0 & 0 \\ 0 & 0 & \gamma & 0 \\ -i\beta\gamma^2 & 0 & 0 & \gamma^2 \end{matrix} & \bigcirc & \bigcirc & \begin{matrix} i\beta\gamma & 0 & 0 & -\beta^2\gamma^2 \\ 0 & i\beta\gamma & 0 & 0 \\ 0 & 0 & i\beta\gamma & 0 \\ \beta^2\gamma^2 & 0 & 0 & i\beta\gamma^2 \end{matrix} \\ \hdashline
\bigcirc & \begin{matrix} \gamma & 0 & 0 & i\beta\gamma \\ 0 & 1 & 0 & 0 \\ 0 & 0 & 1 & 0 \\ -i\beta\gamma & 0 & 0 & \gamma \end{matrix} & \bigcirc & \bigcirc \\ \hdashline
\bigcirc & \bigcirc & \begin{matrix} \gamma & 0 & 0 & i\beta\gamma \\ 0 & 1 & 0 & 0 \\ 0 & 0 & 1 & 0 \\ -i\beta\gamma & 0 & 0 & \gamma \end{matrix} & \bigcirc \\ \hdashline
\begin{matrix} -i\beta\gamma^2 & 0 & 0 & \beta^2\gamma^2 \\ 0 & -i\beta\gamma & 0 & 0 \\ 0 & 0 & -i\beta\gamma & 0 \\ -\beta^2\gamma^2 & 0 & 0 & -i\beta\gamma^2 \end{matrix} & \bigcirc & \bigcirc & \begin{matrix} \gamma^2 & 0 & 0 & i\beta\gamma^2 \\ 0 & \gamma & 0 & 0 \\ 0 & 0 & \gamma & 0 \\ -i\beta\gamma & 0 & 0 & \gamma^2 \end{matrix}
\end{array}\right|
$$

Since, for an antisymmetric tensor, $T_{ij} = -T_{ji}$ and $T_{ii} = 0$, certain rows and columns may be struck out of the direct product transformation matrix $L \times L$. It then becomes

$$
L \times \cdot L = \left|\begin{array}{c:c:c:c}
\begin{matrix} 0 & 0 & 0 & 0 \\ 0 & \gamma & 0 & 0 \\ 0 & 0 & \gamma & 0 \\ 0 & 0 & 0 & \gamma^2 \end{matrix} & \bigcirc & \bigcirc & \begin{matrix} 0 & 0 & 0 & 0 \\ 0 & i\beta\gamma & 0 & 0 \\ 0 & 0 & i\beta\gamma & 0 \\ \beta^2\gamma^2 & 0 & 0 & 0 \end{matrix} \\ \hdashline
\bigcirc & \begin{matrix} \gamma & 0 & 0 & i\beta\gamma \\ 0 & 0 & 0 & 0 \\ 0 & 0 & 1 & 0 \\ -i\beta\gamma & 0 & 0 & \gamma \end{matrix} & \bigcirc & \bigcirc \\ \hdashline
\bigcirc & \bigcirc & \begin{matrix} \gamma & 0 & 0 & i\beta\gamma \\ 0 & 1 & 0 & 0 \\ 0 & 0 & 0 & 0 \\ -i\beta\gamma & 0 & 0 & \gamma \end{matrix} & \bigcirc \\ \hdashline
\begin{matrix} 0 & 0 & 0 & \beta^2\gamma^2 \\ 0 & -i\beta\gamma & 0 & 0 \\ 0 & 0 & -i\beta\gamma & 0 \\ 0 & 0 & 0 & 0 \end{matrix} & \bigcirc & \bigcirc & \begin{matrix} \gamma^2 & 0 & 0 & 0 \\ 0 & \gamma & 0 & 0 \\ 0 & 0 & \gamma & 0 \\ 0 & 0 & 0 & 0 \end{matrix}
\end{array}\right| \tag{15}
$$

Since there are so many zero rows and columns in this matrix, it may be condensed considerably:

$$
L \times \cdot L =
\begin{array}{c|ccc:ccc:ccc:ccc|}
 & (12) & (13) & (14) & (21) & (23) & (24) & (31) & (32) & (34) & (41) & (42) & (43) \\
(12) & \gamma & 0 & 0 & & & & & & & 0 & i\beta\gamma & 0 \\
(13) & 0 & \gamma & 0 & & \bigcirc & & & \bigcirc & & 0 & 0 & i\beta\gamma \\
(14) & 0 & 0 & \gamma^2 & & & & & & & \beta^2\gamma^2 & 0 & 0 \\
\hdashline
(21) & & & & \gamma & 0 & i\beta\gamma & & & & & & \\
(23) & & \bigcirc & & 0 & 1 & 0 & & \bigcirc & & & \bigcirc & \\
(24) & & & & -i\beta\gamma & 0 & \gamma & & & & & & \\
\hdashline
(31) & & & & & & & \gamma & 0 & i\beta\gamma & & & \\
(32) & & \bigcirc & & & \bigcirc & & 0 & 1 & 0 & & \bigcirc & \\
(34) & & & & & & & -i\beta\gamma & 0 & \gamma & & & \\
\hdashline
(41) & 0 & 0 & \beta^2\gamma^2 & & & & & & & \gamma^2 & 0 & 0 \\
(42) & -i\beta\gamma & 0 & 0 & & \bigcirc & & & \bigcirc & & 0 & \gamma & 0 \\
(43) & 0 & -i\beta\gamma & 0 & & & & & & & 0 & 0 & \gamma
\end{array}
\tag{16}
$$

Moreover, since the elements of the tensor must obey $T_{ij} = -T_{ji}$, the following further simplification is possible:

$$
L \times L =
\begin{array}{c}
 \\ (12) \\ (13) \\ (14) \\ \\ (23) \\ (24) \\ (34)
\end{array}
\begin{array}{c}
\begin{array}{cccccc}
(12) & (13) & (14) & (23) & (24) & (34)
\end{array}
\\
\left(\begin{array}{ccc:ccc}
\gamma & 0 & 0 & 0 & -i\beta\gamma & 0 \\
0 & \gamma & 0 & 0 & 0 & -i\beta\gamma \\
0 & 0 & 1 & 0 & 0 & 0 \\
\hdashline
0 & 0 & 0 & 1 & 0 & 0 \\
i\beta\gamma & 0 & 0 & 0 & \gamma & 0 \\
0 & i\beta\gamma & 0 & 0 & 0 & \gamma
\end{array}\right)
\end{array}.
\tag{17}
$$

The second time an index couplet appears in matrix (16) its row or column is deleted in matrix (17).

The restricted Lorentz direct product transformation obtained,

$$L \times L = \begin{array}{c} \\ (12) \\ (13) \\ (14) \\ \\ (23) \\ (24) \\ (34) \end{array} \begin{array}{c} \begin{array}{cccccc} (12) & (13) & (14) & (23) & (24) & (34) \end{array} \\ \left(\begin{array}{cccccc} \gamma & 0 & 0 & 0 & -i\beta\gamma & 0 \\ 0 & \gamma & 0 & 0 & 0 & -i\beta\gamma \\ 0 & 0 & 1 & 0 & 0 & 0 \\ \hdashline 0 & 0 & 0 & 1 & 0 & 0 \\ i\beta\gamma & 0 & 0 & 0 & \gamma & 0 \\ 0 & i\beta\gamma & 0 & 0 & 0 & \gamma \end{array} \right) \end{array}, \tag{18}$$

can be applied to the six electromagnetic field components:

$$\begin{aligned} f_{14} &= -iE_x, & f_{12} &= B_x, \\ f_{24} &= -iE_y, & f_{13} &= -B_y, \\ f_{34} &= -iE_z, \quad \text{and} & f_{23} &= B_z, \end{aligned} \tag{19}$$

considered as a six component vector, to find the transformed electromagnetic field components f' or $\mathbf{E}'$ and $\mathbf{B}'$:

$$\begin{aligned} f'_{14} &= f_{14}, & E'_x &= E_x, \\ f'_{24} &= i\beta\gamma f_{12} + \gamma f_{24}, & E'_y &= \gamma(E_y - \beta B_z), \\ f'_{34} &= i\beta\gamma f_{13} + \gamma f_{34}, & E'_z &= \gamma(E_z + \beta B_y), \\ f'_{12} &= \gamma f_{12} - i\beta\gamma f_{24}, & B'_z &= \gamma(B_z - \beta E_y), \\ f'_{13} &= \gamma f_{13} - i\beta\gamma f_{34}, & B'_y &= \gamma(B_y + \beta E_z), \\ &\text{and} & &\text{and} \\ f'_{23} &= f_{23}, & B'_x &= B_x. \end{aligned} \tag{20}$$

Recall that $\mathbf{E}$ is a real vector and $\mathbf{H}$ is a pseudo or axial vector that can be cast in a 4 by 4 matrix, as follows:

$$f_{\mu\nu} = \begin{pmatrix} 0 & H_z & -H_y & -iE_x \\ -H_z & 0 & H_x & -iE_y \\ H_y & -H_x & 0 & -iE_z \\ iE_x & iE_y & iE_z & 0 \end{pmatrix}. \tag{21}$$

Now consider the well-known form of the electromagnetic force due to Lorentz:

$$\mathbf{F} = q\mathbf{E} + \frac{q}{c}[\mathbf{v} \times \mathbf{H}] = q(\mathbf{E} + [\boldsymbol{\beta} \times \mathbf{H}]), \tag{22}$$

which has the x component

$$F_x = q(E_x + [\boldsymbol{\beta} \times \mathbf{H}]_x) = q(E_x + \beta_y H_z - \beta_z H_y)$$

$$= q(if_{14} + \beta_2 f_{12} + \beta_3 f_{13}) \tag{23}$$

$$= q(f_{12}\beta_2 + f_{13}\beta_3 + f_{14}i) = F_1.$$

Now define a four-dimensional velocity:

$$\mathbb{U} = (v_x, v_y, v_z, ic), \tag{24}$$

and beta:

$$\mathbb{B} = (\beta_x, \beta_y, \beta_z, i). \tag{25}$$

The four-dimensional current is thus

$$J = (j_x, j_y, j_z, i\,\rho c) = (\rho v_x, \rho v_y, \rho v_z, i\,\rho c) \tag{26}$$

or, with q

$$J = (qv_x, qv_y, qv_z, iqc) = (q\beta_x, q\beta_y, q\beta_z, iq). \tag{27}$$

By redefinition:

$$F_1 = (f_{11}j_1 + f_{12}j_2 + f_{13}j_3 + f_{14}j_4) \tag{28}$$

or, better,

$$K_\mu = f_{\mu\nu}j_\nu, \tag{29}$$

since K is commonly used for the four-dimensional Lorentz-Minkowski force. Thus it is evident that the Lorentz force is the interaction of the four-dimensional current with the four-dimensional tensor field:

$$\mathbf{K} = \mathbf{F} \cdot \mathbf{J}. \tag{30}$$

This four-dimensional force has as its last component

$$\begin{aligned} K_4 &= f_{41}j_1 + f_{42}j_2 + f_{43}j_3 + f_{44}j_4 \\ &= iE_x q\beta_x + iE_y q\beta_y + iE_z q\beta_z + 0 \end{aligned} \tag{31}$$

or, better,

$$K_4 = iq\mathbf{E} \cdot \boldsymbol{\beta}, \tag{32}$$

which is a scalar product of ordinary space vectors. This is interpreted physically as the rate of change of energy

According to special relativity, the four-dimensional Minkowski line element or interval between two events in space-time can be expressed by either

$$ds^2 = dx^2 + dy^2 + dz^2 - c^2 dt^2, \tag{33a}$$

$$ds^2 = dx^{12} + dx^{22} + dx^{32} + dx^{42}, \tag{33b}$$

or

$$ds^2 = -dx^{02} + dx^{12} + dx^{22} + dx^{32}, \tag{33c}$$

where the four coordinate axes are labeled x, y, z, or t or more systematically,

$$x' = x, \qquad x^2 = y, \qquad x^3 = z, \qquad x^4 = ict = ix^0 \tag{34}$$

The corresponding Minkowski metrics are given by:

$$g_{ij} = \begin{pmatrix} 1 & 0 & 0 & 0 \\ 0 & 1 & 0 & 0 \\ 0 & 0 & 1 & 0 \\ 0 & 0 & 0 & -c^2 \end{pmatrix}, \qquad \text{for } x, y, z, t; \tag{35a}$$

$$g_{ij} = \begin{pmatrix} 1 & 0 & 0 & 0 \\ 0 & 1 & 0 & 0 \\ 0 & 0 & 1 & 0 \\ 0 & 0 & 0 & 1 \end{pmatrix}, \qquad \text{for } x^1, x^2, x^3, x^4; \tag{35b}$$

and

$$g_{ij} = \begin{pmatrix} -1 & 0 & 0 & 0 \\ 0 & 1 & 0 & 0 \\ 0 & 0 & 1 & 0 \\ 0 & 0 & 0 & 1 \end{pmatrix}, \qquad \text{for } x^0, x^1, x^2, x^3, \tag{35c}$$

which are all quite similar in appearance. Using spherical coordinates, the interval between two events in flat space-time is expressed by

$$ds^2 = dr^2 + r^2 d\theta^2 + r^2 \sin^2\theta\, d\phi^2 - c^2 dt^2. \tag{36}$$

The Minkowski metric for spherical coordinates is

$$g_{ij} = \begin{pmatrix} 1 & 0 & 0 & 0 \\ 0 & r^2 & 0 & 0 \\ 0 & 0 & r^2 \sin^2\theta & 0 \\ 0 & 0 & 0 & -c^2 \end{pmatrix}, \tag{37}$$

which differs rather strikingly in appearance from the matrices (35).

For simplicity, restrict the Lorentz space-time transformation to two of the four dimensions:

$$\mathbf{r}' = \begin{pmatrix} x' \\ ct' \end{pmatrix} = \begin{pmatrix} x_1' \\ x_0' \end{pmatrix}, \quad \mathbf{r} = \begin{pmatrix} x \\ ct \end{pmatrix} = \begin{pmatrix} x_1 \\ x_0 \end{pmatrix}, \quad L = \begin{pmatrix} \gamma & -\beta\gamma \\ -\beta\gamma & \gamma \end{pmatrix}, \tag{38}$$

and

$$\mathbf{r}' = L\mathbf{r}, \tag{39}$$

or

$$\begin{pmatrix} x' \\ ct' \end{pmatrix} = \begin{pmatrix} \gamma & -\beta\gamma \\ -\beta\gamma & \gamma \end{pmatrix} \begin{pmatrix} x \\ ct \end{pmatrix}. \tag{40}$$

Alternatively,

$$\mathbf{r}' = \begin{pmatrix} x' \\ ict' \end{pmatrix} = \begin{pmatrix} x_1' \\ x_4' \end{pmatrix}, \quad \mathbf{r} = \begin{pmatrix} x \\ ict \end{pmatrix} = \begin{pmatrix} x_1 \\ x_4 \end{pmatrix}, \quad L = \begin{pmatrix} \gamma & i\beta\gamma \\ -i\beta\gamma & \gamma \end{pmatrix}, \tag{41}$$

and

$$\mathbf{r}' = L\mathbf{r}, \tag{42}$$

or

$$\begin{pmatrix} x' \\ ict' \end{pmatrix} = \begin{pmatrix} \gamma & i\beta\gamma \\ -i\beta\gamma & \gamma \end{pmatrix} \begin{pmatrix} x \\ ict \end{pmatrix}. \tag{43}$$

Formulations (40) and (43) may be represented as either

$$\mathbf{r}' = \gamma(I - \beta\sigma_x)\mathbf{r} \tag{44}$$

or

$$\mathbf{r}' = \gamma(I - \beta\sigma_y)\mathbf{r}, \tag{45}$$

respectively. Thus in both cases the Lorentz transformation may be written

$$L = \gamma(I - \beta\sigma_x), \tag{46}$$

which is hermitian, or

$$L = \gamma(I - \beta\sigma_y), \tag{47}$$

which is again hermitian. The inverse transformations are given by

$$L^{-1} = \gamma(I + \beta\sigma_x) \tag{48}$$

and

$$L^{-1} = \gamma(I + \beta\sigma_y). \tag{49}$$

It is easy to establish that the Lorentz transformation is hermitian and that it is not unitary. A general rule is: I is unitary-hermitian, and σ is unitary-hermitian, but the sum of two unitary-hermitian matrices usually is only hermitian. To verify the inverse (σ denotes either σ_x or σ_y):

$$\begin{aligned} LL^{-1} &= \gamma(I - \beta\sigma)\gamma(I + \beta\sigma) = \gamma^2(I - \beta\sigma)(I + \beta\sigma) \\ &= \gamma^2(I - \beta^2\sigma^2) = \gamma^2(I - \beta^2 I) = I\gamma^2(1 - \beta^2) = I. \end{aligned} \tag{50}$$

An alternative way of writing equation (40) is as follows:

$$L = \frac{1}{\sqrt{1 - \beta^2}}\{I - \beta\sigma\} = \{\epsilon I - \eta\sigma\}. \tag{51}$$

Orthogonal matrices are special cases of unitary matrices, namely, real unitary matrices. Although there is a common belief that L is a unitary transformation, this belief is faulty, because then the inverse would have to be

$$L = \gamma(I - \beta\sigma), \tag{52}$$

which readily is found to be wrong.

To determine (a) the eigenvalues, (b) the eigenvectors, and (c) the diagonal form of the transformation, observe that since σ_z is the diagonal form of both σ_x and σ_y,

$$L = \gamma(I - \beta\sigma) \tag{53}$$

may be diagonalized to

$$L_D = \gamma(I - \beta\sigma_z), \tag{54}$$

for which the eigenvalues are

$$\lambda = \gamma(1 \pm \beta). \tag{55}$$

A general vector is

$$\mathbf{r} = \begin{pmatrix} x_1 \\ x_0 \end{pmatrix} \quad \text{or} \quad \begin{pmatrix} x_1 \\ x_4 \end{pmatrix}; \tag{56}$$

a unit space vector is

$$\hat{\mathbf{r}} = \begin{pmatrix} 1 \\ 0 \end{pmatrix}; \tag{57}$$

a unit time vector is

$$\hat{\mathbf{t}} = \begin{pmatrix} 0 \\ 1 \end{pmatrix}; \tag{58}$$

and the diagonalized Lorentz matrix is

$$\mathscr{L}_D = \begin{pmatrix} \gamma - \beta\gamma & 0 \\ 0 & \gamma + \beta\gamma \end{pmatrix} = \begin{pmatrix} \dfrac{1 - \beta}{\sqrt{1 - \beta^2}} & 0 \\ 0 & \dfrac{1 + \beta}{\sqrt{1 - \beta^2}} \end{pmatrix}$$
$$= \begin{pmatrix} \sqrt{\dfrac{1 - \beta}{1 + \beta}} & 0 \\ 0 & \sqrt{\dfrac{1 + \beta}{1 - \beta}} \end{pmatrix}. \tag{59}$$

These incomplete thoughts seem to represent a new way of looking at an old subject.

Recall the eigenvectors of σ_x and σ_y:

$$\begin{pmatrix} 0 & 1 \\ 1 & 0 \end{pmatrix} \frac{1}{\sqrt{2}} \begin{pmatrix} 1 \\ 1 \end{pmatrix} = \frac{1}{\sqrt{2}} \begin{pmatrix} 1 \\ 1 \end{pmatrix},$$
$$\begin{pmatrix} 0 & 1 \\ 1 & 0 \end{pmatrix} \frac{1}{\sqrt{2}} \begin{pmatrix} 1 \\ -1 \end{pmatrix} = -\frac{1}{\sqrt{2}} \begin{pmatrix} 1 \\ -1 \end{pmatrix}, \tag{60}$$

and

$$\begin{pmatrix} 0 & -i \\ i & 0 \end{pmatrix} \frac{1}{\sqrt{2}} \begin{pmatrix} 1 \\ i \end{pmatrix} = \frac{1}{\sqrt{2}} \begin{pmatrix} 1 \\ i \end{pmatrix},$$
$$\begin{pmatrix} 0 & -i \\ i & 0 \end{pmatrix} \frac{1}{\sqrt{2}} \begin{pmatrix} 1 \\ -i \end{pmatrix} = -\frac{1}{\sqrt{2}} \begin{pmatrix} 1 \\ -i \end{pmatrix}. \tag{61}$$

These are also eigenvalues of L. Thus

$$L = \gamma(I - \beta\sigma_x), \qquad L = \gamma(I - \beta\sigma_y), \tag{62}$$

and

$$\hat{\mathbf{v}}_\pm = \frac{1}{\sqrt{2}} \begin{pmatrix} 1 \\ \pm 1 \end{pmatrix}, \qquad \hat{\mathbf{u}}_\pm = \frac{1}{\sqrt{2}} \begin{pmatrix} 1 \\ \pm i \end{pmatrix} \tag{63}$$

gives

$$\begin{aligned} \gamma(I - \beta\sigma_x)\hat{\mathbf{v}}_+ &= (\gamma - \beta\gamma)\hat{\mathbf{v}}_+, \\ \gamma(I - \beta\sigma_x)\hat{\mathbf{v}}_- &= (\gamma + \beta\gamma)\hat{\mathbf{v}}_-, \\ \gamma(I - \beta\sigma_y)\hat{\mathbf{u}}_+ &= (\gamma - \beta\gamma)\hat{\mathbf{u}}_+, \\ \gamma(I - \beta\sigma_y)\hat{\mathbf{u}}_- &= (\gamma + \beta\gamma)\hat{\mathbf{u}}_-. \end{aligned} \tag{64}$$

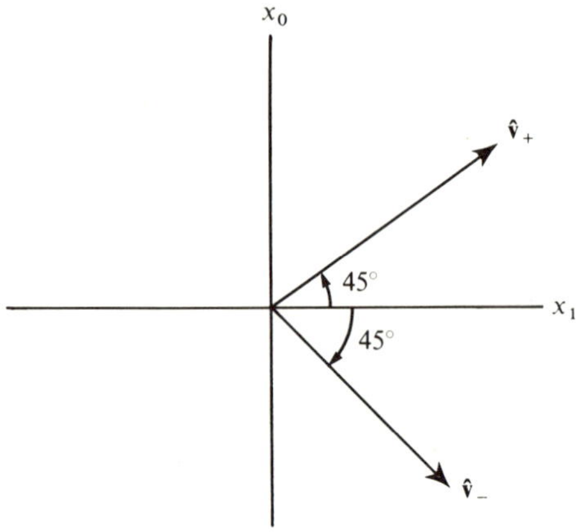

Figure 26. Eigenvectors and light cone directions.

Hence the eigenvectors are seen to be related to the direction of light cone as in Fig. 26.

EXERCISES

2. Find a similarity transformation,

$$uLu^{-1} = L_D,$$

that can be applied to diagonalize L.

3. The eigenvalues of the Lorentz transformation matrix are exactly the factors that appear in the relativistic derivation of the Doppler effect. What other physical phenomena might be related to these eigenvalues?

4. Find the Lorentz transformation between cartesian and spherical coordinates.

5. A tensor of rank three transforms as

$$T'_{ijk} = L_{il} L_{jm} L_{km} T_{lmn}.$$

Show that this may be written as a triple direct product transformation

$$T' = [L \times\cdot (L \times\cdot L)]T = [(L \times\cdot L) \times\cdot L]T.$$

C. Einstein's General Relativity

If a functional relationship between two sets of coordinates

$$X'^i = X'^i(X^j) \qquad \text{or} \qquad X^j = X^j(X'^i) \tag{1}$$

is assumed to exist, but is not yet explicitly specified, then a differential relationship is given by

$$dx'^i = \frac{\partial x'^i}{\partial x^j} dx^j = M^i{}_j dx^j, \tag{2}$$

and a total time derivative is given by

$$\frac{dx'^i}{dt} = \frac{\partial x'^i}{\partial x^j}\frac{dx^j}{dt} + \frac{\partial x'^i}{\partial t}. \tag{3}$$

If X'^i is not an explicit function of time, equation (3) becomes

$$\frac{dx'^i}{dt} = \frac{\partial x'^i}{\partial x^j}\frac{dx^j}{dt} \tag{4}$$

or

$$\dot{X}'^i = M^i{}_j \dot{X}^j, \tag{5}$$

the "standard form for the transformation of a contravariant vector." Typical contravariant vectors are position:

$$\mathbf{r}, \qquad \mathbf{r}', \tag{6}$$

and velocity

$$\mathbf{v} = \dot{\mathbf{r}}, \qquad \mathbf{v}' = \dot{\mathbf{r}}'; \tag{7}$$

others, of course, also exist.

A scalar point function has the derivative

$$\frac{\partial \phi}{\partial x'^i} = \frac{\partial x^j}{\partial x'^i}\frac{\partial \phi}{\partial x^j} \tag{8}$$

or

$$\nabla'_i \phi = \frac{\partial x^j}{\partial x'^i} \nabla_j \phi = M_i{}^j \nabla_j \phi, \tag{9}$$

the standard form for the transformation of a covariant vector. A typical covariant vector is the gradient of a scalar.

Notice the difference between relationship (2) or (5) and (9), the transformation matrices $M^i{}_j$ and $M_i{}^j$ are almost reciprocals. Remember the mnemonic phrase "co-lo-primes-below" to impress the fact that a covariant vector has subscript indices and transforms by a matrix containing primed coordinates in the denominator.

To illustrate that the product of any covariant vector by a contravariant vector is an invariant, from (2) and (8),

$$\begin{aligned} d\phi &= \frac{\partial \phi}{\partial x'^i} dx'^i = \tilde{M}_i^{\ j} \frac{\partial \phi}{\partial x^j} M^i_{\ j} dx^j \\ &= \tilde{M}_i^{\ j} M^i_{\ j} \frac{\partial \phi}{\partial x^j} dx^j = \frac{\partial \phi}{\partial x^j} dx^j, \end{aligned} \tag{10}$$

thus the total differential is an invariant:

$$d\phi = \frac{\partial \phi}{\partial x'^i} dx'^i = \frac{\partial \phi}{\partial x^j} dx^j = \text{invariant.} \tag{11}$$

If $\dot{x}'^i$ is not explicitly dependent on time, the preceding transformation law for velocity,

$$\dot{x}'^i = \frac{\partial x'^i}{\partial x^j} \dot{x}^j + \frac{\partial x'^i}{\partial t} = M^i_{\ j} \dot{x}^j + \frac{\partial x'^i}{\partial t}, \tag{12}$$

reduces to:

$$\dot{x}'^i = M^i_{\ j} \dot{x}^j. \tag{13}$$

Now the second derivative with respect to time may be written

$$\begin{aligned} \ddot{x}'^i &= \frac{d}{dt}\left(\frac{dx'^i}{dt}\right) = \frac{d}{dt}\left(\frac{\partial x'^i}{\partial x^j} \dot{x}^j + \frac{\partial x'^i}{\partial t}\right) \\ &= \frac{d}{dt} M^i_{\ j} \dot{x}^j + M^i_{\ j} \ddot{x}^j + \frac{\partial \dot{x}'^i}{\partial t}, \end{aligned} \tag{14}$$

which may be expanded further and rearranged slightly as follows:

$$\ddot{x}'^i = M^i_{\ j} \ddot{x}^j + \left[\frac{\partial^2 x'^i}{\partial x^j \partial x^k} \dot{x}^k + \frac{\partial^2 x'^i}{\partial x^j \partial t}\right] \dot{x}^j + \frac{\partial}{\partial t}\left[M^i_{\ j} x^j + \frac{\partial x^i}{\partial t}\right], \tag{15}$$

or

$$\ddot{x}'^i = M^i_{\ j} \ddot{x}^j + \left(\frac{\partial^2 x'^i}{\partial x^j \partial x^k}\right) \dot{x}^j \dot{x}^k + 2\left(\frac{\partial^2 x'^i}{\partial x^j \partial t}\right) \dot{x}^j + \frac{\partial^2 x^i}{\partial t^2}. \tag{16}$$

Again, if x'^i is not an explicit function of t,

$$\ddot{x}'^i = \frac{\partial x'^j}{\partial x^j} \ddot{x}^j + \left(\frac{\partial^2 x'^i}{\partial x^j \partial x^k}\right) \dot{x}^j \dot{x}^k, \tag{17}$$

or

$$\ddot{x}'^i = M^i_{\ j} \ddot{x}^j + N_{jk}^{\ \ i} \dot{x}^j \dot{x}^k. \tag{18}$$

Notice that *if* the coordinate transformation relationships $x'^i = x'^i(x^j)$ are linear or affine, then the second derivatives are zero:

$$N_{jk}{}^i \equiv \frac{\partial^2 x'^i}{\partial x^j \partial x^k} = 0, \tag{19}$$

so that equation (18) reduces to

$$\ddot{x}'^i = M^i{}_j \ddot{x}^j, \tag{20}$$

the standard form for the transformation of a contravariant vector. Thus, excluding curvilinear coordinates, acceleration transforms like a contravariant vector.

But, in general, acceleration fails to transform like a vector because of the term containing $N_{jk}{}^i$. Use the ordering convention for the transformations in order to distinguish the prime and unprimed system components:

$$\frac{\partial x'^i}{\partial x^j} = M^i{}_j, \tag{21}$$

$$\frac{\partial x^j}{\partial x'^i} = M_i{}^j, \tag{22}$$

and

$$\frac{\partial^2 x'^\mu}{\partial x^\sigma \partial x^\tau} = N_{\sigma\tau}{}^\mu. \tag{23}$$

Take the generalized coordinates q^i to be the polar coordinates r, θ, defined as follows:

$$q^1 = r = [x^2 + y^2]^{1/2} \tag{24}$$

and

$$q^2 = \theta = \tan^{-1}\frac{y}{x}, \tag{25}$$

with the inverse relationships:

$$x = r\cos\theta \tag{26}$$

and

$$y = r\sin\theta. \tag{27}$$

For the cartesian or x, y, z coordinate system the velocity components are

$$\dot{x} = \dot{r}\cos\theta - r\dot{\theta}\sin\theta \tag{28}$$

and

$$\dot{y} = \dot{r}\sin\theta + r\dot{\theta}\cos\theta. \tag{29}$$

For the polar or r, θ coordinate system, the velocity components are

$$\dot{r} = \frac{\dot{x}x + \dot{y}y}{\sqrt{x^2 + y^2}} \tag{30}$$

and

$$\dot{\theta} = \frac{x\dot{y} - y\dot{x}}{x^2 + y^2}. \tag{31}$$

For the cartesian or x, y coordinate system the acceleration components are

$$\begin{aligned} \ddot{x} &= (\ddot{r} - r\dot{\theta}^2)\cos\theta - (r\ddot{\theta} + 2\dot{r}\dot{\theta})\sin\theta, \\ \ddot{y} &= (\ddot{r} - r\dot{\theta}^2)\sin\theta + (r\ddot{\theta} + 2\dot{r}\dot{\theta})\cos\theta \end{aligned} \tag{32}$$

or

$$\ddot{x} = \ddot{R}\cos\theta - \ddot{\Theta}\sin\theta \tag{33}$$

and

$$\ddot{y} = \ddot{R}\sin\theta + \ddot{\Theta}\cos\theta, \tag{34}$$

where the centripetal contribution,

$$\ddot{R} = (\ddot{r} - r\dot{\theta}^2), \tag{35}$$

may be called the "r-acceleration," and the coriolis contribution,

$$= (r\ddot{\theta} + 2\dot{r}\dot{\theta}), \tag{36}$$

may be called the "θ-acceleration." The $\ddot{R}$, $\ddot{\Theta}$) transforms like a vector.

For the cartesian or x, y coordinate system, the metric tensor

$$g^{mn} = g_{mn} = \begin{pmatrix} 1 & 0 \\ 0 & 1 \end{pmatrix}, \tag{37}$$

For the polar or r, θ coordinate system,

$$\bar{g}_{ij} = \begin{pmatrix} 1 & 0 \\ 0 & r^2 \end{pmatrix} \tag{38}$$

and

$$\bar{g}^{ij} = \begin{pmatrix} 1 & 0 \\ 0 & r^{-2} \end{pmatrix}. \tag{39}$$

Since only those terms involving $\partial\bar{q}_{22}/\partial\bar{q}^1$ do not vanish, the nine Christoffel symbols of the first kind reduce to exactly three, corresponding to just two different values:

$$[\overline{2\,2,1}] = -r \tag{40}$$

and

$$[\overline{2\,1,2}] = [\overline{1\,2,2}] = r. \tag{41}$$

Similarly, there are exactly three Christoffel symbols of the second kind corresponding to just two different values:

$$\left\{\begin{matrix}\bar{1}\\22\end{matrix}\right\} = -r \tag{42}$$

and

$$\left\{\begin{matrix}\bar{2}\\12\end{matrix}\right\} = \left\{\begin{matrix}2\\21\end{matrix}\right\} = \frac{1}{r}. \tag{43}$$

The generalized acceleration component may be written

$$\bar{a}^m = \ddot{q}^m + \left\{\begin{matrix}\bar{m}\\nr\end{matrix}\right\}\dot{q}^n\dot{q}^r. \tag{44}$$

Thus for r,

$$\bar{a}^1 = \ddot{q}^1 + \left\{\begin{matrix}\bar{1}\\nr\end{matrix}\right\}\dot{q}^n\dot{q}^r = \ddot{r} - r\dot{\theta}^2, \tag{45}$$

and for θ,

$$\bar{a}^2 = \ddot{q}^2 + \left\{\begin{matrix}\bar{2}\\nr\end{matrix}\right\}\dot{q}^n\dot{q}^r, \tag{46}$$

$$\bar{a} = \ddot{q} + \left\{\begin{matrix}\bar{2}\\12\end{matrix}\right\}\dot{q}^1\dot{q}^2 + \left\{\begin{matrix}\bar{2}\\21\end{matrix}\right\}\dot{q}^2\dot{q}^1, \tag{47}$$

or

$$\bar{a}^2 = \ddot{\theta} + \frac{2\dot{\theta}\dot{r}}{r} = \frac{1}{r}(r\ddot{\theta} + 2\dot{r}\dot{\theta}). \tag{48}$$

Take the generalized coordinates q^i to be the standard spherical coordinates r, θ, and ϕ defined as follows:

$$\bar{q}^1 = r = (x^2 + y^2 + z^2)^{1/2},$$

$$\bar{q}^2 = \theta = \cot^{-1} \frac{z}{\sqrt{x^2 + z^2}},$$

and (49)

$$\bar{q}^3 = \phi = \tan^{-1} \frac{y}{x},$$

which have the following inverse relationships:

$$x = r \sin\theta \cos\phi,$$

$$y = r \sin\theta \sin\phi,$$

and (50)

$$z = r\cos\theta.$$

The x, y, z components of velocity are

$$\dot{x} = \dot{r} \sin\theta\cos\phi + r\dot{\theta}\cos\theta\cos\phi - r\dot{\phi}\sin\theta\sin\phi,$$

$$\dot{y} = \dot{r} \sin\theta\sin\phi + r\dot{\theta}\cos\theta\sin\phi + r\dot{\phi}\sin\theta\cos\phi,$$

and (51)

$$\dot{z} = \dot{r}\cos\theta - r\dot{\theta}\sin\theta.$$

The r, θ, ϕ components of velocity are*

$$V_r = \dot{r} = V_x \sin\theta\cos\phi + V_y\sin\theta\sin\phi + V_z\cos\theta,$$

$$V_\theta = r\dot{\theta} = V_x\cos\theta\cos\phi + V_y\cos\theta\sin\phi - V_z\sin\theta, \quad (52)$$

and

$$V_\phi = r\sin\theta\dot{\phi} = -V_x\sin\phi + V_y\cos\phi.$$

For a cartesian or x, y, z coordinate system, the line element is

$$ds^2 = dx^2 + dy^2 + dz^2. \quad (53)$$

For a spherical or r, θ, ϕ coordinate system, the line element is

$$ds^2 = dr^2 + r^2 d\theta^2 + r^2 \sin^2\theta\, d\phi^2. \quad (54)$$

For a cartesian coordinate system, x, y, z, the metric tensor is

$$g_{ij} = \begin{pmatrix} 1 & 0 & 0 \\ 0 & 1 & 0 \\ 0 & 0 & 1 \end{pmatrix}. \quad (55)$$

* Notice the similarity between equations (51) and (52) and the transformation of a vector from cartesian to spherical coördinates derived in Chapter 1.

For a spherical coordinate or r, θ, ϕ system, the metric tensor is

$$\bar{g}_{mn} = \begin{pmatrix} 1 & 0 & 0 \\ 0 & r^2 & 0 \\ 0 & 0 & r^2 \sin^2\theta \end{pmatrix}, \tag{56}$$

and the reciprocal metric tensor is

$$\bar{g}^{mn} = \begin{pmatrix} 1 & 0 & 0 \\ 0 & \dfrac{1}{r^2} & 0 \\ 0 & 0 & \dfrac{1}{r^2 \sin^2\theta} \end{pmatrix}. \tag{57}$$

Since there are no ϕ terms in g_{mn}, it must be true that

$$\frac{\partial \bar{g}_{mn}}{\partial \bar{q}^3} \equiv 0. \tag{58}$$

As a consequence, the only nonzero Christoffel symbols are those that involve

$$\frac{\partial \bar{g}_{22}}{\partial \bar{q}^1}, \qquad \frac{\partial \bar{g}_{33}}{\partial \bar{q}^1}, \qquad \text{and} \qquad \frac{\partial \bar{g}_{33}}{\partial \bar{q}^2}. \tag{59}$$

Evaluation of these partial derivatives reveals that there is just one nonzero term for θ:

$$\frac{\partial \bar{g}_{33}}{\partial \bar{q}^2} = 2r^2 \sin\theta \cos\theta; \tag{60}$$

and just two nonzero terms for r:

$$\frac{\partial \bar{g}_{22}}{\partial \bar{q}^1} = 2r \qquad \text{and} \qquad \frac{\partial \bar{g}_{33}}{\partial \bar{q}^1} = 2r \sin^2\theta. \tag{61}$$

Permutation of the nonzero terms produces exactly nine Christoffel symbols, the first of which correspond to the following six distinct expressions:

$$\begin{aligned} &\{\overline{2\,2, 1}\} = -\, r, \qquad \{\overline{1\,2, 2}\} = \{\overline{2\,1, 2}\} = r, \\ &\{\overline{3\,3, 1}\} = -\, r \sin^2\theta, \qquad \{\overline{1\,3, 3}\} = \{\overline{3\,1, 3}\} = r^2 \sin^2\theta, \\ &\{\overline{3\,3, 2}\} = -\, r^2 \sin\theta \cos\theta, \qquad \text{and} \quad \{\overline{2\,3, 3}\} = \{\overline{3\,2, 3}\} = r^2 \sin\theta \cos\theta. \end{aligned} \tag{62}$$

Permutation also produces the related set of exactly nine Christoffel symbols of the second type which correspond to the following six distinct expressions:

$$\begin{Bmatrix} \bar{1} \\ 22 \end{Bmatrix} = -r, \qquad \begin{Bmatrix} \bar{2} \\ 12 \end{Bmatrix} = \begin{Bmatrix} \bar{2} \\ 21 \end{Bmatrix} = \frac{1}{r},$$

$$\begin{Bmatrix} \bar{1} \\ 33 \end{Bmatrix} = -r\sin^2\theta, \qquad \begin{Bmatrix} \bar{3} \\ 13 \end{Bmatrix} = \begin{Bmatrix} \bar{3} \\ 31 \end{Bmatrix} = \frac{1}{r}, \tag{63}$$

$$\begin{Bmatrix} \bar{2} \\ 33 \end{Bmatrix} = -\sin\theta\cos\theta, \qquad \begin{Bmatrix} \bar{3} \\ 23 \end{Bmatrix} = \begin{Bmatrix} \bar{3} \\ 32 \end{Bmatrix} = \cot\theta.$$

The covariant derivative formula yields the acceleration equation:

$$\bar{a}^m = \ddot{\bar{q}}^m + \begin{Bmatrix} m \\ nr \end{Bmatrix} \dot{\bar{q}}^n \dot{\bar{q}}^r, \tag{64}$$

from which it follows that

$$\bar{a}^1 = \ddot{r} - r\dot{\theta}^2 - r\sin^2\theta\dot{\phi}^2,$$

$$\bar{a}^2 = \ddot{\theta} + \frac{2\dot{r}\dot{\theta}}{r} - \sin\theta\cos\theta\dot{\phi}^2, \tag{65}$$

and

$$\bar{a}^3 = \ddot{\phi} + \frac{2\dot{r}\dot{\phi}}{r} + 2\cot\theta\dot{\theta}\dot{\phi}.$$

According to Einstein's gravitational theory, the properties of space are dependent on the amount of matter that is present in a particular locality in the universe. By the principle of equivalence connecting acceleration and force, it is easy to predict that if a large gravitational object, such as the Sun, were to be placed at the center of coordinates, the surrounding space would become curved or non-Euclidian, and there would be a consequent change in the coordinate system, line interval, and metric tensor.

Since a Lorentz transformation normally must involve at least two variables (time and one space coordinate) it is natural to seek a change in both **r** and t. Because the gravitational force for a body at the origin is a radial force with both θ and ϕ perpendicular to the force and therefore almost entirely immune to its effects, the Riemannian line element or general relativistic interval connecting two events in space time is postulated to be

$$ds^2 = e^\mu\, dr^2 + r^2(d\theta^2 + \sin^2\theta\, d\phi^2) - e^\nu\, dt^2, \tag{66}$$

with a metric:

$$g_{ij} = \begin{pmatrix} e^{\mu} & 0 & 0 & 0 \\ 0 & r^2 & 0 & 0 \\ 0 & 0 & r^2 \sin^2\theta & 0 \\ 0 & 0 & 0 & -e^{\nu} \end{pmatrix}, \tag{67}$$

which reduces to the Minkowski flat metric far away from any matter. The determinant of matrix (67)—the so-called "Schwarzschild tensor"—is

$$g = |g_{ij}| = -e^{\mu+\nu} r^4 \sin^2\theta. \tag{68}$$

EXERCISES

12. Compute all of the Christoffel symbols for the Schwarzschild line element

$$(ds)^2 = -e^{\lambda}(dr)^2 - r^2(d\theta)^2 - r^2 \sin^2\theta (d\phi)^2 + e^{\mu}(dt)^2.$$

13. Continue working with the line element and obtain the result that

$$e^{\mu} = e^{-\lambda} = \left[1 - \frac{2m}{r}\right].$$

What is the mass of the Sun in meters, and how small must the Sun's radius be for the Sun to be invisible because of its increased surface gravity?

In ordinary generalized coordinates the equation for a geodesic written for an acceleration based on ordinary time is

$$q^m + \left\{ \begin{matrix} \overline{m} \\ nr \end{matrix} \right\} q^n q^r = 0 \tag{69}$$

$$\frac{d^2\bar{q}^m}{dt^2} + \left\{ \begin{matrix} \overline{m} \\ nr \end{matrix} \right\} \frac{d\bar{q}^n}{dt}\frac{d\bar{q}^r}{dt} = 0. \tag{70}$$

Relativity introduces two important changes in this coordinate system:

1. The number of coordinates is increased from 3 to 4, and the set of cartesian coordinates $x, y, z = x^1, x^2, x^3$ is augmented by either t, $x_0 = ct$, or $x_4 = ict$, as a fourth independent variable.
2. The role played by the nonrelativistic time parameter t is taken over by a new parametric quantity called "interval." Interval, which always appears

as a differential ds, is defined by

$$ds^2 = c^2\,dt^2 - dx^2 - dy^2 - dz^2,$$

$$\frac{ds^2}{dt^2} = c^2 - \frac{dx^2}{dt^2} - \frac{dy^2}{dt^2} - \frac{dz^2}{dt^2},$$

$$\frac{ds^2}{dt^2} = c^2 - v^2, \tag{71}$$

$$ds^2 = c^2\,dt^2 - v^2\,dt^2 = c^2\,dt^2\left(1 - \frac{v^2}{c^2}\right).$$

Thus

$$ds = c\,dt\sqrt{1 - \frac{v^2}{c^2}}. \tag{72}$$

Equation (69) may be rewritten using four-dimensional generalized coordinates, as follows:

$$\frac{d^2\bar{q}^m}{ds^2} + \begin{Bmatrix} m \\ nr \end{Bmatrix} \frac{d\bar{q}^n}{ds}\frac{d\bar{q}^r}{ds} = 0, \tag{73}$$

with $n, m, r = 1, 2, 3, 4$.

D. Geodesics of Motion and Generalized Acceleration

In keeping with the motivating idea of generalized coordinates in Lagrangian and Hamiltonian mechanics—the quest for simplicity—and D'Alembert's principle, which considers forces to be in equilibrium and thereby reduces all problems of dynamics to problems of statics, the concept of a "geodesic" is a great simplification.

Recall the expression for generalized acceleration in terms of covariant derivatives and Christoffel symbols:

$$\bar{a}^m = \dot{\bar{q}}^m + \begin{Bmatrix} m \\ nr \end{Bmatrix} \dot{\bar{q}}^n\dot{\bar{q}}^r. \tag{74}$$

For cartesian coordinates $\bar{q}^i$ the Christoffel symbols are zero and expression (74) reduces to the standard derivative. Moreover, for zero acceleration there will be no force and a particle or object consequently will move in a straight line. This line is a geodesic in flat or Euclidian space.

Now, conceivably, expression (74) may be zero also for a truly generalized coordinate system. Yet the same kind of statement may be made to describe the system. A particle or object will move along a geodesic in a generalized coordinate system provided its generalized acceleration is zero. This does not mean that it moves linearly, because although a geodesic is the shortest distance between two points, in a curved space this shortest distance may not be along a straight line. On the surface of a sphere, for example, which is a curved two-dimensional space, it is a great circle like the Earth's equator.

Therefore the expression for a geodesic of motion in any generalized coordinate system is written

$$\overset{\dot{+}}{q} + \begin{Bmatrix} m \\ nr \end{Bmatrix} \overset{\dot{+}}{q}{}^n \overset{\dot{+}}{q}{}^r = 0. \tag{75}$$

The calculus of variations supplies the classic approach to the problem of finding a geodesic. This approach starts out with the variation of the interval

$$\delta \int ds \tag{76}$$

and finishes by arriving at the Euler-Lagrange equations.

The extremum condition

$$\delta \int ds = 0, \tag{77}$$

may be rewritten as

$$\begin{aligned} \delta \quad ds &= \delta \int \sqrt{g_{ij} dx^i dx^j} \\ &= \delta \int \sqrt{g_{ij} \frac{dx^i}{dt} \frac{dx^j}{dt}}\, dt = 0 \end{aligned} \tag{78}$$

or

$$\delta \int F\, dt = 0, \tag{79}$$

with

$$F = \sqrt{g_{ij} \frac{dx^i}{dt} \frac{dx^j}{dt}} = \dot{S}. \tag{80}$$

Now apply the Euler-Lagrange equation of the calculus of variations to F:

$$\frac{d}{dt}\left(\frac{\partial F}{\partial \dot{x}^k}\right) - \frac{\partial F}{\partial x^k} = 0. \tag{81}$$

Thus

$$\frac{\partial F}{\partial x^k} = \frac{1}{2} \frac{1}{\sqrt{g^{ij} \dot{x}^i \dot{x}^j}} \left(\frac{\partial g_{ij}}{\partial x^k}\right) \frac{dx^i}{dt} \frac{dx^j}{dt} = \frac{\dot{x}^i \dot{x}^j}{2\dot{s}} \left(\frac{\partial g_{ij}}{\partial x^k}\right), \tag{82}$$

and

$$\begin{aligned}\frac{\partial F}{\partial \dot{x}^k} &= \frac{1}{2}\frac{1}{\sqrt{g_{ij}\dot{x}^i\dot{x}^j}}(g_{kj}\dot{x}^j + g_{ik}\dot{x}^i) \\ &= \frac{1}{2\dot{S}}(g_{kj}\dot{x}^j + g_{ik}\dot{x}^i).\end{aligned} \tag{83}$$

Finally

$$\begin{aligned}\frac{d}{dt}\left(\frac{\partial F}{\partial \dot{x}^k}\right) = \frac{1}{2\dot{S}}\left(g_{kj}\ddot{x}^j + g_{ik}\ddot{x}^i + \frac{\partial g_{kj}}{\partial x^i}\dot{x}^i\dot{x}^j + \frac{\partial g_{ik}}{\partial x^j}\dot{x}^i\dot{x}^j\right) \\ - \frac{1}{2\dot{S}^2}\ddot{S}(g_{kj}\dot{x}^j + g_{ik}\dot{x}^i).\end{aligned} \tag{84}$$

Remember that $g_{kj} = g_{jk}$ and $g_{ik} = g_{ki}$. Also, change the dummy variable. Then

$$\frac{d}{dt}\left(\frac{\partial F}{\partial \dot{x}^k}\right) = \frac{1}{\dot{S}}g_{kj}\ddot{x}^j + \frac{1}{2\dot{S}}\left(\frac{\partial g_{kj}}{\partial x^i} + \frac{\partial g_{ik}}{\partial x^j}\right)\dot{x}^i\dot{x}^j - \frac{1}{2}\frac{\ddot{S}}{\dot{S}^2}(g_{kj}\dot{x}^j + g_{ik}\dot{x}^i). \tag{85}$$

Next combine equations (82) and (85):

$$\frac{d}{dt}\left(\frac{\partial F}{\partial \dot{x}^k}\right) - \frac{\partial F}{\partial x^k} = \frac{1}{\dot{S}}\left[g_{ki}\ddot{x}^j + \frac{1}{2}\left(\frac{\partial g_{kj}}{\partial x^i} + \frac{\partial g_{ik}}{\partial x^j} - \frac{\partial g_{ij}}{\partial x^k}\right)\dot{x}^i\dot{x}^j - \frac{\ddot{S}}{\dot{S}}g_{kj}\dot{x}^j\right] = 0; \tag{86}$$

and convert from t to s. Thus $\dot{s} = 1$, $\ddot{s} = 0$, and

$$g_{ki}\frac{d^2x^j}{\partial s^2} + [i\,j, k]\frac{dx^i}{ds}\frac{dx^j}{ds} = 0. \tag{87}$$

Finally change the Christoffel symbol and remove the g_{kj} by multiplying by g^{kl}, then the equation of the geodesic becomes

$$\frac{d^2x^l}{ds^2} + \begin{Bmatrix} l \\ ij \end{Bmatrix}\frac{dx^i}{ds}\frac{dx^j}{ds} = 0. \tag{88}$$

For the transformation of a contravariant acceleration, start with the equation

$$\ddot{\bar{q}}^P = \frac{\partial \bar{q}^P}{\partial q^T}\ddot{q}^T + \frac{\partial^2 \bar{q}^P}{\partial q^m \partial q^n}\dot{q}^m\dot{q}^n. \tag{89}$$

Next insert the expression for the second partial derivative:

$$\ddot{\bar{q}}^P = \frac{\partial \bar{q}^P}{\partial q^T}\ddot{q}^T + \left[\begin{Bmatrix} s \\ mn \end{Bmatrix}\frac{\partial \bar{q}^P}{\partial q^\theta} - \begin{Bmatrix} \bar{P} \\ ij \end{Bmatrix}\frac{\partial \bar{q}^i}{\partial q^m}\frac{\partial \bar{q}^j}{\partial q^n}\right]\dot{q}^m\dot{q}^n. \tag{90}$$

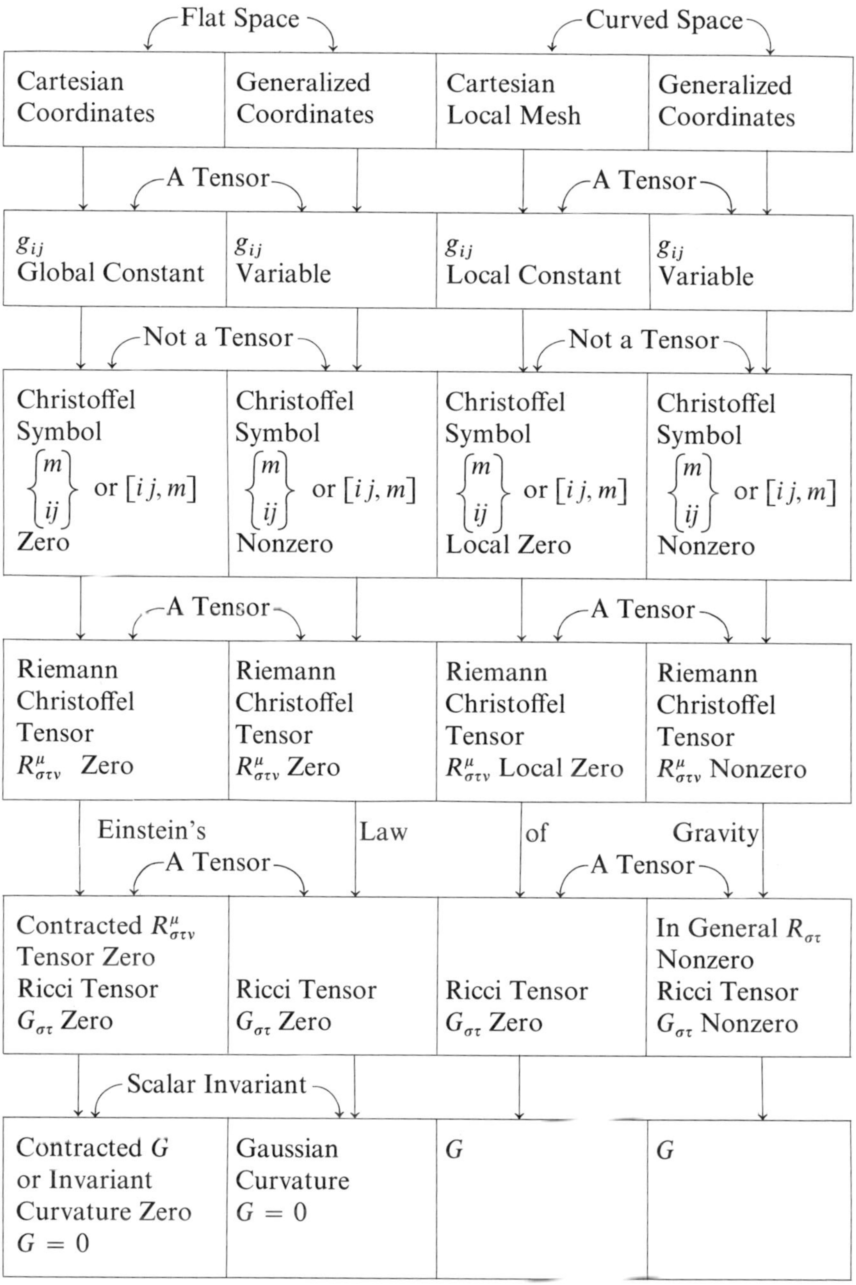

Figure 27. Flowchart for flat and curved spaces showing their metric properties

Rearrange these terms to obtain

$$\left[\ddot{\bar{q}}^P + \left\{\begin{matrix} \bar{P} \\ ij \end{matrix}\right\} \frac{\partial \bar{q}^i}{\partial q^m} \dot{q}^m \frac{\partial \bar{q}^j}{\partial q^n} \dot{q}^n\right] = \left[\ddot{q}^T \frac{\partial \bar{q}^P}{\partial q^T} + \left\{\begin{matrix} s \\ mn \end{matrix}\right\} \dot{q}^m \dot{q}^n \frac{\partial \bar{q}^P}{\partial q^\theta}\right], \quad (91)$$

which reduces to

$$\left[\ddot{\bar{q}}^P + \left\{\begin{matrix} \bar{P} \\ ij \end{matrix}\right\} \dot{\bar{q}}^i \dot{\bar{q}}^j\right] = \left[\ddot{q}^T + \left\{\begin{matrix} T \\ mn \end{matrix}\right\} \dot{q}^m \dot{q}^n\ \frac{\partial \bar{q}^P}{\partial q^T}\right] \quad (92)$$

or

$$\bar{a}^P = a^T \frac{\partial \bar{q}^P}{\partial q^T}. \quad (93)$$

When a Christoffel term is included, the generalized acceleration behaves like a tensor.

Figure 27 represents the most basic facts related to space and gravity.

The five drawings in Fig. 28 provide a qualitative understanding of the meaning of the invariant curvature G.

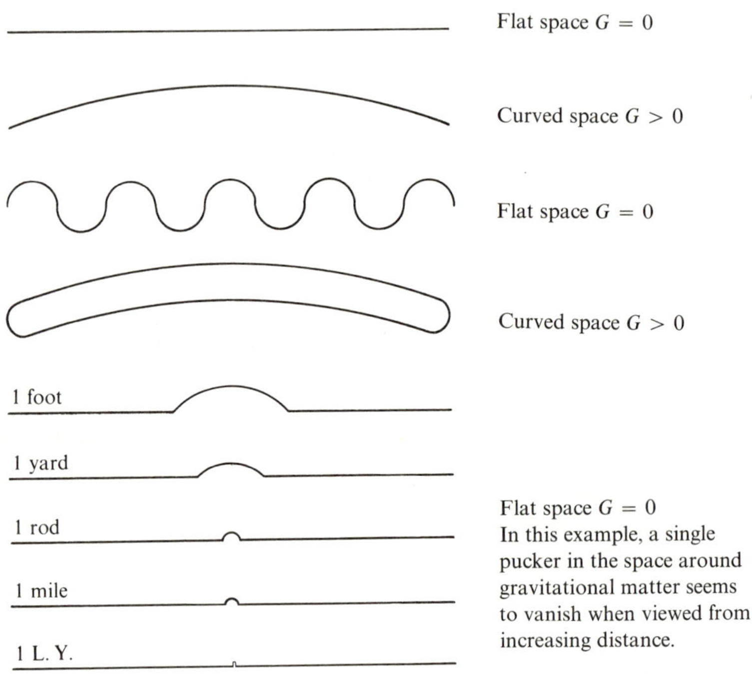

Figure 28. Graphic examples of the invariant curvature G.

6 Applications to Quantum Mechanics

A. Pauli's Spin Matrices and Spinors

A question often raised in the quantum mechanics of second quantization is how to find a matrix a, such that $a^2 = a^{*2} = 0$, $aa^* + a^*a = 1$, and $(a^*a)^2 = a^*a \equiv N$. The matrix that satisfies these conditions is

$$a = \begin{pmatrix} 0 & 0 \\ 1 & 0 \end{pmatrix}, \qquad a^* = \begin{pmatrix} 0 & 1 \\ 0 & 0 \end{pmatrix}. \tag{1}$$

This section is concerned with the relationships between a, a^*, $a^*a \equiv N$, the three nontrivial Pauli spin matrices:

$$\sigma_x = \begin{pmatrix} 0 & 1 \\ 1 & 0 \end{pmatrix}, \tag{2}$$

$$\sigma_y = \begin{pmatrix} 0 & -i \\ i & 0 \end{pmatrix}, \tag{3}$$

and

$$\sigma_z = \begin{pmatrix} 1 & 0 \\ 0 & -1 \end{pmatrix}, \tag{4}$$

and the 2 by 2 identity matrix:

$$I = \begin{pmatrix} 1 & 0 \\ 0 & 1 \end{pmatrix}. \tag{5}$$

Certain linear combinations of the three sigma matrices are deserving of special recognition and names:

$$\frac{\sigma_x + i\sigma_y}{2} \equiv \sigma_+ = \begin{pmatrix} 0 & 1 \\ 0 & 0 \end{pmatrix} = a^*, \tag{6}$$

$$\frac{\sigma_x - i\sigma_y}{2} \equiv \sigma_- = \begin{pmatrix} 0 & 0 \\ 1 & 0 \end{pmatrix} = a, \tag{7}$$

$$\frac{I + \sigma_z}{2} \equiv \Lambda_+ = \begin{pmatrix} 1 & 0 \\ 0 & 0 \end{pmatrix} = a^*a = N, \tag{8}$$

and

$$\frac{I - \sigma_z}{2} \equiv \Lambda_- = \begin{pmatrix} 0 & 0 \\ 0 & 1 \end{pmatrix} = I - N. \tag{9}$$

Other useful relations for these Pauli spin matrices are as follows:

$$\boldsymbol{\sigma} \times \boldsymbol{\sigma} = 2i\boldsymbol{\sigma}, \tag{10}$$

where

$$\boldsymbol{\sigma} = \hat{\mathbf{i}}\sigma_x + \hat{\mathbf{j}}\sigma_y + \hat{\mathbf{k}}\sigma_z, \tag{11}$$

and

$$\sigma_x\sigma_y = i\sigma_z, \tag{12}$$

$$\sigma_y\sigma_z = i\sigma_x, \tag{13}$$

and

$$\sigma_z\sigma_x = i\sigma_y. \tag{14}$$

Relations (12) to (14) can be stated more compactly as follows:

$$\sigma_1\sigma_m = i\epsilon_{lmn}\sigma_n. \tag{15}$$

Additional properties of the sigma matrices are

$$\sigma_x\sigma_y + \sigma_y\sigma_x = 0, \tag{16}$$

$$\sigma_y\sigma_z + \sigma_z\sigma_y = 0, \tag{17}$$

$$\sigma_x\sigma_z + \sigma_z\sigma_x = 0, \tag{18}$$

or

$$\sigma_i\sigma_j + \sigma_j\sigma_i = 2\delta_{ij}; \tag{19}$$

and

$$\sigma_x\sigma_y - \sigma_y\sigma_x = 2i\sigma_z, \tag{20}$$

$$\sigma_y\sigma_z - \sigma_z\sigma_y = 2i\sigma_x, \tag{21}$$

$$\sigma_z\sigma_x - \sigma_x\sigma_z = 2i\sigma_y, \tag{22}$$

or

$$\sigma_1\sigma_m - \sigma_m\sigma_1 = 2i\epsilon_{lmn}\sigma_n. \tag{23}$$

Note also that

$$
\begin{aligned}
&\Lambda_+ + \Lambda_- = 1, && \sigma_+\sigma_- + \sigma_-\sigma_+ = 1, && \sigma_+\sigma_- - \sigma_-\sigma_+ = \sigma_z,\\
&\sigma_+\sigma_- = \Lambda_+, && \sigma_-\sigma_+ = \Lambda_-, && {\sigma_+}^2 = 0, \quad {\sigma_-}^2 = 0,\\
&{\Lambda_+}^2 = \Lambda_+, && {\Lambda_-}^2 = \Lambda_-, && \Lambda_+\Lambda_- = \Lambda_-\Lambda_+ = 0,\\
&\sigma_x\sigma_y\sigma_z = iI && \Lambda_+(\alpha + \beta) = \alpha, && \Lambda_-(\alpha + \beta) = \beta,
\end{aligned}
\tag{24}
$$

and

$$\boldsymbol{\sigma}^2 = {\sigma_x}^2 + {\sigma_y}^2 + {\sigma_z}^2 = 2(\sigma_+\sigma_- + \sigma_-\sigma_+) + {\sigma_z}^2.$$

The matrix

$$\mathbf{S} = \frac{\hbar}{2}\boldsymbol{\sigma} \tag{25}$$

has the same identities as σ. For example

$$\mathbf{S}^2 = 2(S_+S_- + S_-S_+) + {S_z}^2, \tag{26}$$

where

$$S_+ = \frac{S_x + iS_y}{2}, \qquad S_- = \frac{S_x - iS_y}{2}. \tag{27}$$

Now σ_x, σ_y, σ_z, σ_+, σ_-, Λ_+, Λ_-, $\boldsymbol{\sigma}^2$, and $\mathbf{S}^2$ are matrix operators that operate on quantum mechanical operands depending on only two variables. Often these variables can take only two values, 0 or 1, in which case they are called "dichotomic variables." Examples of quantities that are described by functions of only two variables include the angular momentum of an electron in a magnetic field and the existence or nonexistence of a certain particle. (Note that the angular momentum of an electron in a magnetic field can be measured to have only two values, $+\hbar/2$ in the direction of the field or $-\hbar/2$ in the direction opposite to the field.

Such operands affect the following two-element column matrices:

$$\alpha \equiv \begin{pmatrix} 1 \\ 0 \end{pmatrix} \tag{28}$$

and

$$\beta \equiv \begin{pmatrix} 0 \\ 1 \end{pmatrix}, \tag{29}$$

which have the properties

$$\alpha + \beta = \begin{pmatrix} 1 \\ 1 \end{pmatrix}, \tag{30}$$

$$A\alpha + B\beta = \begin{pmatrix} A \\ B \end{pmatrix}, \tag{31}$$

$$\boldsymbol{\alpha}\alpha = \boldsymbol{\beta}\beta = 1, \tag{32}$$

and

$$\boldsymbol{\alpha}\beta = 0. \tag{33}$$

Let

$$\psi(1) = \begin{pmatrix} 1 \\ 0 \end{pmatrix} = |1\rangle \quad \text{and} \quad \psi(0) = \begin{pmatrix} 0 \\ 1 \end{pmatrix} = |0\rangle, \tag{34}$$

where $\psi(1)$ is an eigenfunction representing the existence of a certain particle, say a proton, and $\psi(0)$ is an eigenfunction representing its nonexistence. From the previous discussion, it follows that

$$a\psi(1) = \begin{pmatrix} 0 & 0 \\ 1 & 0 \end{pmatrix}\begin{pmatrix} 1 \\ 0 \end{pmatrix} = \begin{pmatrix} 0 \\ 1 \end{pmatrix} = \psi(0), \tag{35}$$

$$a\psi(0) = \begin{pmatrix} 0 & 0 \\ 1 & 0 \end{pmatrix}\begin{pmatrix} 0 \\ 1 \end{pmatrix} = \begin{pmatrix} 0 \\ 0 \end{pmatrix} = 0, \tag{36}$$

$$a^*\psi(1) = \begin{pmatrix} 0 & 1 \\ 0 & 0 \end{pmatrix}\begin{pmatrix} 1 \\ 0 \end{pmatrix} = \begin{pmatrix} 0 \\ 0 \end{pmatrix} = 0, \tag{37}$$

$$a^*\psi(0) = \begin{pmatrix} 0 & 1 \\ 0 & 0 \end{pmatrix}\begin{pmatrix} 0 \\ 1 \end{pmatrix} = \begin{pmatrix} 1 \\ 0 \end{pmatrix} = \psi(1), \tag{38}$$

or, more compactly,

$$a^*\psi(n) = (1 - n)\psi(1 - n) \tag{39}$$

and

$$a\psi(n) = n\psi(1 - n), \tag{40}$$

where $n = 0, 1$. These often are referred to as the "destruction operator" a and as the "creation operator" a^*

For a many-particle system

$$\Psi(ni) = \Psi(n_1, n_2, n_3, \cdots) = |n_1, n_2, n_3, \cdots\rangle \tag{41}$$

where $n_i = 0, 1$ and $0 \sim \begin{pmatrix} 0 \\ 1 \end{pmatrix}$, $1 \sim \begin{pmatrix} 1 \\ 0 \end{pmatrix}$. For example, at a prescribed time

$$\Psi(n_i) = \left| \begin{pmatrix} 0 \\ 1 \end{pmatrix}_1 \begin{pmatrix} 0 \\ 1 \end{pmatrix}_2 \begin{pmatrix} 1 \\ 0 \end{pmatrix}_3 \cdots \begin{pmatrix} 1 \\ 0 \end{pmatrix}_k \cdots \right\rangle. \tag{42}$$

If either a_k or $a_k{}^*$ operates on Ψ, then only $(\ \ )_k$ is affected.

Next define $n \equiv a^*a = \begin{pmatrix} 1 & 0 \\ 0 & 0 \end{pmatrix}$ and $N_k \equiv a_k{}^*a_k = \begin{pmatrix} 1 & 0 \\ 0 & 0 \end{pmatrix}_k$, for which

$$N^2 = a^*aa^*a = a^*(1 - a^*a)a = a^*a = N, \tag{43}$$

since

$$a^2 = 0 \qquad \text{and} \qquad aa^* + a^*a = I. \tag{44}$$

If N is diagonalized and nondegenerate, then $N = \begin{pmatrix} \lambda_1 & 0 \\ 0 & \lambda_2 \end{pmatrix}$, and $N^2 = N$, which implies that $\lambda_i{}^2 = \lambda_i$ or that the eigenvalues are 0 and 1. This fact can be seen more easily from the determinant $|N - \lambda I| = \begin{pmatrix} 1 - \lambda & 0 \\ 0 & -\lambda \end{pmatrix}$. Since N has these eigenvalues, it follows that

$$N\psi(n) = n\psi(n), \tag{45}$$

or

$$N\psi(1) = \begin{pmatrix} 1 \\ 0 \end{pmatrix} = 1\psi(1), \tag{46}$$

and

$$N\psi(0) = \begin{pmatrix} 0 \\ 0 \end{pmatrix} = 0\psi(0). \tag{47}$$

A second application of α and β arises in the Pauli treatment of spin $\frac{1}{2}$ particles. For such particles, only two values of angular momentum, $+\hbar/2$ in the direction of the magnetic field (spin up) and $-\hbar/2$ in the opposite direction, (spin down) can be measured. The spin part of the wavefunction then might be of the form

$$(A\alpha + B\beta) = \begin{pmatrix} A \\ B \end{pmatrix}. \tag{48}$$

Now

$$\psi^*(s)\psi(s) = (A^*B^*)\begin{pmatrix} A \\ B \end{pmatrix} = A^2 + B^2, \tag{49}$$

where $|A|^2$ and $|B|^2$ are the probabilities that the particle is spin up or spin down, respectively.

A third application of α and β arises in the discussion of isotopic spin. Let $P = \begin{pmatrix} 1 \\ 0 \end{pmatrix}$ and $N = \begin{pmatrix} 0 \\ 1 \end{pmatrix}$, where P represents a proton or positive charge and N represents a neutron or zero charge. Now a and a^* can operate on P and N just as they did on $\psi(1) + \psi(0)$. Perhaps it should be noted that the selection of P and N is arbitrary and $P = \begin{pmatrix} 1 \\ 0 \end{pmatrix}$ and $N = \begin{pmatrix} 0 \\ 1 \end{pmatrix}$ appear in the representation used by high-energy physicists and that $P = \begin{pmatrix} 0 \\ 1 \end{pmatrix}$, $N = \begin{pmatrix} 1 \\ 0 \end{pmatrix}$ appear in the representation used by low-energy physicists. Also, in the treatment of isotopic spin,

$$\boldsymbol{\sigma} \to \boldsymbol{\tau},$$

$$\tau_x = \begin{pmatrix} 0 & 1 \\ 1 & 0 \end{pmatrix},$$

$$\tau_y = \begin{pmatrix} 0 & -i \\ i & 0 \end{pmatrix},$$

and

$$\tau_z = \begin{pmatrix} 1 & 0 \\ 0 & -1 \end{pmatrix},$$

also

$$\sigma_+ \Rightarrow \tau_+ \quad \text{and} \quad \sigma_- \Rightarrow \tau_-.$$

Finally, note that the operator a cannot be diagonalized, since

$$a^2 = a^{*2} = \begin{pmatrix} 0 & 0 \\ 0 & 0 \end{pmatrix} = 0,$$

which means that the squares of each of its eigenvalues is zero, and also that a, a^*, and N are identically zero.

EXERCISES

8. Determine the effect of previously discussed operators when applied to α and β by evaluating:

(a)	(b)	(c)	(d)
$\sigma_z\alpha$	$\sigma_z\beta$	$\Lambda_+\alpha$	$\Lambda_+\beta$
$\sigma_+\alpha$	$\sigma_+\beta$	$\Lambda_-\alpha$	$\Lambda_-\beta$
$\sigma_-\alpha$	$\sigma_-\beta$		

9. Evaluate $\boldsymbol{\sigma\sigma}^* - I$, where $\boldsymbol{\sigma}$ denotes the column vector

$$\begin{pmatrix} \sigma_x \\ \sigma_y \\ \sigma_z \end{pmatrix}.$$

What can be said about the resulting tensor?

10. Alladi Ramakrishnan has said: "Everything seems to flow from the traility of the Pauli matrices." Write a brief essay tracing the development of this theme in his writings.

$$|S| = \sqrt{\tfrac{1}{2}(1 + \tfrac{1}{2})} = \frac{\sqrt{3}}{2} = 0.833$$

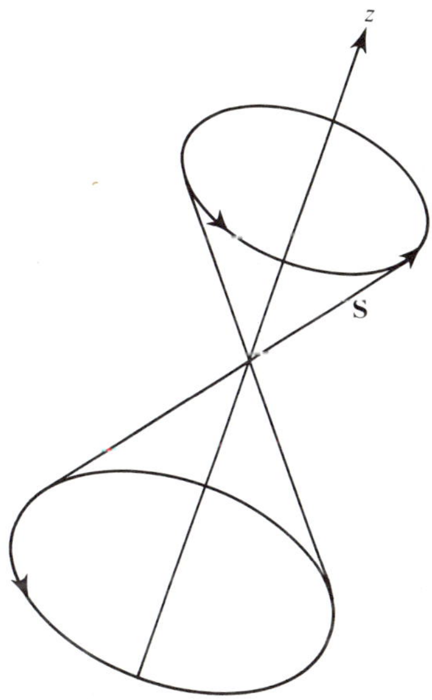

Figure 1. Representation of the spin vector.

From a vectorial point of view electron spin $\mathbf{S}$ has a magnitude $|\mathbf{S}| = \sqrt{S(S+1)}$. As shown in Fig. 1, the spin vector precesses about the z axis with a projection on the z axis of $\pm\frac{1}{2}$. In the N representation, which finds application in second quantization where radiant energy is absorbed and emitted, the symbol $\Psi(0)$ denotes a wave function with no elementary particle present and the symbol $\Psi(1)$ denotes a wave function with an elementary particle present. The elementary particle in question, of course, is typically an electron that appears or disappears at a certain point in time and space. In the N representation, these wave functions are

$$\Psi(0) = \begin{pmatrix} 1 \\ 0 \end{pmatrix} \tag{50}$$

and

$$\Psi(1) = \begin{pmatrix} 0 \\ 1 \end{pmatrix}, \tag{51}$$

which are represented by the same matrices that are used in another connection to represent isospin, as follows:

$$\rho = \begin{pmatrix} 1 \\ 0 \end{pmatrix} \tag{54}$$

and

$$\eta = \begin{pmatrix} 0 \\ 1 \end{pmatrix}. \tag{53}$$

To keep from adding to the confusion in actual calculations, it is often necessary to attach subscripts to certain quantities, as

$$(\Psi_\sigma \Psi_\tau)_A, \tag{54a}$$

to represent spin and isospin of particle A; or

$$\Psi_\sigma \Psi_\tau \Psi_\rho, \tag{54b}$$

to represent spin, isospin, and possibly parity of a particle. Also familiar in the N representation are the so-called creation operator:

$$a^* = \begin{pmatrix} 0 & 0 \\ 1 & 0 \end{pmatrix}, \tag{55a}$$

and annihilation operator:

$$a = \begin{pmatrix} 0 & 1 \\ 0 & 0 \end{pmatrix}. \tag{55b}$$

Formally the operators a and a^* are the same as the shifting or spin flipping operators σ_+ and σ_-. They have the following effect on the wave functions (50) and (51):

$$a^*\psi(0) = \psi(1), \tag{56}$$

and

$$a\psi(1) = \psi(0). \tag{57}$$

Also, the operator N, used to measure the population, has the same form,

$$N = \begin{pmatrix} 0 & 0 \\ 0 & 1 \end{pmatrix}, \tag{58}$$

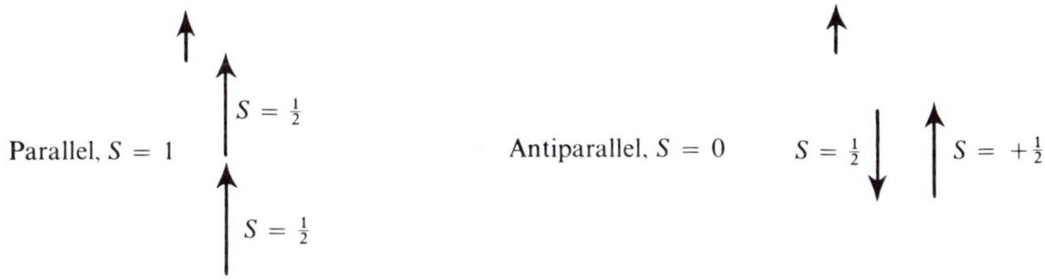

Figure 2. The vector addition of two spins.

as one of the projection operators:

$$\Lambda_+ = \left(\frac{1 + \sigma_z}{2}\right) = \begin{pmatrix} 1 & 0 \\ 0 & 0 \end{pmatrix} \quad \text{and} \quad \Lambda_- = \frac{1 - \sigma_z}{2} = \begin{pmatrix} 0 & 0 \\ 0 & 1 \end{pmatrix}. \tag{59}$$

The matrix N has eigenvalues 1 and 0; hence

$$N\psi(0) = 0\psi(0) \quad \text{and} \quad N\psi(1) = 1\psi(1). \tag{60}$$

EXERCISE 11

11. Show the following:
(*a*) $a^2 = 0$ and $(a^*)^2 = 0$.
(*b*) $(aa^* + a^*a) = 1$.
(*c*) $N = a^*a$.
(*d*) $a^*a^*aa = N(1 - N) = 0$.

The electron spin vector precesses about the z axis in such a way that its projection on the z axis is either $+\frac{1}{2}$ or $-\frac{1}{2}$. For two electrons, spin vectors add with directions either parallel or antiparallel as shown in Fig. 2. Actually, as shown in Fig. 3, there are four possibilities. These arise because $S = 1$ can have any one of three projections on the z axis and is called the "triplet" state (for the helium atom, the triplet state is called "orthohelium"), and $S = 0$ can have only one projection and is called the "singlet" state (for

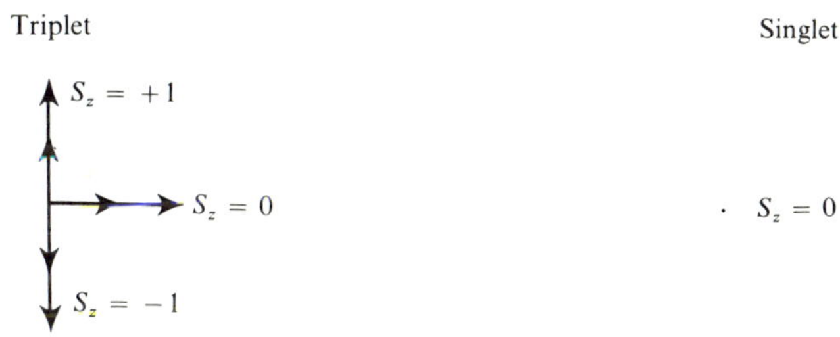

Figure 3. The allowed projection states for two spins.

helium it is "parahelium"). Designate "spin up" by the Greek letter alpha and "spin down" by the Greek letter beta; designate the individual electrons by the Arabic figures 2, 3, and so forth. Thus:

$$\text{Electron 1, spin up is denoted } \alpha(1) = \begin{pmatrix} 1 \\ 0 \end{pmatrix}_1 .$$

$$\text{Electron 2, spin down is denoted } \beta(2) = \begin{pmatrix} 0 \\ 1 \end{pmatrix}_2 .$$

With these definitions the singlet and triplet states can be expressed as follows:

Triplet State	Singlet State
$\alpha(1)\alpha(2)\ \uparrow\uparrow$	
$\frac{1}{\sqrt{2}}[\alpha(1)\beta(2) + \alpha(2)\beta(1)]\ \uparrow\downarrow + \downarrow\uparrow$	$\frac{1}{\sqrt{2}}[\alpha(1)\beta(2) - \alpha(2)\beta(1)]\ \uparrow\downarrow - \uparrow\downarrow$
$\beta(1)\beta(2)\ \downarrow\downarrow$	

From the standpoint of the Schroedinger equation, the energy of a system depends only upon the principal quantum number n and the orbital angular momentum quantum number l, that is,

$$E = E(n, l).$$

Since there is no dependency on the magnetic quantum number m or the spin quantum number s, all wave functions with the same n and l have exactly the same energy and for this reason are said to be "degenerate." If a magnetic field is applied, however, the degeneracy is removed.

The sigma vector product,

$$\boldsymbol{\sigma}(1) \cdot \boldsymbol{\sigma}(2), \tag{61}$$

refers in operator notation to the spin coupling of electrons 1 and 2. To evaluate the eigenvalues of this spin coupling for the three triplet states and the singlet state of the operator, keep in mind that the scalar or dot product of two sigma vectors is an operator that operates on a product (or sum of products) of spin up and/or spin down matrices, as follows:

$$\boldsymbol{\sigma}(1) \cdot \boldsymbol{\sigma}(2) = \sigma_x(1)\sigma_x(2) + \sigma_y(1)\sigma_y(2) + \sigma_z(1)\sigma_z(2). \tag{62}$$

Moreover, $\boldsymbol{\sigma}(1)$ can operate only on $\alpha(1)$ or $\beta(1)$, and $\boldsymbol{\sigma}(2)$ can operate only on $\alpha(2)$ or $\beta(2)$.

Given

$$\sigma_x = \begin{pmatrix} 0 & 1 \\ 1 & 0 \end{pmatrix}, \quad \sigma_y = \begin{pmatrix} 0 & -i \\ i & 0 \end{pmatrix}, \quad \sigma_z = \begin{pmatrix} 1 & 0 \\ 0 & -1 \end{pmatrix},$$
$$\alpha = \begin{pmatrix} 1 \\ 0 \end{pmatrix}, \quad \beta = \begin{pmatrix} 0 \\ 1 \end{pmatrix} \tag{63}$$

it is possible to evaluate some partial operations:

$$\sigma_x\alpha = \begin{pmatrix} 0 & 1 \\ 1 & 0 \end{pmatrix}\begin{pmatrix} 1 \\ 0 \end{pmatrix} = \begin{pmatrix} 0 \\ 1 \end{pmatrix} = \beta,$$
$$\sigma_y\alpha = \begin{pmatrix} 0 & -i \\ i & 0 \end{pmatrix}\begin{pmatrix} 1 \\ 0 \end{pmatrix} = \begin{pmatrix} 0 \\ i \end{pmatrix} = i\beta, \tag{64}$$
$$\sigma_z\alpha = \begin{pmatrix} 1 & 0 \\ 0 & -1 \end{pmatrix}\begin{pmatrix} 1 \\ 0 \end{pmatrix} = \begin{pmatrix} 1 \\ 0 \end{pmatrix} = \alpha;$$

and:

$$\sigma_x\beta = \begin{pmatrix} 0 & 1 \\ 1 & 0 \end{pmatrix}\begin{pmatrix} 0 \\ 1 \end{pmatrix} = \begin{pmatrix} 1 \\ 0 \end{pmatrix} = \alpha,$$
$$\sigma_y\beta = \begin{pmatrix} 0 & -i \\ i & 0 \end{pmatrix}\begin{pmatrix} 0 \\ 1 \end{pmatrix} = \begin{pmatrix} -i \\ 0 \end{pmatrix} = -i\alpha, \tag{65}$$
$$\sigma_z\beta = \begin{pmatrix} 1 & 0 \\ 0 & -1 \end{pmatrix}\begin{pmatrix} 0 \\ 1 \end{pmatrix} = \begin{pmatrix} 0 \\ -1 \end{pmatrix} = -\beta;$$

and:

$$\begin{aligned} &\sigma_x(1)\sigma_x(2)\alpha(1)\alpha(2) = \beta(1)\beta(2), && \sigma_x(1)\sigma_x(2)\beta(1)\beta(2) = \alpha(1)\alpha(2), \\ &\sigma_y(1)\sigma_y(2)\alpha(1)\alpha(2) = -\beta(1)\beta(2), && \sigma_y(1)\sigma_y(2)\beta(1)\beta(2) = -\alpha(1)\alpha(2), \\ &\sigma_z(1)\sigma_z(2)\alpha(1)\alpha(2) = \alpha(1)\alpha(2), && \sigma_z(1)\sigma_z(2)\beta(1)\beta(2) = \beta(1)\beta(2). \end{aligned} \tag{66}$$

Many other products can be formed.

It is somewhat surprising to learn that all triplet states have the same eigenvalue, 1, and the singlet state has an eigenvalue, −3.

$$\begin{array}{llll} \alpha(1)\alpha(2) & \frac{1}{\sqrt{2}}(\alpha(1)\beta(2) + \alpha(2)\beta(1)) & \beta(1)\beta(2) & \text{triplet} \\ & \frac{1}{\sqrt{2}}(\alpha(1)\beta(2) - \alpha(2)\beta(1)) & & \text{singlet} \end{array} \tag{67}$$

$$\begin{aligned}[\boldsymbol{\sigma}(1)\cdot\boldsymbol{\sigma}(2)]\alpha(1)\alpha(2) &= [\sigma_x(1)\sigma_x(2) + \sigma_y(1)\sigma_y(2) + \sigma_z(1)\sigma_z(2)]\alpha(1)\alpha(2) \\ &= \beta(1)\beta(2) - \beta(1)\beta(2) + \alpha(1)\alpha(2) \\ &= [+1]\alpha(1)\alpha(2).\end{aligned} \tag{68}$$

Also

$$\begin{aligned}&[\boldsymbol{\sigma}(1)\cdot\boldsymbol{\sigma}(2)]\frac{1}{\sqrt{2}}(\alpha(1)\beta(2) + \alpha(2)\beta(1)) \\ &\quad = [\sigma_x(1)\sigma_x(2) + \sigma_y(1)\sigma_y(2) + \sigma_z(1)\sigma_z(2)]\frac{1}{\sqrt{2}}(\alpha(1)\beta(2) + \alpha(2)\beta(1)) \\ &\quad = \frac{1}{\sqrt{2}}(\beta(1)\alpha(2) + \alpha(1)\beta(2) + i\beta(1)(-i)\alpha(2) + i\beta(2)(-i)\alpha(1) \\ &\qquad + \alpha(1)(-1)\beta(2) + \alpha(2)(-1)\beta(1)) \\ &\quad = [+1]\frac{1}{\sqrt{2}}(\alpha(1)\beta(2) + \alpha(2)\beta(1)).\end{aligned} \tag{69}$$

Further,

$$\begin{aligned}[\boldsymbol{\sigma}(1)\cdot\boldsymbol{\sigma}(2)]\beta(1)\beta(2) &= [\sigma_x(1)\sigma_x(2) + \sigma_y(1)\sigma_y(2) + \sigma_z(1)\sigma_z(2)]\beta(1)\beta(2) \\ &= \alpha(1)\alpha(2) - \alpha(1)\alpha(2) + \beta(1)\beta(2) \\ &= [+1]\beta(1)\beta(2).\end{aligned} \tag{70}$$

And

$$\begin{aligned}&[\boldsymbol{\sigma}(1)\cdot\boldsymbol{\sigma}(2)]\frac{1}{\sqrt{2}}(\alpha(1)\beta(2) - \alpha(2)\beta(1)) \\ &\quad = [\sigma_x(1)\sigma_x(2) + \sigma_y(1)\sigma_y(2) + \sigma_z(1)\sigma_z(2)]\frac{1}{\sqrt{2}}(\alpha(1)\beta(2) - \alpha(2)\beta(1)) \\ &\quad = \frac{1}{\sqrt{2}}(-\beta(2)\alpha(1) + \beta(1)\alpha(2) - \beta(2)\alpha(1) + \beta(1)\alpha(2) \\ &\qquad - \alpha(1)\beta(2) + \alpha(2)\beta(1)) \\ &\quad = [-3]\frac{1}{\sqrt{2}}(\alpha(1)\beta(2) - \alpha(2)\beta(1)).\end{aligned} \tag{71}$$

The following remarks pertain to the transformation of σ_y to diagonal form. Since

$$|\sigma_y - \lambda I| = 0 \tag{72}$$

has the solutions

$$\lambda = \pm 1, \tag{73}$$

the Hermitian matrix

$$\sigma_y = \begin{pmatrix} 0 & -i \\ i & 0 \end{pmatrix} \tag{74}$$

has the diagonal form

$$\begin{pmatrix} 1 & 0 \\ 0 & -1 \end{pmatrix}. \tag{75}$$

Thus

$$U^{-1}\sigma_y U = \begin{pmatrix} 1 & 0 \\ 0 & -1 \end{pmatrix} \tag{76}$$

or

$$\sigma_y U = U\begin{pmatrix} 1 & 0 \\ 0 & -1 \end{pmatrix}. \tag{77}$$

To solve for the matrix U, proceed as follows:

$$\sigma_y U = \begin{pmatrix} 0 & -i \\ i & 0 \end{pmatrix}\begin{pmatrix} a & b \\ c & d \end{pmatrix} = \begin{pmatrix} a & b \\ c & d \end{pmatrix}\begin{pmatrix} 1 & 0 \\ 0 & -1 \end{pmatrix}, \tag{78}$$

where

$$\begin{pmatrix} a & b \\ c & d \end{pmatrix} \tag{79}$$

represents the unknown matrix U. Then

$$\begin{pmatrix} -ic & -id \\ ia & ib \end{pmatrix} = \begin{pmatrix} a & -b \\ c & -d \end{pmatrix}, \tag{80}$$

which yields

$$c = ia \qquad \text{and} \quad d = -ib. \tag{81}$$

Thus

$$U = \begin{pmatrix} a & b \\ ia & -ib \end{pmatrix}. \tag{82}$$

The unitary property $U^\dagger = U^{-1}$ gives

$$\begin{pmatrix} a^* & -ia^* \\ b^* & ib^* \end{pmatrix} \begin{pmatrix} a & b \\ ia & -ib \end{pmatrix} = \begin{pmatrix} 1 & 0 \\ 0 & 1 \end{pmatrix} \tag{83}$$

or

$$\begin{pmatrix} 2a^2 & 0 \\ 0 & 2b^2 \end{pmatrix} = \begin{pmatrix} 1 & 0 \\ 0 & 1 \end{pmatrix}. \tag{84}$$

A usable solution is

$$a = b = \frac{1}{\sqrt{2}}, \tag{85}$$

and then

$$U = \begin{pmatrix} \frac{1}{\sqrt{2}} & \frac{1}{\sqrt{2}} \\ i\frac{1}{\sqrt{2}} & -i\frac{1}{\sqrt{2}} \end{pmatrix}$$

$$\text{and} \tag{86}$$

$$U^{-1} = \begin{pmatrix} \frac{1}{\sqrt{2}} & -i\frac{1}{\sqrt{2}} \\ \frac{1}{\sqrt{2}} & i\frac{1}{\sqrt{2}} \end{pmatrix}.$$

Rather than taking the usable solution $a = b = 1/\sqrt{2}$, proceed more generally; let

$$a = \frac{1}{\sqrt{2}} e^{i\phi} \quad \text{and} \quad b = \frac{1}{\sqrt{2}} e^{i\theta}. \tag{87}$$

Then choose the phase angles θ and ϕ. With this approach,

$$U = \begin{pmatrix} \frac{1}{\sqrt{2}} e^{i\phi} & \frac{1}{\sqrt{2}} e^{i\theta} \\ \frac{i}{\sqrt{2}} e^{i\phi} & \frac{-i}{\sqrt{2}} e^{i\theta} \end{pmatrix}. \tag{88}$$

and if $\theta = 0$ and $\phi = 0$,

$$U = \begin{pmatrix} \frac{1}{\sqrt{2}} & \frac{1}{\sqrt{2}} \\ \frac{i}{\sqrt{2}} & \frac{-i}{\sqrt{2}} \end{pmatrix} \tag{89}$$

as before. But if $\theta = 3\pi/2$ and $\phi = 0$, then

$$a = \frac{1}{\sqrt{2}} \quad \text{and} \quad b = \frac{-i}{\sqrt{2}}; \tag{90}$$

then:

$$U = \begin{pmatrix} \frac{1}{\sqrt{2}} & \frac{-i}{\sqrt{2}} \\ \frac{i}{\sqrt{2}} & -\frac{1}{\sqrt{2}} \end{pmatrix}, \tag{91}$$

which is both hermitian and unitary, or hermitary.

An alternate approach first finds U^{-1} and then forms the product $U^{-1}\sigma_y U$. The formula for an inverse matrix is

$$U^{-1} = \frac{1}{\det U} M \tag{92}$$

where M is the matrix of the signed cofactors of U^T.
On this basis,

$$\det U = \begin{vmatrix} a & b \\ c & d \end{vmatrix} = (ad - bc) \tag{93}$$

and

$$M = \begin{pmatrix} d & -b \\ -c & a \end{pmatrix}. \tag{94}$$

Consequently,

$$U^{-1} = \begin{pmatrix} \frac{d}{ad - bc} & \frac{-b}{ad - bc} \\ \frac{-c}{ad - bc} & \frac{a}{ad - bc} \end{pmatrix} \tag{95}$$

and

$$U^{-1}\sigma_y U = \begin{pmatrix} \dfrac{d}{ad - bc} & \dfrac{-b}{ad - bc} \\ \dfrac{-c}{ad - bc} & \dfrac{a}{ad - bc} \end{pmatrix} \begin{pmatrix} 0 & -i \\ i & 0 \end{pmatrix} \begin{pmatrix} a & b \\ c & d \end{pmatrix}. \tag{96}$$

Continuing,

$$U^{-1}\sigma_y U = \begin{pmatrix} \dfrac{-ib}{ad - bc} & \dfrac{-id}{ad - bc} \\ \dfrac{ia}{ad - bc} & \dfrac{ic}{ad - bc} \end{pmatrix} \begin{pmatrix} a & b \\ c & d \end{pmatrix}$$

$$= \begin{pmatrix} \left(\dfrac{-iab}{ad - bc} + \dfrac{-icd}{ad - bc}\right) & \left(\dfrac{-ib^2}{ad - bc} + \dfrac{-id^2}{ad - bc}\right) \\ \dfrac{ia^2}{ad - bc} + \dfrac{ic^2}{ad - bc} & \dfrac{iab}{ad - bc} + \dfrac{icd}{ad - bc} \end{pmatrix}. \tag{97}$$

This matrix must be equated to the diagonal form and the equations solved for the relations between a, b, c, and d as before. This approach is obviously more tedious than the previous one.

The Pauli spin matrices satisfy the following quadruple cross-product:

$$(\boldsymbol{\sigma} \times \mathbf{A}) \times (\boldsymbol{\sigma} \times \mathbf{B}) = \mathbf{A} \times \mathbf{B} - i(\mathbf{A}\mathbf{B}^T + \mathbf{B}\mathbf{A}^T)\boldsymbol{\sigma}, \tag{98}$$

where **A** and **B** are two arbitrary three-dimensional vectors. To establish this relationship, let

$$\mathbf{A} = \begin{vmatrix} a \\ b \\ c \end{vmatrix}, \quad \mathbf{B} = \begin{vmatrix} d \\ e \\ f \end{vmatrix}, \quad \text{and} \quad \boldsymbol{\sigma} = \begin{vmatrix} \sigma_x \\ \sigma_y \\ \sigma_z \end{vmatrix} \equiv \begin{vmatrix} x \\ y \\ z \end{vmatrix}. \tag{99}$$

Recall from Exercise 9 the sigma-matrix multiplication table:

	x	y	z
x	I	$+iz$	$-iy$
y	$-iz$	I	$+ix$
z	$+iy$	$-ix$	I

;

also:

$$\boldsymbol{\sigma} \times \mathbf{A} = \det\begin{pmatrix} \hat{\mathbf{i}} & \hat{\mathbf{j}} & \hat{\mathbf{k}} \\ x & y & z \\ a & b & c \end{pmatrix} = \begin{pmatrix} yc - zb \\ za - xc \\ xb - ya \end{pmatrix},$$

$$\boldsymbol{\sigma} \times \mathbf{B} = \det\begin{pmatrix} \hat{\mathbf{i}} & \hat{\mathbf{j}} & \hat{\mathbf{k}} \\ x & y & z \\ d & e & f \end{pmatrix} = \begin{pmatrix} yf - ze \\ zd - xf \\ xe - yd \end{pmatrix}, \tag{100}$$

and

$$\mathbf{A} \times \mathbf{B} = \det\begin{pmatrix} \hat{\mathbf{i}} & \hat{\mathbf{j}} & \hat{\mathbf{k}} \\ a & b & c \\ d & e & f \end{pmatrix} = \begin{pmatrix} bf - ce \\ cd - af \\ ae - bd \end{pmatrix}. \tag{101}$$

These results may now be used to establish the required result:

$(\boldsymbol{\sigma} \times \mathbf{A}) \times (\boldsymbol{\sigma} \times \mathbf{B})$

$$= \det\begin{pmatrix} \hat{\mathbf{i}} & \hat{\mathbf{j}} & \hat{\mathbf{k}} \\ yc - zb & +za - xc & xb - ya \\ yf - ze & +zd - xf & xe - yd \end{pmatrix}$$

$$= \begin{pmatrix}
+zaxe & -zayd & -xcxe & +xcyd & +xbxf & -xbzd & -yaxf & +yazd \\
\diagdown\diagup & \diagdown\diagup & \diagdown\diagup & \diagdown\diagup & \diagdown\diagup & \diagdown\diagup & \diagdown\diagup & \diagdown\diagup \\
+iy & -ix & I & +iz & I & -iy & -iz & +ix \\
-ycxe & +ycyd & +zbxe & -zbyd & +xbyf & -xbze & -yayf & +yaze \\
\diagdown\diagup & \diagdown\diagup & \diagdown\diagup & \diagdown\diagup & \diagdown\diagup & \diagdown\diagup & \diagdown\diagup & \diagdown\diagup \\
-iz & I & +iy & -ix & +iz & -iy & I & +ix \\
+yczd & -ycxf & +zbxf & -zbzd & +xcyf & -xcze & +zaze & -zayf \\
\diagdown\diagup & \diagdown\diagup & \diagdown\diagup & \diagdown\diagup & \diagdown\diagup & \diagdown\diagup & \diagdown\diagup & \diagdown\diagup \\
+ix & -ic & +iy & I & +iz & -iy & I & -ix
\end{pmatrix}$$

$$= \begin{pmatrix} bf - ce & 2ad & ae + bd & af + cd \\ +cd - bf & ae + bd & 2be & bf + ce \\ ae - bd & af + cd & bf + ce & 2cf \end{pmatrix} \begin{pmatrix} I \\ +ix \\ +iy \\ +iz \end{pmatrix} \tag{102}$$

$$= \begin{pmatrix} bf - ce \\ +cd - af \\ ae - bd \end{pmatrix} + \begin{pmatrix} 2ad & ae + bd & af + cd \\ ae + bd & 2be & bf + ce \\ af + cd & bf + ce & 2cf \end{pmatrix} \begin{pmatrix} +ix \\ +iy \\ +iz \end{pmatrix}$$

$$= \mathbf{A} \times \mathbf{B} + (+i) \left[\begin{pmatrix} ad & ae & af \\ bd & be & bf \\ cd & ce & cf \end{pmatrix} + \begin{pmatrix} da & db & dc \\ ea & eb & ec \\ fa & fb & fc \end{pmatrix} \right] \begin{pmatrix} x \\ y \\ z \end{pmatrix}$$

$$= \mathbf{A} \times \mathbf{B} + i \left[\begin{pmatrix} a \\ b \\ c \end{pmatrix} (d, e, f) + \begin{pmatrix} d \\ e \\ f \end{pmatrix} (a. b, c) \right] \begin{pmatrix} \sigma_x \\ \sigma_y \\ \sigma_z \end{pmatrix}$$

$$= \mathbf{A} \times \mathbf{B} + i[\mathbf{AB}^T + \mathbf{BA}^T]\boldsymbol{\sigma},$$

which prominently displays the "alternator" $[AB^T + (AB^T)^T]$ of AB^T.

EXERCISE 12

Given that **A** and **B** are two noncommuting vectors, prove the following:
(*a*) $(\boldsymbol{\sigma} \cdot \mathbf{A})(\boldsymbol{\sigma} \cdot \mathbf{B}) = \mathbf{A} \cdot \mathbf{B} - i\boldsymbol{\sigma} \cdot (\mathbf{A} \times \mathbf{B})$.
(*b*) $(\boldsymbol{\sigma} \times \mathbf{A}) \cdot (\boldsymbol{\sigma} \times \mathbf{B}) = 2\mathbf{A} \cdot \mathbf{B} + i\boldsymbol{\sigma} \cdot (\mathbf{A} \times \mathbf{B})$.
(*c*) $(\boldsymbol{\sigma} \times \mathbf{A}) \times (\boldsymbol{\sigma} \times \mathbf{B}) = \mathbf{A} \times \mathbf{B} + i[\mathbf{A}(\mathbf{B} \cdot \boldsymbol{\sigma}) + \mathbf{B}(\mathbf{A} \cdot \boldsymbol{\sigma})]$.

B. Dirac's Relativistic Wave Equation

In 1927 Dirac created the relativistic wave equation that bears his name. Because it has been found to be highly successful in the description of electrons and other particles that possess spin $\frac{1}{2}$, a detailed description of this equation

$$\left[\gamma_\mu\left(P_\mu - \frac{e}{c} A_\mu\right) - im_0 c\right]\Psi = 0, \tag{1}$$

and its solution can be found in several books.*

*W. Pauli, *Algemeinen Prinzipen der Quantenmechanik*, H.d.P., Vol. V/1, J. Springer, 1958; A. Sommerfeld, *Wellenmechanik*, Ungar. Eng. translation, Bell, England; H. Bethe and Salpeter, *Quantum Mechanics of One and Two Electron Systems*, H.d.P., J. Springer; and J. A. Eisele, *Modern Quantum Mechanics with Applications to Elementary Particle Physics*, New York: Wiley, 1969.

In this section the complete set of sixteen 4 by 4 Dirac matrices is expressed as various direct products of the three 2 by 2 Pauli spin matrices and the unit matrix,

$$\sigma_1 = \begin{pmatrix} 0 & 1 \\ 1 & 0 \end{pmatrix}, \quad \sigma_2 = \begin{pmatrix} 0 & -i \\ i & 0 \end{pmatrix}, \quad \sigma_3 = \begin{pmatrix} 1 & 0 \\ 0 & -1 \end{pmatrix},$$

and

$$\sigma_4 = \begin{pmatrix} 1 & 0 \\ 0 & 1 \end{pmatrix}, \tag{2}$$

according to the rule

$$\Gamma_i = \sigma_j \times\cdot\ \sigma_k. \tag{3}$$

Table 1 shows all 16 Γ matrices and their direct product equivalents.

Table 1 The Dirac Matrices in Direct Product Form

Scalar

$$I = \begin{pmatrix} I & 0 \\ 0 & I \end{pmatrix} = \sigma_4 \times\cdot\ \sigma_4$$

Vector

$$\gamma_1 = \begin{pmatrix} 0 & -i\sigma_x \\ i\sigma_x & 0 \end{pmatrix} = \sigma_2 \times\cdot\ \sigma_1$$

$$\gamma_2 = \begin{pmatrix} 0 & -i\sigma_y \\ i\sigma_y & 0 \end{pmatrix} = \sigma_2 \times\cdot\ \sigma_2$$

$$\gamma_3 = \begin{pmatrix} 0 & -i\sigma_z \\ i\sigma_z & 0 \end{pmatrix} = \sigma_2 \times\cdot\ \sigma_3$$

$$\gamma_4 = \begin{pmatrix} I & 0 \\ 0 & -I \end{pmatrix} = \sigma_3 \times\cdot\ \sigma_4$$

Tensor

$$-i\gamma_2\gamma_3 = \begin{pmatrix} \sigma_x & 0 \\ 0 & \sigma_x \end{pmatrix} = \sigma_4 \times\cdot\ \sigma_1$$

$$-i\gamma_3\gamma_1 = \begin{pmatrix} \sigma_y & 0 \\ 0 & \sigma_y \end{pmatrix} = \sigma_4 \times\cdot\ \sigma_2$$

$$-i\gamma_1\gamma_2 = \begin{pmatrix} \sigma_z & 0 \\ 0 & \sigma_z \end{pmatrix} = \sigma_4 \times\cdot\ \sigma_3$$

Table 1 *continued*

$$-i\gamma_1\gamma_4 = \begin{pmatrix} 0 & \sigma_x \\ \sigma_x & 0 \end{pmatrix} = \sigma_1 \times\cdot\ \sigma_1$$

$$-i\gamma_2\gamma_4 = \begin{pmatrix} 0 & \sigma_y \\ \sigma_y & 0 \end{pmatrix} = \sigma_1 \times\cdot\ \sigma_2$$

$$-i\gamma_3\gamma_4 = \begin{pmatrix} 0 & \sigma_z \\ \sigma_z & 0 \end{pmatrix} = \sigma_1 \times\cdot\ \sigma_3$$

Axial vector

$$-i\gamma_2\gamma_3\gamma_4 = -\ i\gamma_1\gamma_5 = \begin{pmatrix} \sigma_x & 0 \\ 0 & -\sigma_x \end{pmatrix} = \sigma_3 \times\cdot\ \sigma_1$$

$$-i\gamma_3\gamma_1\gamma_4 = -\ i\gamma_2\gamma_5 = \begin{pmatrix} \sigma_y & 0 \\ 0 & -\sigma_y \end{pmatrix} = \sigma_3 \times\cdot\ \sigma_2$$

$$-i\gamma_1\gamma_2\gamma_4 = -\ i\gamma_3\gamma_5 = \begin{pmatrix} \sigma_z & 0 \\ 0 & -\sigma_z \end{pmatrix} = \sigma_3 \times\cdot\ \sigma_3$$

$$-i\gamma_1\gamma_2\gamma_3 = +\ i\gamma_4\gamma_5 = \begin{pmatrix} 0 & -iI \\ iI & 0 \end{pmatrix} = \sigma_2 \times\cdot\ \sigma_4$$

Pseudoscalar

$$\gamma_1\gamma_2\gamma_3\gamma_4 \equiv \gamma_5 = \begin{pmatrix} 0 & -I \\ I & 0 \end{pmatrix} = -\ \sigma_1 \times\cdot\ \sigma_4$$

Notice that the Dirac equation (1) appears symmetric in tensor notation with the one scalar matrix I and four vector matrices γ_μ represented:

$$\Big\{\gamma_1\Big(P_1 - \frac{e}{c}A_1\Big) + \gamma_2\Big(P_2 - \frac{e}{c}A_2\Big) + \gamma_3\Big(P_3 - \frac{e}{c}A_3\Big) + \gamma_4\Big(P_4 - \frac{e}{c}A_4\Big) - im_0cI\Big\}\Psi = 0. \quad (4)$$

If represented in direct product form using the Pauli matrices, however, the symmetry is lost:

$$\Big\{\sigma_2 \times\cdot\ \sigma_1\Big(P_1 - \frac{e}{c}A_1\Big) + \sigma_2 \times\cdot \sigma_2\Big(P_3 - \frac{e}{c}A_2\Big) + \sigma_2 \times\cdot\ \sigma_3\Big(P_3 - \frac{e}{c}A_3\Big) + \sigma_3 \times\ \sigma_4\Big(P_4 - \frac{e}{c}A_4\Big) + \sigma_4 \times\cdot\ \sigma_4(-im_0c)\Big\}\Psi = 0 \quad (5)$$

Pauli has shown that the relativistic and gauge invariance of the Dirac equation is not lost with the addition to equation (1) of six tensor terms proportional to the three electric and three magnetic field components:

$$\gamma_\mu \gamma_\nu f_{\mu\nu}. \tag{6}$$

The occurrence of such terms is usually reserved for the "squared" form of the Dirac equation in connection with the appearance of intrinsic Pauli spin of the electron. There is no reason forbidding the appearance of spin terms in the linear form, and it has been pointed out by Bethe and Salpeter that these correspond to higher order correction terms. It is possible to restore the complete symmetry of the Dirac equation when written in direct product form by adding the "missing" axial vector terms $\gamma_\mu\gamma_5$ and the pseudoscalar term γ_5 to form the 16-term version:

$$\left[\sum_{i=1}^{4} \sigma_i \times\cdot \sum_{j=1}^{4} \sigma_j C_{ij}\right]:\Psi = 0 \tag{7}$$

or

$$\sigma_i C_{ij} \Psi \sigma_j{}^T = 0, \tag{8}$$

where the triproduct has been introduced. Hence the four-component Dirac solution becomes

$$\Psi = \begin{pmatrix} \psi_1 & \psi_2 \\ \psi_3 & \psi_4 \end{pmatrix}. \tag{9}$$

EXERCISES

13. What values of C_{ij} are appropriate to the conventional five-term form of the Dirac equation?

14. What values of C_{ij} are appropriate to the augmented version of the Dirac equation according to Pauli, Bethe, and Salpeter?

15. What values of C_{ij} are possible for the complete 16-term form of the Dirac equation? Is it possible to have axial vector momentum and potential terms and a pseudoscalar mass term in addition to the vector and scalar terms?

16. The complete Pauli transformation:

$$\sigma_w A \sigma_w{}^T + \sigma_x A \sigma_x{}^T + \sigma_y A \sigma_y{}^T + \sigma_z A \sigma_z{}^T$$

may be written as follows:

$$[\sigma_w \cdot\times \sigma_w + \sigma_x \cdot\times \sigma_x + \sigma_y \cdot\times \sigma_y + \sigma_z \cdot\times \sigma_z]:A.$$

$\downarrow$ $I \times I$ Identifier $\qquad$ $\downarrow$ $K \times K$ Transpose cispose

Therefore

$$\left[\begin{pmatrix}1&0&0&0\\0&1&0&0\\0&0&1&0\\0&0&0&1\end{pmatrix}+\begin{pmatrix}0&0&0&1\\0&0&1&0\\0&1&0&0\\1&0&0&0\end{pmatrix}+\begin{pmatrix}0&0&0&-1\\0&0&1&0\\0&1&0&0\\-1&0&0&0\end{pmatrix}\right.$$

$$\left.+\begin{pmatrix}1&0&0&0\\0&-1&0&0\\0&0&-1&0\\0&0&0&1\end{pmatrix}\right]:A,$$

becomes

$$\left[\begin{pmatrix}2&0&0&0\\0&0&2&0\\0&2&0&0\\0&0&0&2\end{pmatrix}=2I\,\dot{\times}\,I=2\text{ times transposer}\right]:A=2A^T.$$

Evaluate the complete Dirac transformation:

$$\Gamma_i\,A\,\Gamma_i^T.$$

Bibliography

Abramowitz and Stegun, *Handbook of Mathematical Functions*, National Bureau of Standards Applied Math, Series 55, 1969.

Bellman, *Introduction to Matrix Analysis*, McGraw-Hill, New York, 1960.

Bocher, *Introduction to Higher Algebra*, Macmillan, New York, 1947.

Churchill, *Fourier Series and Boundary Volume Problems*, McGraw-Hill, New York, 1941.

Churchill, *Modern Operational Mathematics in Engineering*, McGraw-Hill, New York.

Churchill, *Introduction to Complex Variables*, McGraw-Hill, New York, 1944.

Collatz, *Eigenwert-Probleme*, Chelsea, New York, 1948.

Conkwrite, *Differential Equations*, Macmillan, 1949.

Cartan, *The Theory of Spinors*, M.I.T. Press, Cambridge, Mass., 1966.

Courant and Hilbert, *Methods of Mathematical Physics*, Interscience, New York, 2 vols., 1955, 1962.

Dwight, *Tables of Integrals and Other Mathematical Data*, Macmillan.

Eisenhart, *Riemanian Geometry*, Princeton University Press, 1950.

Frank and von Mises, *Differential und Integralgleichungen der Mechanik und Physik*, Rosenberg, New York, 1943.

Franklin, *Methods of Advanced Calculus*, McGraw-Hill, New York, 1944.

Franklin, *Matrix theory*, Prentice-Hall, Englewood Cliffs, N.J., 1968.

Frazer, Duncan and Collar, *Elementary Matrices*, Cambridge University Press, 1952

Gradshteyn and Ryzhik, *Tables of Integrals, Series, and Products*, Academic Press, New York, 1962.

Graustein, *Introduction to Higher Geometry*, Macmillan, New York, 1930.

Hahn, *Elementary Matrix Algebra*, 2nd Ed., Macmillan, New York, 1967.

Handbook of Mathematical Tables, 2nd Ed., The Chemical Rubber Co., 1964.

Hildebrand, *Methods of Applied Mathematics*, Prentice-Hall, Englewood Cliffs, N.J., 1952.

Hopf, *Introduction to the Differential Equations of Physics*, Dover, New York, 1948.

Householder, *The Theory of Matrices in Numerical Analysis*, Blaisdell Publishing Co., 1964.

Householder, *Principles of Numerical Analysis*, McGraw-Hill, New York, 1953.

Jahnke and Emde, *Tables of Functions with Formulae and Curves*, Dover, 1945.

Kamke, *Differentialgeichungen Lösungemethoden und Lösungen*, Chelsea.

Kaplan, *Advanced Calculus*, Addison-Wesley, Reading, Mass., 1952.

Lass, *Vector and Tensor Analysis*, McGraw-Hill, New York, 1950.

Levi-Civita, *Absolute Differential Calculus*, Blackie and Sons Limited, London, 1947.

MacDuffee, *Matrix Theory*, Chelsea Press, New York, 1946

MacRobert, *Spherical Harmonics*, Dover, New York, 1948.

Madelung, *Mathematischen Hilfsmittle des Physiken*, Dover, New York, 1943.

Magnus and Oberhettinger, *Special Functions of Mathematical Physics*, Chelsea, New York, 1949.

Marcus and Minc, *A Survey of Matrix Theory and Matrix Inequalities*, Allyn Bacon, 1964.

Margenau and Murphy, *The Mathematics of Physics and Chemistry*, Von Nostrand, Princeton, N.J., 1951.

Mathews and Walker, *Mathematical Methods of Physics*, Benjamin, New York, 1965.

Menzel, *Fundamental Formulas of Physics*, Dover, New York, 1960.

Morse and Feshbach, *Methods of Theoretical Physics*, McGraw-Hill, New York, 1953.

Murnaghan, *Applied Mathematics*, Wiley, New York, 1948.

Nering, *Linear Algebra and Matrix Theory*, John Wiley, New York, 1964.

Patterson, *Projective Geometry*, Wiley, New York, 1946.

Peirce, A., *Short Table of Integrals*, Ginn and Company, New York, 1929.

Phillips, *Vector Analysis*, Wiley, New York, 1933.

Pierpont, *Functions of a Complex Variable*, Ginn and Company, New York, 1914.

Rainich, *Mathematics of Relativity*, Wiley, New York, 1950.

Schneider, *Recent Advances in Matrix Theory*, University of Wisconsin Press, Madison, 1964.

Schreier and Sperner, *Modern Algebra and Matrix Theory*, Chelsea, New York, 1951.

Sneddon, *Fourier Transforms*, McGraw-Hill, New York, 1951.

Sobolev, *Partial Differential Equations of Mathematical Physics*, Addison-Wesley, Reading, Mass., 1964.

Sokolnikoff, *Tensor Analysis*, New York, 1964.

Titchmarch, *Eigenfunction Expansions*, Oxford University Press, 1950.

Todd, *Survey of Numerical Analysis*, McGraw-Hill, New York, 1962.

Watson, *Theory of Bessel Functions*, Cambridge University Press, 1952.

Webster, *Partial Differential Equations of Mathematical Physics*, Hafner, New York, 1950.

Westlake, *A Handbook of Numerical Matrix Inversion and Solution of Linear Equations*, John Wiley, New York, 1968.

Whittaker and Watson, *Modern Analysis*, Cambridge University Press, 1952.

Wilkinson, *The Algebraic Eigenvalue Problem*, Clarendon Press, Oxford, 1965.

Wylie, *Advanced Engineering Mathematics*, McGraw-Hill, New York, 1951.

Zassenhause, *The Theory of Groups*, Chelsea, New York, 1949.

Index